WE ARE FROM THE STARS

EXPLORING SCIENCE

LAYMAN'S GUIDE OF:

FROM: *WHEN IT ALL STARTED*

TO: *WHERE IT MAY GO*

TOM DEAL

Note for Librarians: A cataloguing record for this book is available from Library and Archives Canada at www.collectionscanada.ca/amicus/index-e.html
ISBN 1-4120-5244-0

PUBLISHING™

Offices in Canada, USA, Ireland and UK

This book was published *on-demand* in cooperation with Trafford Publishing. On-demand publishing is a unique process and service of making a book available for retail sale to the public taking advantage of on-demand manufacturing and Internet marketing. On-demand publishing includes promotions, retail sales, manufacturing, order fulfilment, accounting and collecting royalties on behalf of the author.

Book sales for North America and international:
Trafford Publishing, 6E–2333 Government St.,
Victoria, BC v8t 4p4 CANADA
phone 250 383 6864 (toll-free 1 888 232 4444)
fax 250 383 6804; email to orders@trafford.com
Book sales in Europe:
Trafford Publishing (uk) Limited, 9 Park End Street, 2nd Floor
Oxford, UK ox1 1hh UNITED KINGDOM
phone 44 (0)1865 722 113 (local rate 0845 230 9601)
facsimile 44 (0)1865 722 868; info.uk@trafford.com
Order online at:
trafford.com/05-0139

10 9 8 7 6 5

CONTENTS

ACKNOWLEDGEMENTS

To a captivating Connie and our daughters - Dianne, Linda and Patricia; also their husbands Jim, George and Gary - for allowing me the freedom to research and develop this book. To Mark Mason, my eagle-eyed proofreader, who suggested the corrections needed to properly present this book for publication.

SUGGESTIONS FOR READING

This is not a textbook but was developed in a similar manner. All the topics are in sequence - some readers may move from topic to topic randomly and enjoy the book. However, it is hoped that readers will pursue their reading from the beginning through to the last page. This is the direction the book was written - presenting the subjects in an order that helps develop the subjects that follow. Definitions are explained as subjects are encountered - the choice is yours.

The subject matter starts with the beginning of everything - the big bang - and ends with the explanation of black holes. This book will incorporate the system used in many scientific books by capitalizing the objects of the solar system - our planet Earth, earth (dirt); our Moon, other moons; our Sun, other suns; our Universe, other universes

The purpose of this book is **not** to make everyone want to become a scientist (a few will do). The primary object is to provide information in several fields that require **facts** and **verification.** Other fields also require these same components - I present the fields familiar to me. The idea is to encourage each reader to question the **how** and **why** of everything until **verification** is made. You will then have the **facts**. You now understand **why** you would not buy the Brooklyn Bridge or oceanfront property in Arizona. **Research** and **verification** are the foundations of science - these should also be applied to any and all aspects of life.

DEDICATION

This book is dedicated to my five grandchildren: Kelsey, Ben, Tessa, Alyssa and Carley. Each generation arrives into a busy world and their observations and the teachings of the times guide their learning. The perspective these five children encountered of society and the sciences entering the twenty-first century was unbelievable compared to the realities their ancestors met as each of them entered their new centuries. The discoveries in science during their ancestor's lifetimes, especially the last four hundred years, have been proven reliable and are used today in our everyday life without question.

Education works only if the recipient really wants it. I want all to enjoy, question and marvel at things that are now known and are being discovered. Four hundred years ago many people believed they lived on a stationary Earth at the center of the Universe as the Sun and stars went around them. Before the seventeenth century, science was observational. Only after the seventeenth century started did it become experimental. Before I was born, the following discoveries occurred:

1609 Galileo Galilei - observational astronomy
1609 Johannes Kepler - planetary motion and laws
1666 Isaac Newton - laws of gravity and motion
1718 Edmund Halley - comet orbits and laws
1879 Albert Michelson - determined the speed of light
1898 Marie Curie - radioactivity
1900 Max Planck - quantum theory
1905 Albert Einstein - theory of relativity
1913 Niels Bohr - atomic structure
1922 Edwin Hubble - expanding Universe

During my lifetime we have learned how tremendous these discoveries have been by the many applications achieved. These applications have permitted further advances at a rate that is sometimes unbelievable.

PREFACE

Scientists have the opportunity to view certain subjects in more depth than most non-scientists. This is accomplished through the use of an intricate language necessary to accomplish their goals - mathematics. This language of numbers, coordinates, derivatives, integrals, functions, extensions, exponentials and many other processes allows the exploration of the known and unknown in a systematic and precise manner.

The distinguished geophysicist David Greggs' hero was William Thompson, better known as Lord Kelvin. On the door of Greggs' office was a quotation from Lord Kelvin:

> "I often say that when you can measure what you are speaking about and express it in numbers, you know something about it; but when you cannot express it in numbers, your knowledge is of a meager and unsatisfactory kind; it may be the beginning of knowledge, but you have scarcely, in your thoughts, advanced to the state of Science, whatever the matter may be."

One outlook on physicists: A theory is proposed, a prediction is made on the basis of that theory and the prediction is *verified* by observation with repeated experimentations, lending weight and credibility to the theory. Scientists cannot always explain why things are as they are, but they can tell you *how* and *why* they work.

Astrophysicists and astronomers place their minds in our solar system, our galaxy and beyond to view and explore. Nuclear physicists and particle physicists place their minds inside the nucleus of the atom and then inside the proton and the neutron to ask: "Is it possible for me to go further?" Molecular biologists are exploring nature and the human body for the causes of and the prevention of diseases, DNA corrections and genetically building replacement parts for our bodies. All this study and research of the disciplines of science is providing the human race with more accurate information as to "from whence we come" to "whither we go" (Petrarch 1304-1374).

We are thankful for the diversity of the human race. Wouldn't it be a problem if everyone wanted to be a truck driver? Who would build the trucks, who would manufacture or grow the products to be carried in these trucks, who would sell these items, who would ... ? During the development of our solar system and the accumulation of the material that formed the planets, Earth happened to emerge under conditions that eventually led to the beginning of and development of life. Had Earth formed closer to the Sun there would have been no water - it would have boiled away. Had Earth formed further from the Sun, it would have been too cold for bacteria to form and there would have been no life there either.

Returning to the first sentence in this PREFACE, I wish to amend this to *include* all non-scientists due to the approach taken in this book. A formula, which may be half as long as this printed line, speaks directly to the scientist in the time it takes to read the formula. The explanation to the layman could require an hour of time or a chapter in a book. The intent of this book is to allow the layman to understand the basics of many sciences using words - not formulas. Simple formulas will be shown in examples to clarify and verify information. More complete data on many subjects can be found in the APPENDIX. Diagrams are used when possible because "one picture is worth a thousand words." I want young people to be able to read this book and understand the contents enough to encourage them to want to know more - create an appetite for knowledge.

Fields you will visit include: cosmology, astronomy, astrophysics, atomic physics, nuclear physics, particle physics, quantum physics, relativity, geology, geophysics and molecular biology. You will traverse from the beginning of our Universe, proceed to the present and be given a peek into what is now known of the future of our planet and the rest of the Universe. I hope you enjoy this most wonderful journey.

HISTORY

We are living our history. Our descendants will live their history. The means are now available to accurately record these happenings for the peoples of later times. Volumes have been written over the past few thousand years, much of it well after the events occurred, as interpreted by the writers themselves. History and ideas are usually slanted, not always intentional, by the writers and recorders as they preserve information for posterity as they see it. You must know who wrote what and why. Technology now permits the separation of fact from fiction; truth from myth, verifying that truth is often stranger than fiction. We humans are an interesting species. The following parlor talk is for your enjoyment.

Wheeled vehicles appeared in the form of animal-drawn war chariots soon after 3500 BC. The potter's wheel, invented during this same time, has been called the "first really mechanical device."

From the earliest records as far back as 2750 BC, a victory in warfare was usually followed by a fairly wholesale slaughter of prisoners. Those who were spared were carried off to be slaves of their conquerors or were held for ransom. The capture of a town meant its looting and destruction and the elimination of the remaining population. "Remember to rape and pillage BEFORE you burn!"

If history is defined as an "honest attempt first to find out what happened, then explain why it happened," Herodotus of Halicarnassus (484-425 BC) deserves to be called the "Father of History." As he stated, "My duty is to report what has been said, but I do not have to believe it."
The first truly scientific historian was Thucydides (460-400 BC), who believed that his history would become "an everlasting possession" for those who desire a clear picture of what has happened and what is likely to be repeated.

The Babylonians achieved little that today can accurately be called science. They did observe nature and collect data, which is the first requirement of science. To explain natural phenomena, they did not go beyond the formulation of myths that explained things in terms of the unpredictable whims of the gods. The Sun, Moon and five visible planets were thought to be gods who were able to influence human lives, and their movements were charted. From these observations, the pseudoscience of astrology originated.

Hippocrates, the "Father of Medicine," founded a school in 420 BC in which he emphasized the value of observation and the careful interpretation of symptoms. He was firmly convinced that disease resulted from natural, not supernatural, causes. He also gave medicine a sense of service to humanity with the Hippocratic Oath, still in use today.

An example of the origination of a myth in Greece by Phillip II; also used years later by Alexander the Great:

The ancient Greeks and Persians had a long history of warfare. The Persians invaded Greece from the west in 480 BC and burned Athens. The Greeks eventually drove the Persians from Athens and continued until they had driven the Persians out of Greece. The final victory of the Greeks occurred by the defeat of the Persian fleet in the Bay of Salamis. This caused Xerxes and his forces to return to Asia. This victory resulted in much political oratory in Athens.

As Athens progressed into the fourth century BC, the greatness of Athens' past was paramount in the current thinking of the people. Around 380 BC many decrees were invented and quoted as if passed in the great days of 480 BC. Phillip II, Emperor of Greece and the father of Alexander the Great, used these decrees with great brilliance. War was declared on the barbarians "on behalf of the Greeks to punish the Persian outrages to Greek temples in 480 BC." The temples that Persia had sacked were in Athens, but the empire had grown such a slogan of revenge on Persia for the sacking of ALL Greek temples - some which had never existed. Philip II's myth of his grand design was conceived to appeal to the people of Athens. The appeal was not without effect elsewhere. One small Boetian city called its role on the march "a revenge for its ancestors' sake against the barbarians." Another city damned its traitors as "foes of Phillip II and the Greek war." Around 336 BC, forty-four years later, Alexander the Great continued this myth during his conquests.

The Alexandria Library of Egypt, had it remained undamaged by wars and religious zealots, would have given us the knowledge of the Egyptians, Greeks, Romans and those little known civilizations that rose and fell beyond the shores of the Mediterranean. The library was founded in 295 BC and contained over five hundred thousand volumes, or rolls, and its annex in the Temple of Serapes contained an additional forty-three thousand volumes.

The destruction by fire of the collection of books at Alexandria was an incalculable loss and certainly one of the greatest catastrophes in the history of scholarship. In 48 BC Julius Caesar arrived in Alexandria in pursuit of his military rival Pompey, who sought refuge in the town. Caesar sided with Cleopatra in her civil war against her brother Ptolemy XIII. Responding to a sudden attack by Ptolemy's forces in 47 BC, Caesar ordered his soldiers to hurl flaming torches at enemy ships in the harbor. Fanned by the Mediterranean breezes, the flames leaped from the ships to the buildings onshore. Some of the buildings which were destroyed were the library's warehouses. Partial restitution for the losses of the library was made by Cleopatra (69-30 BC), the last of the Ptolemies, who founded a new collection. To this reportedly was added the library at Pergamum, presented to Cleopatra by Caesar's protégé Mark Anthony. The next burning of the contents of the library occurred in 272 as Roman troops of Emperor Lucius Domitius Aeroeliones suppressed an uprising in bitter fighting. The entire collection was destroyed in 391 by a mob of infuriated Christians led by Roman Emperor Theodosius I the Great in their war on paganism. A later collection was destroyed in 640 by an army of Muslims which effectively eliminated the library. Caliph Dinar I, when asked by the Muslim conqueror what to do with a half-million scrolls, answered, "As for the books you mention, if what is written in them agrees with the Book of God, they are not required; if it disagrees, they are not desired. Destroy them, therefore." For selfish objectives great knowledge was lost forever as each religion attempted to conform history to its single outlook.

Why the ancient Greek philosophers - in defiance of universal human behavior - chose to cast aside greed, ambition, superstition and impulses to devote themselves to strenuous inquiry and dispassionate debate about the nature of man and his world is a mystery, but their thinking shaped the development of Western Civilization.

The reason the ancient Greeks embellished their accurate knowledge of the apparent motions and positions of the celestial bodies with mythological tales is not clear. It may be that the authors of these tales looked upon the ordinary Greek citizens as children who had to be entertained by myths if they were to be taught the truths about the Sun, Moon, planets and stars. In these mythological stories the Greek writers presented certain philosophical ideas mixed with pure fancy. Cosmogony, which deals with the origin of the cosmos, and cosmology, dealing with the evolution of the cosmos, were presented as poetic revelations.

In the Later Han Dynasty (23-220), a scholar wrote the *History of the Han* (the Earlier Han Dynasty that existed from 202 BC-8), and thereafter it was customary for each dynasty to write the official history of its immediate predecessor. The Chinese believed that the success and failures of the past provided guidance for one's own time and the future. As stated in the *Historical Records,* "Events of the past, if not forgotten, are teachings about the future."

The real difference in Middle Age writings was the almost total lack of reference to observation or experiment. They followed the medieval Aristotelian ideal of using pure logic to arrive at conclusions - conclusions that were then held to be self-evident products of reason itself. Alternatively, they sought to solve problems by appeals to the authority contained within accepted writings and texts.

One old story illustrates the attitude toward observation. There was a debate concerning the number of teeth in a horse's mouth. One by one the scholars got up and cited their sources. One quoted Aristotle; another, one of the church fathers; and so on. Eventually a very junior member of the group rose and pointed out that there was a horse outside and everyone could go out and count its teeth. At this suggestion, the brothers "fell upon him, smote him hip and thigh, and cast him from the company of educated men."

The modern world, so far as mental outlook is concerned, began in the seventeenth century.

Bertrand Russell

The idea of a canal across Central America had been the dream of many during the nineteenth century. The project had many difficulties and was eventually taken over by President Theodore Roosevelt. Politics was naturally involved and Roosevelt had no qualms about building the canal. The following conversation was "reported" to have occurred:

Once Roosevelt asked his adviser Elihu Root: "Have I answered the charges; have I defended myself?" Root grinned: "You certainly have, Mr. President. You have shown that you were accused of seduction and you have conclusively proved that you were guilty of rape."

The Panama Canal was opened in 1914.

President Franklin Roosevelt was a man of few words. During World War II he ordered General Dwight Eisenhower to "assemble a force for the invasion of Europe."

For a successful technology, reality must take precedence over public relations, for nature cannot be fooled.
--- On the "O" ring failure on the space shuttle *Challenger*---

Richard Feynman

... the beauty of the *real* world - the scientific world ...

SCIENTIFIC HISTORY

The skies of our ancestors appeared to be very low as they looked at the stars. When the ancient Sumerian, Chinese and Korean astronomers climbed the steps of their stone ziggurats to study the stars, they had reason to assume that they obtained a better view because they were closer to these stars. The Egyptians regarded the sky as a tent canopy supported by the mountains that marked the "Four Corners" of the Earth. Because the mountains were not all that high, neither were the heavens. The Greek Sun was so near that Icarus had achieved an altitude of only a few thousand feet when the heat from the Sun melted the wax in his wings causing him to fall into the Aegean Sea.

Stonehenge is one of thousands of old, time-reckoning machines in which all the moving parts were located IN THE SKY. The Great Pyramid of Gizeh (Giza) was aligned to the pole star of its time - Thuban - and it was possible to read the seasons from the position of the pyramid's shadow. The Mayans of ancient Yucatan inscribed stone monuments with formulas useful in predicting solar eclipses and the rising of Venus as the "Morning Star." The twenty-eight poles of the Cheyenne and Sioux medicine lodges were used to record the days of the lunar month.

Someone had to develop a concept to improve on the idea of the Earth being flat - the Sun rising on one side and setting on the other - always hoping the Sun would remember to return and WHERE! It was assumed that the Moon was female (always changing her looks and appearing only as she desired). It never occurred to the ancients that the Moon was hidden behind dense clouds during bad weather. Something had to be devised to relieve these people of the worry of these many problems. The rejection of mythological explanations and the use of reason to explain natural phenomena has been called the "Greek miracle."

The most famous man to have lived at Croton, Italy, was named Pythagoras (582-497 BC), who moved to South Italy from his birthplace of Samos, Greece, in 531 BC. He believed that only a numerical explanation could explain the Universe; nature could best be interpreted in quantifiable terms. Mathematical science was not new but Pythagoras elevated the subject to universal status. He developed many arithmetic sequences including the fact that adding the squares of the legs of a right triangle equals the square of the hypotenuse. He contributed to geometry by formulating many of its definitions. Pythagoras was one of the first to believe that the Earth and the Universe are spherical in shape and that the Sun, Moon and the planets have a motion of their own. He also believed that the Earth rotated, providing day and night. He determined Venus to be both the morning and evening star and also discovered that the planets move in separate orbits, tilted at different angles to the plane of the Sun - the ecliptic. It is not known for sure if it was Pythagoras or a later group of Pythagorean astronomers who removed Earth from its lofty position as the center of all action and made it a planet like all the others. This created the concept of a *heliocentric solar system* in which all objects are viewed or measured from the center of the Sun. To these astronomers this replaced the existing *geocentric system* in which all objects are viewed or measured from the center of Earth. These ideas had to wait two thousand years for Nicholas Copernicus.

Parmenides (539-469 BC) regarded the Universe as a series of concentric bands composed of the two primary elements - fire, which is gentle, thin and homogenous; and night, which is dark, thick and heavy - and of mixtures of them, the whole system being directed by an unnamed goddess established at its center.

Plato (427-347 BC) knew the Athenian philosopher Socrates (470-399 BC) from his boyhood. Plato founded the *Academy* in 387 BC at Athens as an institute for the systematic pursuit of philosophical and scientific research. It became the recognized authority in both mathematics and jurisprudence. He presided over it until he died at Chalcis on the island of Euboea. His nephew, Speusippus, succeeded him as head of the *Academy*. The *Academy* was closed in 529 by the Christian Emperor Justinian in his zeal for Christian orthodoxy. The *Academy* existed for nine hundred and sixteen years.

Eudoxus (400-347 BC) of Cnidus, Greece entered Plato's *Academy* in 385 BC to complete his studies of geometry - *geo-metry*, the "measurement of the Earth." To the simple spherical Universe that had been proposed by Parmenides a century earlier, Eudoxus simply added more spheres. He constructed a model of twenty-seven spheres to explain the motion of the stars. The motion of each planet was explained as resulting from the motion of four spheres fixed one within the other so that each interior sphere revolved around its own axis while participating in the motions of the external spheres revolving around different axes. The Sun and the Moon required three spheres each with the fixed stars requiring only one each. By simple manipulation of these twenty-seven spheres he could show at any time the position of other objects relative to Earth. This was convenient, but did not explain the independent movings of each object. Eudoxus divided the sky into degrees of longitude and latitude, gave a better approximation of the length of the solar year and improved the calendar. He estimated the Earth's circumference at four hundred thousand stadia. Eculid later incorporated many discoveries of Eudoxus into his work.

Aristotle (384-322 BC) was born at Stagira, a Greek colonial town on the northwestern shores of the Aegean Sea. He attended Plato's *Academy* at Athens from age seventeen to thirty-seven (367-347 BC) where he worked side by side with Plato. He had Alexander as a student while at the Macedonian Court in Pella from age thirty-seven to forty-nine (347-335 BC). He was head of the *Peripatetic School* in the Lyceum in Athens from age forty-nine to sixty-two (335-322 BC). He based his model of the Universe on the spheres of Eudoxus. Spheres were added and parameters were adjusted with the result that the Universe now supported fifty-five clearly marked spheres. Beyond its outermost sphere, Aristotle argued that nothing could exist, not even space. At the center sat a firmly fixed Earth.

Aristarchus of Samos, Greece (Third Century BC) was noted for being the first to maintain that the Earth orbits the Sun - the heliocentric solar system that had been postulated by Pythagorean astronomers almost three centuries earlier. A quotation in the *Arenarius* of Archimedes from another work of Aristarchus proves that he anticipated the great discovery of Copernicus. He is said to also have invented a hemispherical sundial. Due to Aristotle's stature and insistence that Earth occupied the center location, not the Sun, the heliocentric system lay dormant until the 1500's when Nicholas Copernicus reintroduced it - a delay of almost eighteen hundred years.

Archimedes (287-212 BC) was born at Syracuse, in Sicily. He was the son of Pheidias, an astronomer. Archimedes also believed that the Sun was the center of the solar system, if not the whole Universe, and that the stars were really "far" away. He also finalized the formulas determining the volumes of the cylinder, cone and spheres, describing their relationship to PI (π). Some of his treatises were: *On the Sphere and Cylinder; Measurements of a Circle; On Conoids and Spheroids; On Spirals; On the Equilibrium of Planes; Quadrature of the Parabola; On Floating Bodies; The Sand Reckoner; The Method Treating of Mechanical Problems; Book of Lemmas.*

The fifty-five sphere model of Aristotle dominated thinking for over five hundred years until someone else with authority appeared on the scene - Claudius Ptolemy (not of the royal Ptolemies). He was born in the Second Century in Ptolemais on the Nile and funding for his astronomical studies came from the Ptolemaic Dynasty of Egypt through the Museum of Alexandria. He was a hard working astronomer charting stars from an observatory at Canopus, a city named for a star, situated fifteen miles east of Alexandria. The title of his principal work was *He Mathematical Syntaxis (The Mathematical Collection).* It was later known as *Ho megas astronomas (The Great Astronomer).* Arabian astronomers named it *Almagest,* Arabic for *"The Greatest."* It is divided into thirteen books. Ptolemy also produced *Geographike Huphegesis (Guide to Geography)* which was divided into seven books. Ptolemy's system of the Sun, planets and stars was different - it had lost nearly all the symmetry of Aristotle's spheres, but it worked, more or less, with Earth still at the center. All the planets out to Saturn were now known. Although no one was really "right" about the structure of the Universe, knowledge was progressing with many of the mysteries being explained scientifically.

Mikolai Kopernik was born in Torun, Poland on February 19, 1473 - the intellectual revolutionary we know today as Nicholas Copernicus. He is the man who brought about Ptolemy's downfall with his publication *De revolutionibus orbium coelestium (On the Revolutions of the Heavenly Spheres),* the book that would set the Earth into motion round the Sun with the other planets out to Saturn in their proper orbits. The proper sequence was now in place for further advancement. He died in Frauenburg, Poland May 24, 1543.

Galileo Galilei was born at Pisa, Italy on February 15, 1564. He was a mathematician, astronomer and physicist who made three significant contributions to the founding of modern scientific thought. (1) As the first man to use the telescope - his *optik* tube - to study the skies, Galileo amassed evidence that proved the Earth revolves around the Sun and is not the center of the Universe. (2) Galileo informally stated the principles later incorporated in Newton"s first two laws. Because of his pioneer work in gravitation and motion and in the combining of mathematical analysis with experimentation, Galileo is often referred to as the "father of modern mechanics and experimental physics" - he was the world's first *physicist.* (3) Perhaps the most far-reaching of Galileo's achievements was his re-establishment of mathematical rationalism as opposed to Aristotle's logico-verbal approach, and his insistence that the *"Book of Nature* is written in mathematical characters." Through this he was able to found the modern experimental method. Galileo sent a letter to Johannes Kepler on April 4, 1597 concerning the Copernican theory. They were cognizant of each other's works. His first decisive astronomical observations were published in 1610 in *Sidereus Nuncius.* In his *Letters on the Sunspots,* printed at Rome in 1613, he took up a more definite position on the Copernican theory. Movement of the spots across the face of the Sun, Galileo maintained, proved Copernicus was right and that Ptolemy was wrong. *Dialogo des Massimi Sistems (Dialogue of the Two Chief World Systems)* came out in 1632 as a literary and philosophical masterpiece. In 1634 he completed his *Discorsi e Dimostrazion Mathematiche intorno due nuove scienze ... (Dialogue on Two New Sciences).* As the first person to apply the telescope to a study of the skies, Galileo, in late 1609 and early 1610, announced a series of astronomical studies. He found the surface of the Moon was irregular and not smooth; he observed that the Milky Way was composed of a collection of distant stars; he discovered the moons of Jupiter. He also observed spots on the Sun, the phases of Venus and the three forms of Saturn (observation of the rings from different attitudes of Saturn - the rings perpendicular to the observer, the rings in line with the observer and the rings at an angle to the observer). Galileo died at Arcetri, Italy on January 8, 1642.

Tycho Brahe (1546-1601), a Danish astronomer, was the greatest naked-eye observer of all time. From his Uraniborg Observatory in Denmark, he conducted observations of the motions of the planets for over twenty years. These observations, due to their great accuracy, exceeded all previous observational accomplishments. He moved to Prague in 1597 and later became associated with Johannes Kepler.

Johannes Kepler was born on December 27, 1571 in Weil, Germany. He was an astronomer, whose epoch in the history of the physical sciences ranks between those of Copernicus and Newton, symbolizing a transition from ancient to modern times. Kepler joined Tycho Brahe in Prague in January 1600. This memorable conjunction of the two leading astronomers of their time - the aging great observer and the brilliant young theoritician - lasted only twenty-two months before Tycho's death on October 24, 1601. Kepler published *Astronomia Nova* in 1609, the two laws of planetary orbits about the Sun. The publication of *Astronomia Nova* and the almost simultaeous invention of the telescope make the year 1609-1610 also a dividing line between ancient and modern times. Kepler produced *De Harmonice Munds*, his third planetary law, in 1619 at Linz, Austria. He died in Regensburg, Germany on November 15, 1630.

Isaac Newton, English physical scientist and mathematician, one of the greatest figures in the history of science, was born at Woolsthorpe, near Grantham in Lincolnshire, on December 25, 1642. Late in 1665 Newton developed what is now called the Binomial Theorem, and soon after this the method of fluxions, an early form of the differential calculus. In May of 1666 he introduced the inverse of fluxions, or the principle of integral calculus. Other observations he made at this time were: An analysis by experiment of the composition of white light and the nature of colors; the discovery of the gravitational force holding the Moon in its orbit. Newton succeeded to the Lucasian Chair of Mathematics at Cambridge at the age of twenty-six. His *Philosophiae Naturalis Principa Mathematica (Mathematical Principles of Natural Philosophy)* was published in the summer of 1687. This explained why the planets orbited the Sun, why ocean tides occurred and dispelled the mysteries of the era as explained through the knowledge of gravity. He also developed the corpuscular theory of light. Newton died on March 20, 1727, and was buried in Westminster Abbey on March 28, 1727.

PHYSICS

The word "physics" is derived from the Greek physis - nature.

This book must start somewhere and this is just as good a place as any. The concept is to start at the beginning and proceed in an orderly manner to the conclusion. That is the intent of this book - it may appear to be a "random walk" at times, but the projected goal is always in sight.

The technology explosion we are now enjoying was born in the pain of World War II. I received my Bachelor of Science in Electrical Engineering from the University of Arkansas in 1952 and then completed my graduate studies in nuclear physics and advanced mathematics. This was during the time physics courses were classified as CLASSICAL and the concept of MODERN physics was being introduced. The CLASSICAL physics were reduced to the basics for MODERN physics today. Physics is a science that requires careful measurements. The early Greek philosophers were called *physikoi* (physicists) because their main interest was in investigating the physical world.

Classical physics is divided into the following: Mechanics; Heat; Sound; Electricity and Magnetism; and Light. Modern physics is divided into: Atomic, Molecular, Electron; Nuclear; Particle; Solid State; and Fluid, Plasma.

We must give credit where credit is due. One poll lists the following as the:

WORLD'S TOP TEN SCIENTISTS

Isaac Newton	1666
Albert Einstein	1905
Galileo Galilei	1609
Charles Darwin	1859
Aristotle	340 BC
Euclid	300 BC
James C. Maxwell	1865
Louis Pasteur	1880
Thomas Edison	1880
Nicholas Copernicus	1514

Entering the technical field at the appropriate time, all I had to do was "keep up" with each advancement as it occurred over the last half of the twentieth century. Geology books had to be rewritten due to the discovery that radioactivity had a half-life and now past dates and ages could be determined. Geological deposits that had been estimated to be the "unbelievable" age of one million years were found to be almost TWO BILLION years old! Astronomy realized it possessed more tools at this time and great strides were begun. The sciences proceeded to further divisions. The following is a more accurate description of the fields, using size as the determinant.

OBJECT	DIMENSION	SCIENCE FIELD
Elementary particle	10^{-13} cm or less	Particle physics
Atomic nucleus	10^{-12} cm	Nuclear physics
Atom	10^{-8} cm	Atomic physics
Molecule	10^{-7} cm	Chemistry
Giant molecule	10^{-5} cm	Biochemistry
Planet Earth	10^{9} cm	Geology
Star	10^{9}-10^{14} cm	Astrophysics
Galaxy	10^{22} cm	Astronomy
Galactic cluster - Universe	10^{25}-10^{28} cm	Cosmology

NOTE: Exponentials are the small numbers always to the upper right of the BIG numbers. This identifies the number of places to move the decimal point. The minus (-) sign means to move the decimal point to the left. No sign means positive (+) and the decimal point is moved to the right as required by the exponential. 10^{-8} cm is the short version of 1.0×10^{-8} cm. 10^{9} cm is the short version of 1.0×10^{9} cm.

10^{-8} cm = 0.00000001 centimeter

10^{9} cm = 1,000,000,000 centimeters

Cosmology, which deals with the evolution of the cosmos, is best defined as proficiency in it and many of the sciences listed above it. I try to show the reader how each is dependent on one or several of the others - it is so intertwined, but similar. It has been a great half-century of learning. Many discoveries and theories occurred with several theories being rejected or replaced as new results were confirmed.

One of the great conflict of theories that had been argued for years was resolved scientifically in 1964. Until this time, two theories were in competition in explaining the expanding Universe Edwin Hubble had discovered. The "big bang" theory was first proposed by a Belgian priest and mathematician named George Lemaitre. In 1927, he wrote a paper that developed a mathematical superstructure connecting the observed redshifts, Edwin Hubble's observations of galactic recession from each other, with the expanding Universe of general relativity developed by Albert Einstein. Later, the astrophysicist Fred Hoyle designated this creation by an intentionally ugly name that stuck - the BIG BANG. He then began work on another theory that would hopefully be correct - the "steady state."

In 1948, Thomas Gold, Herman Bondi and Fred Hoyle proposed that the Universe was infinitely old and generally unchanging - there had been no creation event, no high-density infancy from which the Universe had evolved. To explain why the galaxies are not infinitely far apart in an infinitely old, expanding Universe, the theory proposed that hydrogen atoms materialized spontaneously, out of empty space, and proceeded to condense into new stars and galaxies. These new galaxies filled in the space between existing galaxies to maintain the "steady state."

This dispute was permanently resolved in 1964 in favor of the "big bang" theory when Arno Penzias and Robert Wilson, scientists for Bell Labs, discovered that microwave reception from *any* direction at *all* times from space was constant. They later received the Nobel Prize for their discovery. The "steady state" theory was eliminated. Further verification was accomplished in 1992 by astrophysicist George Smoot at the Lawrence Livermore Laboratory in California. Data from the Cosmic Background Explorer (COBE) satellite measuring microwave radiation in space discovered wrinkles, or ripples, that occurred some three hundred thousand years after the "big bang." Another verification predicted by relativity.

This conflict and others were fascinating to follow, study and observe their resolutions. All did not resolve in the manner anticipated, but they were resolved. A scientist will answer a question in ONE of three ways: YES, NO or I DON'T KNOW. This is the reason for the rapid advances witnessed. If a theory is proven wrong, improved or replaced, so be it - the facts support the decision. These advances have resulted in two observations: The more we learned, the simpler and more logical things became; and, the more we learned, previously unknown avenues were made available for exploration.

Another observation on physicists: Theoretical physicists use blackboards, computers and messy desks to develop theories. The experimental physicists appear with sledgehammers, dynamite, bulldozers, big laboratories, little laboratories - anything needed to tear objects apart to verify their theories. These are the people with the cyclotrons and super colliders which smash atoms to pieces to research and advance their theories.

The seeds of this book may have been sown near the end of my sophomore year in the spring of 1950 when I was asked by the Physics Department to prepare a presentation on the workings of the atomic bomb - the fission process. This was to be presented to civic groups, educators, students and others of the general public. The presentation was to contain no formulas or highly technical descriptions - reduce the data and information into everyday understandable language for the layman. The Cold War was underway and every audience welcomed this understanding of these "mysteries."

Just a few years later it became apparent that the pace of future developments would expand at a fantastic rate. I became involved with the cutting edge of technology in the new developments in radar, communications, electronic countermeasures and other highly sophisticated areas as an Airborne Combat Information Center (CIC) Officer in the U. S. Navy operating Airborne Warning and Control (AWACS) planes in the heat of this Cold War. Our technology was continuously expanding and the efficiency of presenting this information increased the more we reduced these technical terms into easier understanding models.

The transistor had been developed in 1948 and was introduced in 1952, dooming the era of vacuum tubes. The age of integrated circuits had been born and micro-miniaturization was on the horizon. This was the "electronic explosion" that reshaped the world. Quantum mechanics could now be fully understood and used with unlimited advances in science now possible. The limiting factor of these advancements was that new scientific discoveries outpaced the ability of manufacturing to produce profitable products before new and improved products were discovered - newly manufactured products were already obsolete. Investment requirements in new plants and equipment were much higher due to the complicated processes and tolerances that were now required - a new era in manufacturing was being created. One of my adjustments was having my sliderule replaced by scientific calculators and computers; oh well - -

SCIENTISTS OF NOTE

Ralph Alpher - proved the big bang mathematically
Gerd Binnig - superconductivity, STM, "feel" atoms
Hans Bethe - how a star functions, energy production in a star
Niels Bohr - structure of the atom, quantum mechanics
Tycho Brahe - astronomer, planetary orbits
Marie Curie - radioactivity
Louis DeBroglie - quantum mechanics
Hans Dehmelt - isolated single atomic particles
Paul Dirac - mechanics of the atom
Arthur Eddington - nuclear physics energy - emf
Enrico Fermi - fission, nuclear physics, particle physics
Joseph Fraunhofer - glass, lens
Richard Feynman - electromagnetism as a quantum theory
George Gamow - nuclear physics, big bang, expanding Universe
Sheldon Glashow - unification of electromagnetic and weak force
Alan Gurth - big bang, inflation hypothesis
Edmund Halley - cometary orbits and periods
Stephen Hawking - black holes
Werner Heisenberg - uncertainty principle
William Herschel - discovered Uranus in 1781
Robert Hooke - study of solid materials
Edwin Hubble - astronomer, expanding Universe
James Hutton - physician, geologist - found Earth is alive
Werner Israel - relativity
Gustav Kirchoff - voltage, current calculations
Lord Kelvin - see William Thompson
Gottfried Wilhelm Leibniz - mathematics, calculus
Sir Joseph Norman Lockyer - discovered helium
Murray Gell-Mann - quarks
Robert Millikan - light properties, charge of the electron
Wolfgang Pauli - exclusion principle
Roger Penrose - mathematics, black hole
Arno Penzias - Bell Labs, microwaves from space, verified expanding Universe
Max Planck - quantum mechanics, quantum theory
Heinrich Rohrer - superconductivity, STM, "feel" atoms
Ernest Rutherford - radioactivity, mass of the nucleus

Carlo Rubbia - European Center for Nuclear Research (CERN)
Meghnad Saha - spectrum of the Sun
Abdus Salam - unification of electromagnetic and weak forces
J. J. Thompson - electron properties, detected the electron
William Thompson (Lord Kelvin) - thermodynamics
Simon van de Meer - CERN, antiproton
Stephen Weinberg - unification of electromagnetic and weak forces
Robert Wilson - Bell Labs, microwaves from space, verifying expanding Universe
Chen Ning Yang - nonconservation of parity
Peter Zeeman - splitting energy levels and spectra in a magnetic field

All the Nobel recipients and candidates in the sciences and related fields are also included. The greatest were the original thinkers - the observers and experimenters who discovered the impossible - Heinrich Hertz discovered that waves were created by a spark of electricity that was later developed into radio, television and satellite communications. Others could develop the idea into wonderful results, but without the original discovery, the results would have been impossible.

AND a group of Electrical Engineers that placed many of these original discoveries into practical applications:

George Westinghouse (1846-1914)
Nikola Tesla (1856-1943)
Charles Steinmetz (1865-1923)

A scientist can discover a new star, but he cannot make one. He would have to ask an engineer to do that.

Gordon L Gregg

MATHEMATICS

Equations are more important to me, because politics is for the present, but an equation is something for eternity.

Albert Einstein

The fear of mathematics experienced by most people is responsible for their lack of interest in the physical sciences. It is a barrier that effectively cuts them off from a full appreciation of scientific discoveries and prevents them from enjoying vast areas of nature that have been revealed through painstaking research. For, as Roger Bacon appreciated: "Mathematics is the door and key to the sciences . . . For the things of this world cannot be made known without the knowledge of mathematics."

The Sumerians, and the Babylonians who succeeded them, were the first to make significant advances in mathematics and astronomy. By 1800 BC they had developed a number system based on 60. It can be evenly divided by 2, 3, 4, 5, 6, 10, 12, 15 and 30. Ancients had trouble with fractions. Building on the work of the Sumerians, Babylonians made advances in arithmetic, geometry and algebra. They compiled tables for multiplication and division and for square and cube roots to simplify computations with both whole numbers and fractions. They knew how to solve linear and quadratic equations. Their knowledge of geometry included the theorem later formulated by Pythagoras - the square of the hypotenuse of a right-angle triangle is equal to the sum of the squares of the other two sides. The Egyptians were less skilled in mathematics than were the Mesopotamians. Their arithmetic was limited to addition and subtraction, which also served them when they needed to multiply and divide. They only understood simple algebra but had considerable knowledge of practical geometry. The zero (0) first appeared in the numbering sequence in 683 in Kampuchea and Sumatra in Southeast Asia. All previous numbering systems had begun at one (1).

The Arabs added to algebra by contributing quadratic and biquadratic equations in 820 and to trigonometry by inventing the sine, cosine and tangent. They were also involved in mathematical geography, astronomical observations, latitude and longitude, optics, chemistry and pharmacy.

Many years ago Richard Courant wrote: "The river of mathematics, if separated from physics, might break up into many separate little rivulets and finally dry up altogether. What has happened is rather different. It is as if the various streams of mathematics have overflowed their banks, run together, and flooded a vast plain, so that we see countless currents, separating and merging, some of them

23

quite shallow and aimless. Those channels that are still deep and swift-flowing are easy to lose in the general chaos."

Mathematics today is in a constant flux of development. New requirements are needed - space flight, orbitals, satellite intercepts of the planets and comets, computer miniaturization and many others. Likewise, many requirements are not needed - the aerodynamics of Zeppelins. Due to the requirements of engineering, physics, astronomy, satellite and space flights, all the forms of mathematics are now programmed for computers, solving equations much faster - ~~billions~~ trillions of times faster.

Mathematic eras:

Egyptian, Sumerian	3000-1600 BC
Babylonian	1700- 300 BC
Greek	600- 200 BC
Greco-Roman	150- 525
Arabic	750-1450
Western	1100-1600
Modern	1600-Present

A few types of mathematics being used today:

Arithmetic	Descriptive Geometry
Algebra	Vector Analysis
Geometry	Polar Coordinates
Trigonometry	Rectangular Coordinates
Analytical Geometry	Statistical Mechanics
Differential Calculus	Celestial Mechanics
Integral Calculus	Quantum Mechanics
Partial Differential Equations	Quantum Electrodynamics
Statics	Quantum Chromodynamics
Dynamics	Molecular Dynamics

HISTORY OF MATHEMATICS

2200 BC	Mathematical tables at Nippur
1650 BC	Rhind papyrus - numbering problems
600 BC	Thales - beginning of deductive geometry
540 BC	Pythagoras - arithmetic, geometry
380 BC	Plato
340 BC	Aristotle
300 BC	Euclid - systematization of deductive geometry
225 BC	Appollonius - conic sections
	Archimedes - circle and sphere, area of parabolic segment, infinite series, mechanics, hydrostatics
150	Ptolemy - trigonometry, planetary motion
250	Diophantus - theory of numbers
300	Pappus - collections and commentaries, cross ratio
820	al Khowarrizmi - algebra
1100	Omar Khayyam - cubic equations, calendric problems
1150	Bhashara - algebra
1202	Fibonacci - arithmetic, algebra, geometry
1545	Tartaglia, Cardano, Ferrari - algebraic equations of a higher degree
1580	Viete - theory of equations
1600	Harriot - algebraic symbolisms
1610	Kepler - polyhedra, planetary motion
1614	Napier - logarithms
1635	Fermat - number theory, maxima and minima
1637	Descartes - analytic geometry, theory of equations
1650	Pascal - conics, probability theory
1680	Newton - calculus, theory of equations, gravity, planetary motion, infinite series, hydrostatics, dynamics
1682	Leibniz - calculus
1700	Bernoulli - calculus, probability
1750	Euler - calculus, complex variables, applied mathematics
1780	Lagrange - differential equations, calculus of variations
1805	Laplace - differential equations, planetary theory, probability
1820	Gauss - number theory, differential geometry, algebra, astronomy
1825	Bolyai, Lobatchevksy - non-Euclidean geometry
1854	Riemann - integration theory, complex variables, geometry
1880	Cantor - theory of infinite sets

1890	Weierstrass - real and complex analysis	
1895	Poincare - topology, differential equations	
1899	Hilbert - integral equations, foundations of mathematics	
1907	Brouwer - topology constructivism	
1910	Russell, Whitehead - mathematical logic	

CHINESE MATHEMATICS

INVENTION	YEAR	WESTERN DEVELOPMENT
Decimal system	1400 BC	2300 years later
A place for zero (0)	400 BC	1100 " "
Negative numbers	100 BC	600 " "
Extraction of higher roots and solutions of higher numerical equations	100 BC	600 " "
Decimal fractions	100 BC	1600 " "
Using algebra in geometry	300	1000 " "
A refined value of PI (π)	300	1200 " "
"Pascals" triangle of binomial coefficients	1100	427 " "

ROMAN NUMERALS

I	1
V	5
X	10
L	50
C	100
D	500
M	1,000

BASICS

MANKIND DEVELOPED FOUR ITEMS FOR

HIS DAILY USE:

THREE OF THESE ARE:

TIME

MEASUREMENT

WRITTEN LANGUAGE

INTRODUCTION

Before we can begin this journey, we must realize that when mankind arrived upon Earth's landscape, there was very little difference between humans and the surrounding animal world. Communication between these early people undoubtedly began with hand signals and certain grunts. As speech developed in each area, the communication was greatly simplified as long as those in communication with each other could be in the range of their voices or signals. Later, marks on trees or stacks of rocks designating trails for these people to follow (the first means to convey a message without using the voice) eventually led to the first signs in mud that later hardened, preserving the message. This may have been the beginning of writing - spelling would come later after the development of the languages and the alphabets.

As communications improved, it was realized that not enough information was being conveyed. Meetings were impossible to arrange because there was no WHEN or WHERE. Telling someone that they would meet in two sunrises was no good - where were they to meet? This led to adding "... on the second hill toward the sunrise by the big tree." Now a position was established - time and distance (two sunrises and big tree).

The reason these subjects are placed at the beginning of this book is because TIME, MEASUREMENT and the WRITTEN LANGUAGE are required to understand facts. This is merely a history of each.

TIME

When did time start? What would today's date be if we had no Moon?
What would today's date be if we had two Moons?
Time is nothing more than distance divided by velocity. $D = RT$, $T = D/R$

ARROWS OF TIME

There are at least three different arrows of time - something that distinguishes the past from the future, giving *direction* to time. First, there is the thermodynamic arrow of time, the *direction* of time in which disorder or entropy increases (one can go readily from the cup on the table in the past to the broken cup on the floor in the future, but not the other way around).

Then there is the psychological order of time. This is the *direction* in which we feel as time passes, the *direction* in which we remember the past but not the future.

Finally there is the cosmological order of time. This is the *direction* of time in which the Universe is expanding rather than contracting. The laws of science do not distinguish between the forward and backward *direction* of time.

Stephen Hawking
A Brief History of Time

Time as an activity appears to be involved in each experience we encounter. Let us examine the flow of time. If time is flowing, how fast does it move? Does time pass? What does it pass? There is no instrument which can record the flow of time, or measure its rate of passage. A clock measures *intervals* of time, not the speed of time. A clock would have the same input as a ruler when comparing a ruler to a speedometer. There is no speedometer for time. Time doesn't flow at all; it merely *is*. As a clock ticks, the moment passes and another comes into existence, a process that we *call* the flow of time. A world that has no conscious observers would not have a river of time - the flow would not exist. The river of time appears to flow only when conscious observers are present, apparently begun by the Sumerians several thousand years ago.

When we state that something is located to the *left* of this or to the *right* of that, we are defining *directions*, not places. When we say *past* and *future* we are defining *directions*, not moments. There is apparently no *present* because this instance of *now* is already in the *past*.

Earth bases its clocks on solar time. This is the time it takes Earth to rotate and have the Sun located overhead at noon each day - the twenty-four hour *solar day*. The solar day (noon to noon) on the Moon would be around twenty-nine Earth days; Mercury's solar day would be one hundred seventy-six Earth-days (two of Mercury's years); Venus' solar day would be one hundred sixteen Earth-days (a little over half of one Venus year).

If you stand in one spot and look at a star at midnight, it will appear in the same position at an earlier time the next night. This is due to Earth's movement around the Sun during its orbit. This amount is 0.985 degrees each day (360 degrees per year divided by 365.25 days per year). When you see this star the next night, it will appear 3 minutes and 56 seconds earlier (0.985 degrees per day times 4 minutes per degree equals 3.94 minutes per day shorter [1440 minutes per day divided by 360 degrees per day equals 4 minutes per degree]). This is the *sidereal day* which is only 23 hours 56 minutes 4 seconds (24 hours minus 3 minutes 56 seconds) due to the movement of Earth along its orbit of the Sun. The *solar day* is based on the Sun being overhead. The *sidereal day* is based on a star being overhead. The *sidereal year* is based on Earth's complete cycle of the constellations.

Tropical time, the time we use for our calendar, is based on the Sun's movement relative to the equator, the Vernal Equinox. Atomic time is based on quantum changes in the state of atoms. Different types of time are required for various scientific and technical purposes. Celestial navigation and geodesy require mean solar time, and the study of planetary and satellite motions require sidereal time.

To indicate time and to *measure time intervals*, various types of clocks have been developed. Galileo discovered, around 1583, that the time of the swing of a pendulum was nearly independent of the amplitude of the swing. Christian Huygens contributed to the theory of the pendulum and had the first pendulum clock constructed in 1656. Numerous improvements were subsequently incorporated and pendulum clocks reached a high state of precision from 1900 to 1925. The rates of the best pendulum clocks are constant to about one thousandth of a second. The quartz crystal clocks were the next step in providing a more accurate precision due to the constant frequency - several million cycles per second. Quartz crystals change frequency with age.

For centuries, Earth was our timekeeper. The Sun rose and set, and the day was broken into hours, minutes and seconds based on Earth's rotation. This is the reason astronomical observatories, such as the one in Greenwich, England, kept official time. Until the twentieth century, pendulum clocks were calibrated against the rotation of Earth by taking astronomical measurements. But as clocks grew more precise, they exposed the idiosyncracies of our planet. It wobbles, it oscillates, and it undergoes slight shifts in shape, all of which affect its rotation. As a standard of accuracy, the pendulum gave way in the 1940's to electrically induced vibrations in quartz crystals, which in turn gave way in the 1950's to measurements of atomic activity. And, in effect, Earth gave way to the atom as a gauge of time. Instead of using a definition of the second based on Earth's rotation, scientists began to search for one based on frequencies generated by certain atoms - particularly cesium - as they changed from one atomic state to another. Atomic frequencies, unlike the frequency of a pendulum's swing, have the virtue of being the same anywhere in the Universe. In 1967, the international definition of the second shifted to an atomic standard. The Bureau of International desPois et Mesures (BIM) near Paris now defines a second as "the duration of 9,192,631,770 periods of the radiation corresponding to transition between the two hyperfine levels of the ground state of the cesium 133 atom." It is accurate to 0.0000000000000015 (1.5×10^{-15}). If it were to run for twenty million years, it would neither lose nor gain a second.

When we say *time*, we are using the shortened version of what is known as "Earth-time." Since this is the only measure discovered and used by mankind to denote when an event occurred or will occur, it is constant on Earth whether we are on the surface, in the air or under the sea. It is a concept unique to Earth.

Said ibn Yuf - Saadia Gaon (882-942) stated: "If the world were uncreated, then time would be infinite. Infinite time cannot be transversed. Hence, the present moment couldn't have come to be. But the present moment clearly exists. Hence, the world had a beginning." He was born in the Faiyum district of Egypt and moved to Babylonia in 922.

From the practical point of view, time must have started when a caveman suddenly realized that sunrise and sunset were repetitive. The first statement on "time" may have been to a fellow caveman: "I'll meet you at the big rock when the 'big light' is born again - bring your club!" At this momentous meeting a religion may have also been invented at this "time" by the "big rock" which then became "holy" because it must have witnessed this miracle since the beginning of "time."

Sumerians established the second as a unit of time with sixty seconds in a minute. Sixty minutes were allocated to an hour, 60 x 60 = 3600 seconds to each hour - this is 1/60 of 1/60 of 1/24 of a day measured from noon to noon. They divided a circle into 360 equal units - 60 x 6 = 360.

One of the predictions of general relativity is that time should appear to run slower near a massive body like the Earth. This is because there is a relation between the energy of light and its frequency - the greater the energy, the higher the frequency. As light travels upward from the Earth, the gravitational field of Earth causes a loss of energy, which results in a reduction of frequency. To someone high up it would appear that everything down below was taking longer to happen. This prediction was tested in 1962 using a pair of very accurate clocks mounted at the top and bottom of a water tower. The clock at the bottom, nearer the Earth, ran slower in exact agreement with general relativity. The difference in the speed of clocks at different heights above the Earth is very important with the advent of the very accurate navigation systems based on signals from satellites. If the predictions of general relativity were ignored, the position calculated would be wrong by several miles for the Ground Positioning System (GPS).

According to the theory of relativity there is no absolute time, no universal time. Each observer has his own measure of time. The time for someone on one star will be different from someone on a distant star because of the different gravitational fields of the stars - stars are not the same size. The stronger the gravity, the more pronounced the timewarp. If gravity were to reach a critical value, time would appear to grind to a halt to an outside observer. This could happen to one observing a black hole.

The day is to some extent a natural division of time. Its subdivision into a number of equal intervals of twenty-four hours is a late development and is purely artificial. In most primitive societies the day was recognized only as an alternation between light and darkness. The artificial division of the day into hours became necessary for commercial reasons as trade became important. When the concept of hours was first used, these intervals were unequal in duration. Day and night were divided into equal periods causing the periods in the northern hemisphere to be longer during the day and shorter during the night in the summer months. The reverse would be true during the winter months. The invention of the sundial helped alleviate this troublesome problem during the day. Aristarchus and the Chaldean astronomer Berossus each produced a sundial during the third century BC. The arrival of the mechanical clock clarified these problems. High noon became the observer's meridian; AM (ante meridian) was before meridian of high noon; PM (post meridian) was after the meridian of high noon. Time to all other species, if noted, would be through the change of seasons or the aging process.

The week is an entirely artificial division of time and cannot be correlated with any astronomical or natural phenomena except that it is a closed cycle of days. This period first appeared as the time between trading days for primitive peoples. For some African tribes this period was four days; the ancient Assyrians used a six day interval; the ancient Romans had an eight day period and the Incas had a ten day period. The present seven day week may have been derived from the use of the Assyrian division with one day added for religious purposes.

The Old Kingdom of Egypt (2613-2181 BC) which ushered in the Pyramid Age, produced the world's first known solar calendar, the direct ancestor of our own. The very early Egyptian calendar was on the lunar model as evidenced by the crescent-moon hieroglyph for "month." In order to plan their farming operations in accordance with the annual flooding of the Nile, the Egyptians kept records and discovered that the average period between flooding was 365 days. The solar year had twelve months of thirty days each and at the end of the last month five additional days were added - coinciding with the annual flooding of the Nile. This allowed the discrepancy of being short one-quarter of a day each year to accumulate into a noticeable error. For a long time the priests of Egypt prevented any change in the calendar to adjust to a widening gap created by a quarter of a day added each year. In 238 BC King Ptolemy III decreed the addition of one day every four years to correct the error. They also noted that the Nile flood coincided with the annual appearance of the Dog Star - Sirius - on the eastern horizon at dawn and they soon associated the two phenomena.

The original Latin calendar (about eighth century BC) contained ten months of 304 days - five months having thirty-one days each, four months having thirty days each and one month having twenty-nine days. The year began on March 1 giving the months names - October for the eighth month as an example - with the day beginning at the midnight hour. The Roman calendar experienced several changes. King Numa Pompelius added two months, Januarius and Febuarius prior to 700 BC. King Numa's calendar had seven months of twenty-nine days, four months of thirty-one days and Febuarius with twenty-eight days for a total of 355 days. About 451 BC the months were rearranged to the present order with the year beginning in Januarius.

By 46 BC, when Julius Caesar was made Pontifex Maximus, calendar dates had come to be out of step with natural events. After many modifications and corrections, Caesar caused the year to consist of 365 days with an additional day added every four years - what King Ptolemy III had decreed two hundred years earlier. Julius Caesar had now created the Julian calendar.

Monk Dionysius Exiguus introduced a Catholic system of dating events beginning the year he calculated Christ was born using BC (Before Christ) and AD (anno Domini). This year became the year 532 AD on the Julian calendar. Non-Christians use BCE (Before Common Era) and CE (Common Era). The geological calendars use BP (before present). This book will use the term BC denoting BEFORE CALENDAR. Those dates with no notation will be considered to be within the present calendar system.

The Julian calendar, being 365 days 6 hours long, exceeded the solar year by eleven minutes and fourteen seconds. This caused dates of physical phenomena to occur earlier and earlier. The season dates assumed a growing discrepancy, which became annoying. On February 24, 1582 Pope Gregory III caused ten days to be dropped from the Julian calendar creating the Gregorian calendar - the one now in use. Also, only the century years divisible by 400 were to be leap years. This was called the Catholic calendar and was adopted by France and other Catholic countries. Protestant countries were slow to adopt this version. Great Britain and the American Colonies adopted it in 1752. Russia switched from the Julian calendar to the Gregorian calendar in November 1917, officially changed on January 31, 1918.

The Gregorian calendar is not absolutely correct, the calendar being twenty-six seconds longer than the tropical year. This difference will not amount to one day until 3,323 years have elapsed - the year 4905 (year 1582 + 3323 = 4905). This calendar has perhaps a dozen defects of differing degrees of seriousness. In general, the chief weakness is in the uneven division into half years, quarter years, months and weeks. The quarters now contain ninety, ninety-one or ninety-two days.

It is apparent that each civilization had its own method of recording days, months and years. Radioactive dating can determine past dates with predicted accuracy today.

AZTECS - They knew astronomy and had calculated the precise length of the 365.25 day year. Glyphs (pictures) were used for calendar years, with dots from one to thirteen representing actual years. Every fifty-two years a "century" was turned. The first year in the new century was always designated Two Reed. Two House, marked by two dots, is the founding year of Tenochtitlan in 1325. King Nezahualpilli of Texoco was described by Friar Torquemada as being a great astronomer.

CHINA - Calendar with months based on the phases of the Moon, and years on the position of the Sun. It begins at 2637 BC and counts years in cycles of sixty.

Neolithic Period	5000 BC
Shang Dynasty	1700-1050 BC
Western Zhon	1050- 221 BC
Qin Dynasty	221 BC- 220

EGYPTIAN - Roots of Egyptian civilization trail back to around 9000 BC. The First Dynasty arose around 3000 BC or earlier. Time was measured in Dynastic Periods. Kingdoms, or Periods, were used to apply dates to events from 2920 BC until 332 BC.

EUROPE - Referred to time using Ages.

Ice Age/Old Stone Age	2,000,000 years ago-12,000 BC
Middle and New Stone Age	13,000-2000 BC
Bronze Age	2000- 800 BC
Iron Age	800 BC- 0

GREEKS - Referred to time using Periods.

Beginning of Mycanean Civilization	1600 BC
Early Period	1050-750 BC
Archaic Period	750-500 BC
Classical Period	500-323 BC
Hellenistic Period	323- 31 BC

Alexander the Great (356-323 BC) - his body lay in state until the fourth century, almost six hundred years. He died June 10, 323 BC.

ISRAELITES -

Canaanite Cultures	3000-1200 BC
Period of Judges/United Monarchy	1200- 920 BC
Divided Monarchy	920- 586 BC
Exile and Return	586- 323 BC
Greek and Early Roman	326- 4 BC
Early Roman	4- 135

The Jewish calendar dates from 3761 - the Jewish date of creation. Hillel II established this in the fourth century on a nineteen year cycle adjusted to the solar cycle.

MAYA - Interaction of two calendars using glyphs (pictures). Date One Imix Four Vayeb. Solar year of 365 days with eighteen to twenty day months with five extra days, a period called Uayeb. It was a fifty-two year period for the two calendars to overlap. Maya writings often combine ideographs with phonetic syllables to enhance the significance of a word.

Preclassic Period	1500- 250 BC
Early Classic Period	250BC- 600
Late Classic Period	600- 900
Post Classic Period	900-1500

By 1000 BC Maya astronomers had determined that the year was not 365 days long, but 365.24. Not until 1582, almost twenty-six hundred years later, did European astronomers come to this determination. The Maya had also determined that Venus required 583.92 days to rotate almost exactly five times between one closest approach to Earth and the next. Also during this early period the Maya had devised a multi-part numbering system which enabled them to calculate with great accuracy the exact dates going back ten thousand years or more and equally distant into the future.

The Maya had the ability to forecast eclipses of the Sun. Even then they could forecast that a total eclipse of the Sun would occur on Sunday, March 29, 1987. Their meticulous observations had for centuries been recorded in tables inscribed on papyrus-like sheets and jealously guarded by the priests who perfected them by making minute adjustments. These calculations were complete into the twenty-second century (we are now only in the twenty-first century). Their system allowed them to look either forward or backward for thousands of years to verify or predict.

The priests kept this information within their circle to maintain power over rulers that did not please them. They would threaten to blot out the Sun if such and such was not done as they requested. The rulers usually refused until the eclipse began. The priests would then raise their demands before restoring the Sun. The rulers usually graciously relented and the Sun was allowed to reappear. It is no wonder that the priesthood became elated when an eclipse was due in their area.

On July 12, 1562 Diege de Landa, Fourth Catholic Bishop of Yucatan gathered all known copies of the scrolls containing the history and astronomical observations of the Maya civilization and burned them in a large bonfire. Only three in all Mayaland survived and it is from them that the history of this great civilization was learned. The burnings were to protect Catholicism from Maya "heresy" or any other religious influence the natives may want to use or reintroduce. The Mayan history was intentionally destroyed - all paganism and its history were to be eradicated by the ruling Catholic religion.

MUSLIMS - In 622 the Muslim (Islamic) calendar was formed on the lunar cycle. Each month begins with a new moon.

NATIVE AMERICANS - Anasazi astronomers constructed a unique Sun calendar. They studied the Moon, stars and planets to discern universal patterns of movement and incorporated the knowledge into religious rites and mythology. Their device marked solar time for some seven hundred years until the late 1980's.

OLMEC - Civilization founded in Mexico around 1200 BC and existed until around 400 BC. They were the first people of North America to build religious centers with pyramidal structures.

PERSIANS -

Pre-Achaemenid Period	4000-500 BC
Achaemenid Period	550-300 BC
Seleucid and Parthian Period	300-224 BC
Sassanian Period	224 BC-642

SUMERIANS -

Ubaid Period	5900-4000 BC
Urek Period	4000-3000 BC
Early Dynastic Period	3000-2350 BC
Akkadian Period	2350-2150 BC
Neo-Sumerian Period	2150-2000 BC

The year 2000 in the Western World, using the Gregorian calendar, is:

YEAR		
	5760	Hebrew
	5119	Mayan
	4697	Chinese
	2543	Buddhist
	1993	Ethiopian
	1921	Hindu
	1420	Muslim

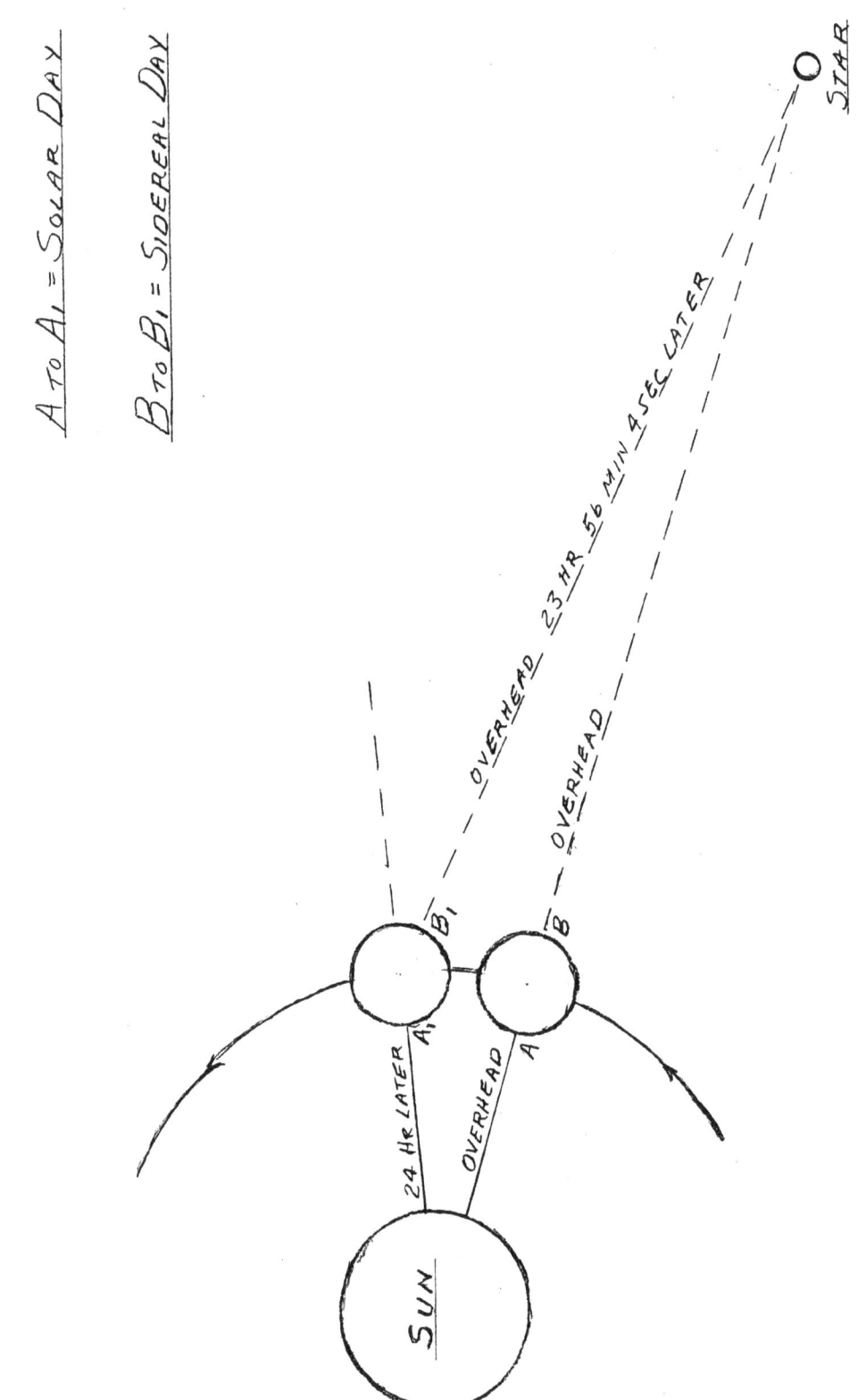

A to A₁ = Solar Day

B to B₁ = Sidereal Day

STAR

OVERHEAD   23 HR 56 MIN 4 SEC LATER

OVERHEAD

B₁

B

24 HR LATER

OVERHEAD

A₁

A

SUN

38

MEASUREMENT

Pounds, miles and Fahrenheit, or kilograms, kilometers and Celsius?

The development of weights and measures has been taking place since the time of their prehistoric origin. It seems likely that units of measure first used in prehistory were those of length and that they derived from parts of the human body; the length of the foot, the width of the palm and the length of the forearm.

In early historic times relatively precise linear units were needed to make astronomic and geodetic measurements. One of the earliest linear units was the foot, first the length of any human foot and later the length of a specific foot. By gradual evolution it became the foot as used by Egyptians and Greeks. The Romans brought it to Europe and Britain where it was modified with time. The legal definition of the FOOT was described in a book published in Germany in 1522:

> "Stand at the door of a church on a Sunday morning and
> bid sixteen men to stop, tall ones and short ones, as they happen
> to pass out when the service is finished; then make them put their
> left feet one behind the other, and the length thus obtained shall
> be a right and lawful *rood* to measure and survey the lane with,
> and the sixteenth part of it shall be a right and lawful FOOT."

(*Rood* - a unit of length varying locally from five and one-half to eight yards.)

The inch was originally a thumb's breadth. In the Roman duodecimal system it was defined as one-twelfth of a foot and was also introduced to Britain during the Roman occupation. By tradition, Henry I is supposed to have decreed that the yard should be the distance from the tip of his nose to the end of his thumb. The mile of 5,280 feet was established probably in the reign of Henry VII and was made a statute by Elizabeth I.

The metric system is the international decimal system of weights and measures based on the meter and the kilogram. The meter, the unit of length, was to be the one ten-millionth part of the meridianial quadrant of the Earth. Geodetic measurements for this purpose were made on a arc from Dunkerque, France to Mont-Jouy, near Barcelona, Spain. The metric system was established in France in the 1790's - length = 3.2811 feet (39.37 inches). The original distance was calculated to be one ten-millionth (10^{-7}) of the distance from the North Pole to the Equator. Later accuracy in measuring this distance from the pole showed an error of less than one-fifth of one percent. The actual measurement was 18,500 meters too long. Subtracting this error, the measurement would have given the length to equal 3.275 feet (39.30 inches). A kilometer is equal to one thousand meters.

TIME can be measured more accurately than LENGTH. The meter is defined as the distance light travels in 0.000000003335640951982 seconds (3.335640951982 x 10^{-9} seconds) - a little over three billionths of a second - as measured by a cesium clock. This is one 299,792,458th part of a second. Light travels at 299,792,458 meters per second.

The gram, the unit of mass, was to be equal to the mass of a cubic centimeter (1 centimeter equals 0.01 meter) of pure water at the temperature of its maximum density - four degrees Celsius. A kilogram is equal to one thousand grams.

Galileo formulated the first formula for both science and the world and it concerned time and distance. In his gravity experiments he determined that the time for an object to fall a given distance was proportional to the square root of that distance; $d = 16t^2$.

d = distance = $16t^2$ feet $\qquad\qquad d = 16t^2$

v = velocity = $32t$ feet/second $\qquad d' = 32t = v = dx/dt$

a = acceleration = 32 feet/second/second
$\qquad\qquad\qquad$ = 32 feet/second^2 $\qquad v' = 32 = a = dv/dt = (d/dt)(dx/dt)$
$\qquad\qquad\qquad\qquad\qquad\qquad\qquad\qquad = d^2x/dt^2$

WRITTEN LANGUAGE

The Greeks called Egyptian papyrus rolls BIBLIA (books) because Byblos was the shipping point for this widely used writing material.

Writing, in the widest sense, is a system of human intercommunication by means of visible conventional markings. Writing is expressed not by objects themselves but by markings on objects. The Sumerians took the important step leading to fully developed writing. The development of writing is a prerequisite to civilization.

The Sumerians were the first civilized inhabitants of ancient Babylonia (modern Iraq). Perhaps as early as 5000 BC the Sumerians came from the east or descended from the mountains of Elam to the swampy plain (ancient Sumer) at the head of the Persian Gulf. They drained the swamps, instituted flood control and established agriculture on a permanent basis. With the development of trade with surrounding areas - Iran, Elam, Assyria, India and the Mediterranean coast - the Sumerian settlements grew into prosperous city-states, which by 3500 BC possessed a mature civilization characterized by urban life, metal working, textile manufacture, monumental architecture and an efficient system of writing (cuneiform). Due to the prevalent use of clay tablets as writing material, the linear strokes of writing acquired a wedge-shaped appearance by being pressed into the soft clay with the slanted edge of the stylus.

It is probable that the example of Sumerian writing stimulated the Egyptians to develop a script of their own. Sumerian clay tablets, the world's first writing, were pictures of concrete things such as a person, a sheep, a star or a measure of grain. These pictures, pictographs, later were replaced by phonetic (or syllabic) writing when the scribes realized that a sign could represent a sound as well as an object or idea. By 2800 BC, the use of syllabic writing had reduced the number of signs from nearly two thousand to six hundred. It was not until early in the Old Kingdom of Egypt (2500 BC) that the Egyptian hieroglyphs which represented objects, ideas and syllables took the further step of using the alphabetic characters for twenty-four consonant sound. This influenced their Semitic neighbors in Syria to produce an alphabet that, in its Phoenician form, became the forerunner of our own.

There are seven ancient Oriental systems of writing, of which the oldest is Sumerian, first used in southern Mesopotamia around 3100 BC. From there the main principles of Sumerian writing may have spread eastward, first to the neighboring Proto-Elamites and then to the Proto-Indians in the valley of the Indus.

One of the Near East writings may, in turn, have been the stimulus leading to the creation of Chinese writing. Around 3000 BC the Sumerian influence presumably worked its way westward to Egypt. Egyptian influence, in turn, spread toward the Aegean where about 2000 BC, Cretan writing originated, and a few centuries later Hittite hieroglyphic writing in Anatolia (present day Turkey).

The Phoenician consonantal script provided the new topological pattern on which the Ugaritic and Old Persian systems were constructed, keeping only the outer likeness of the wedge form. The Phoenician alphabet of twenty-two consonant symbols is related to the thirty-character alphabet of Ugarit, a Canaanite city that was destroyed about 1200 BC. The Greeks added the vowel signs to complete the alphabet.

During the last twenty-five hundred years alphabets have spread to the furthest corners of the Earth, but the principles of writing have not changed. Hundreds of alphabets throughout the world, different as they may be in outer form, use the principles first and last established in Greek writing.

THINGS OUT THERE

The Universe has its center everywhere and its edge nowhere.

Nicholas of Cusa in the fifteenth century

One important feature we have learned about nature is that it performs its "miracles" minimally, using only as much technology as it needs.

Before there was intelligent life, the Universe did not "really" exist.

Eugene Wigner

Is our Universe only a grain of sand on someone else's beach? Are the grains of sand on our beaches someone else's Universes?

CALCULATING SPACE DISTANCES

The nearest star to our Sun is Proxima Centauri, one of a multiple star system located in the Centaurus constellation, located some 4.4 light years distant, over 25,865,896,510,080 miles - almost twenty-six trillion miles. To travel to this star at one million miles per hour, 278 miles each second, 8,766,000,000 miles each year - almost nine billion miles each year - would take over 2950.7 years, almost three thousand years. The speed of light is 670 times faster.

Each of us has looked out toward the stars. When we were children we looked because there were so many and they were very pretty. Later we looked for the Big Dipper and the Little Dipper. We could even find Polaris, the North Star. They were "way up there" so we went our way and they continued on their way.

How little did we know that these bright objects we observed in the night sky were really "going on their way." Some of these objects are racing away from us and each other at tremendous speeds in space. The nearest star to us, our Sun, is only a medium-sized star located about halfway out to the edge of our galaxy - the Milky Way. When we look inward toward the center of the Milky Way we see stars - billions of stars. When we look in the opposite direction toward the outer edge of our galaxy we see billions more. We are part of the same system as these stars and are traveling along with them. We remain relatively the same distance from each other as time passes. When we look "up" or "down", the billions of bright objects we see are also the stars of our galaxy - and other GALAXIES that appear to us as ordinary individual stars. These galaxies are what are moving away from us and each other at millions of miles per hour. We are LOOKING BACK IN TIME each night we look outward.

Turning on the light in our living room appears as an instantaneous event to each of us. This is because we see visible light (a string of photons) that is emitted from a light bulb (excited electrons). It is instantaneous to our senses because the light switch is ten feet from the light - which also places us ten feet from the light. Light and electricity travel very fast. Light travels at 186,282 miles per second (300,000 kilometers per second). Electricity travels much slower - one-third the speed of light - 62,094 miles per second (100,000 kilometers per second). These speeds are instantaneous to our senses. It seems that the light we switch on appears immediately, but there is a time interval involved. It takes time for the electric current to travel the ten feet to the light bulb and time for the light (photons) to reach our eyes.

The elapsed time for this action is expressed by the formula of TIME equals

DISTANCE TRAVELED divided by the SPEED of the object moving - electrons or photons. To solve for how much time was required for the light to travel the ten feet from the light bulb to our eyes, we divide ten feet by 186,282 miles per second (983,571,068 feet per second - almost one billion feet per second) and arrive at an elapsed time of 0.00000001 seconds - ten billionths of a second (10^{-8} seconds). The electrons traveled the ten foot distance from the switch to the light in thirty billionths of a second for a total elapsed time of forty billionths of a second from the time we turned the switch on until the light (photons) reached our eyes. Time is required for electricity and light to leave their source and travel to some destination.

As long as the light is on, photons continue to travel to our eyes. Consider the photon as a waveform particle. When the light goes off, this string of photons ceases at the source but the ones emitted continue to travel to our eyes. Visualize a stream of water from a firehose. When the water is suddenly cut off, the stream of water continues until the end of this stream of water reaches its destination. If this is a long stream of water, it takes a longer time before the end of the stream reaches its destination. Light behaves accordingly except for the fact that it will travel forever in a straight line. Therefore, it takes light (a string of many photons) time to go from one place to another.

Another example is our own star - our Sun. Light emitted from the Sun arrives on Earth eight minutes and twenty seconds later. In distances this close we say we are ninety-three million miles away or we could say we are eight minutes and twenty seconds apart. Multiplying eight minutes and twenty seconds (five hundred seconds) times the speed of light (186,282 miles per second), we arrive at the distance of a little over ninety-three million miles from the Sun to Earth. Proxima Centauri is the solar system's nearest star. It must be located in the Milky Way galaxy to be this close and the distance is so far that it is not measured in miles or kilometers. The simplest method to measure distances this great or greater is to use the distance traveled by light, using the speed of light and time traveled to determine the distance. These objects require the term LIGHT YEARS. We obtain this distance by multiplying the speed of light (186,282 miles per second) TIMES sixty seconds per minute TIMES sixty minutes per hour TIMES twenty-four hours per day TIMES 365.25 days per year. This is 5,878,612,843,200 miles per year - five trillion eight hundred seventy-eight billion six hundred twelve million eight hundred forty-three thousand two hundred miles per year. A light year is almost six trillion miles long. How far, in miles, is our nearest star, Proxima Centauri, which is located 4.4 light years away? Multiplying the light year distance by 4.4 equals 25,865,896,510,080 miles - almost twenty-six trillion miles. Remembering that the light we see at night as we look at Alpha Centauri (one of Proxima Centauri's companions) is traveling to us in streams of photons, the ones that reach our eyes left this star 4.4 years (four years and five months) ago. We assume this stream is continuing. If this star suddenly went "out" this instant, it would be 4.4 years before the last photon (light) emitted would reach Earth. Only 4.4 years from this date would we know that this star had ceased emitting photons - it went "out." WE ARE LOOKING BACK IN TIME!

So the light year (LY) is our unit of measure for the distance to stars, galaxies and other events in the Universe. Our galaxy - the Milky Way - has been determined to have a diameter of one hundred thousand light years. It rotates at a rate of two hundred forty million years per cycle. The solar system is moving at five hundred thousand miles per hour within this system. Earth's orbital speed around the Sun is about 66,588 miles per hour rotating on its axis at 1,037 miles per hour at the equator. We are traveling about six hundred thousand miles per hour as we journey along in the solar system as it journeys in the Milky Way.

Our galaxy is rotating at a minimum of one million miles per hour with the Andromeda galaxy, two million two hundred thousand light years away. Also included are the Large Magellanic Cloud, one hundred sixty-three thousand light years away, and the Small Magellanic Cloud located two hundred thousand light years away. Many other galaxies are also involved. We are part of a galactic cluster. The Milky Way and the Andromeda are racing toward each other at three hundred kilometers per second (671,100 miles per hour) and will pass harmlessly through each other two billion two hundred million years in the future. Some distant galaxies are presently in this process.

BIG BANG

The opening sentence of this book should be:
"Once upon a time - very long ago - we began."

The "big bang" is the origin of SPACE as well as of matter and energy. There was no *pre-existing void* in which the big bang happened. At the big bang itself, the Universe had zero size and was infinitely hot. From the instant of the big bang until 10^{-43} seconds had elapsed, the superstring theory is the best we have to imagine what occurred during this span of time. All particles, including quarks, are pictured as having tiny stringlike structures buried inside them, with different particles corresponding to different types of vibrations of strings.

At 10^{-43} seconds *Gravity* was separated due to the formation of the particles. Gravity came into its own because of the very nature that the particles had included *mass* in their formation and gravitational forces were born out of necessity. Now that *Gravity* was available, the accumulation of the particles into groups could officially begin. From this time until 10^{-33} seconds, the Universe was going through a "freezing" (expansion) session labeled INFLATION as it increased in size 10^{50} times - from the size of an elementary particle to the size of a grapefruit. Inflation caused the Universe to forget its initial state. The *Strong Force* was frozen away. When the period of inflation was over, the big bang had begun and, as they say, the rest is history. *The big bang describes how our Universe is evolving, not how it began.*

At 10^{-11} seconds the *Electromagnetic Force* separated from the *Weak Force* supplying the *FOUR FORCES* that are still in use today governing all matter. One millisecond (10^{-3} second) after the big bang, the Universe was 1.8 trillion degrees. At about ten milliseconds (10^{-2} seconds) from the beginning, quarks combined to make particles. As the Universe expanded, the temperature of the radiation decreased. One second after the big bang, the temperature would have fallen to about ten billion degrees. This is about a thousand times the temperature at the center of the Sun and temperatures reached in hydrogen bomb explosions. At this time the Universe would have contained mostly photons, electrons, neutrinos, their antiparticles, together with some protons and neutrons.

About one hundred seconds after the big bang, the temperature would have fallen to one billion degrees, the temperature inside the hottest stars. At this temperature protons and neutrons would no longer have sufficient energy to escape the attraction of the *Strong Nuclear Force*, and would have started to combine to produce the nuclei of atoms of deutrium (heavy hydrogen) which contains one proton and one neutron. About a quarter of the protons and neutrons would have been converted into helium nuclei (two protons and two neutrons) along with a small amount of heavy hydrogen and other elements. The remaining neutrons would have decayed into protons, which are nuclei of ordinary hydrogen, and electrons to complete the formation of hydrogen atoms. About three minutes into the life of the Universe, the temperature dropped to the point where nuclei could survive collisions and proceed to add more particles to become a larger, more complex nucleus.

Within only a few hours of the big bang, the production of helium and other elements would have stopped. The expansion of the Universe continued for the next million years or so until the temperature had dropped to a few thousand degrees. In regions that were more dense than average, the expansion would have slowed down by the extra gravitational attraction. As the collapsing region became smaller, it would spin faster and faster and in time form disk-like rotating gases from which the galaxies were born. After a few hundred thousand years the temperature of the Universe had become low enough so that if an electron fell into orbit around a nucleus, later collisions would not cause them to separate. Once an atom was formed, it would now stay together.

The expansion was now in full force from birth and in its infancy began the accumulation process of the three forces of *Strong, Weak* and *Electromagnetic* through the use of the weakest of the *Four Forces - Gravity*. Atoms were accumulated into molecules, molecules into matter, matter into gaseous galactic clouds, then to stars, into a process that continues today. As time went on, the hydrogen and helium gas in the galaxies broke up into smaller clouds that would collapse under their own gravity. This continuing collapse caused the temperature of the gas to increase to the point that nuclear fusion was initiated. This is the birth of a star.

The total energy of the Universe was zero before it started and is still zero today. What happens is that the positive energy locked up in mass is cancelled by the negative energy of the gravitational field. "The Universe is simply one of those things that happens from time to time."

Ralph A. Alpher revealed mathematically how the Universe began in a dissertation presented in 1948. It was titled *The Origin of the Chemical Elements* and was dated February 18, 1948. Robert Wilson and Arno Penzias of Bell Labs made the initial verification of the big bang and the expanding Universe in 1964 in the United States.

Space is not absolutely cold - the temperature of the Universe appears to be three Kelvin (-270° Celsius). This is the exact temperature the Universe should be if it all began some thirteen billion years ago with a big bang whose temperature then was more than one hundred thousand billion billion billion degrees (10^{32} degrees).

The direction the big bang chose in its expansion has been studied thoroughly in attempting to discover what choice was taken. One of the choices could have been what is now known as an "open" universe. This universe is shaped like a saddle and keeps on expanding forever, but slows as time passes. Another choice is called the "closed" universe, a spherical shape, in which the expansion slows and eventually reverses, collapsing in a cosmic crunch. The actions for this "closed" version would have been caused by gravitational forces. A "flat" universe exists when the expansion eventually slows very nearly to a stop but never actually reverses. In the very late 1990's new facts became available due to observations using some of our latest technology.

The new finding revealed that the edges of space are rushing away from one another at an ever-increasing rate. The Universe, it seems, is not just growing, but growing faster and faster. The bigger the Universe gets, the faster it grows. This observation was observed and verified by two separate groups while studying a very distant supernova. The light from this supernova, and others at great distances away, is stretched out less than was predicted given the current rate of expansion. This shows that the Universe expanded more slowly in the past that it does now. Expansion was not slowing down as expected, it was speeding up.

Today's cosmologists are calling this force *dark energy* - *dark* because it may be impossible to detect and *energy* because it is not matter (the only other option). "The discovery of an accelerating Universe was simultaneously the biggest surprise and the most anticipated discovery in astronomy," said Michael Turner, University of Chicago cosmologist. It put *dark energy* on the map. If the Universe is made up of a repulsive energy rather than matter, then its ultimate fate is not described by its shape after all. Remember that matter and energy are interchangeable - $E = MC^2$.

SEQUENCE OF EVENTS

TIME (SECONDS)	EVENT
0	Beginning of time
10^{-43}	Normal gravity - Planck time
10^{-38}	Inflation - grand repulsion
10^{-35}	Hot big bang - quark era
10^{-33}	Strong force separated
10^{-11}	Weak force weakened - emf separated
10^{-6}	Proton era - hadronic era
10^{-4}	Electron era - leptonic era
1	Gamma ray era - neutrinos loose Contents of the known Universe - volume is three light years wide
3 Minutes	Helium making (20% He nuclei; 80% H nuclei)
10 Years	Galaxies seeded
10^4 Years	Matter predominant
3×10^5 Years	Atoms forming; one billionth present volume of space
10^6 Years	Transparent Universe
5×10^9 Years	Acceleration begins

RADIATION ERA	TIME	DENSITY (KG/M³)	TEMP K	EVENT
	0	INFINITY		
Planck				Quantum gravity.
	10^{-43} sec	10^{95}	10^{32}	
GUT				Strong, weak, emf united.
	10^{-35} sec	10^{75}	10^{27}	
Hadron				Heavy and light particles obtain thermal equilibrium.
	10^{-4} sec	10^{16}	10^{12}	
Lepton				Light particles obtain thermal equilibrium.
	10^{2} sec	10^{4}	10^{9}	
Nuclear				Deutrium and helium formed by fusion.
	3×10^{10} sec (1,000 years) (10^{3} years)	10^{-13}	6×10^{4}	

MATTER ERA	TIME	DENSITY (KG/M³)	TEMP K	EVENT
	10^{3} years	10^{-13}	6×10^{4}	
Atomic				Matter begins domination Atoms form. Electromagnetic radiation decouples.
	10^{6} years	10^{-19}	10^{3}	
Galactic				Galaxies and large scale structures form.
	10^{9} years	3×10^{-25}	10	
Stellar				All galaxies have formed. Stars continue to form.
	10^{10} years	10^{-26}	3	

FOUR FORCES

Force-carryng particles are grouped into four categories according to the strength of the force and the particles with which they interact.

GRAVITATIONAL FORCE - universal. Weakest of the four forces. Acts over long distances. It is always attractive. Spin 2. Gravitons.

ELECTROMAGNETIC FORCE - 10^{24} stronger than the gravitational force. Causes the electrons to orbit the nucleus of the atoms just as the gravitational attraction causes the Earth to orbit the Sun.

NUCLEAR FORCE - WEAK - responsible for radioactivity.

NUCLEAR FORCE - STRONG - holds the quarks together in the proton and neutron, and holds the protons and neutrons together in the nucleus of the atom. Spin 1. Particle called GLUON, which only interacts with itself and quarks. At high energies the strong force becomes much weaker, and the quarks and gluons behave almost like free particles.

If the *strong force* is set at one (1), the *electromagnetic force (emf)* would be set at 10^{-2} or 0.01. The *weak force* would be 10^{-13} and the *gravitational force* would be 10^{-38}. The *strong* and *weak forces* have limited ranges. The *electromagnetic* and *gravitational forces* have no limits as they obey the inverse square law.

FIRST MILLION YEARS OF THE UNIVERSE

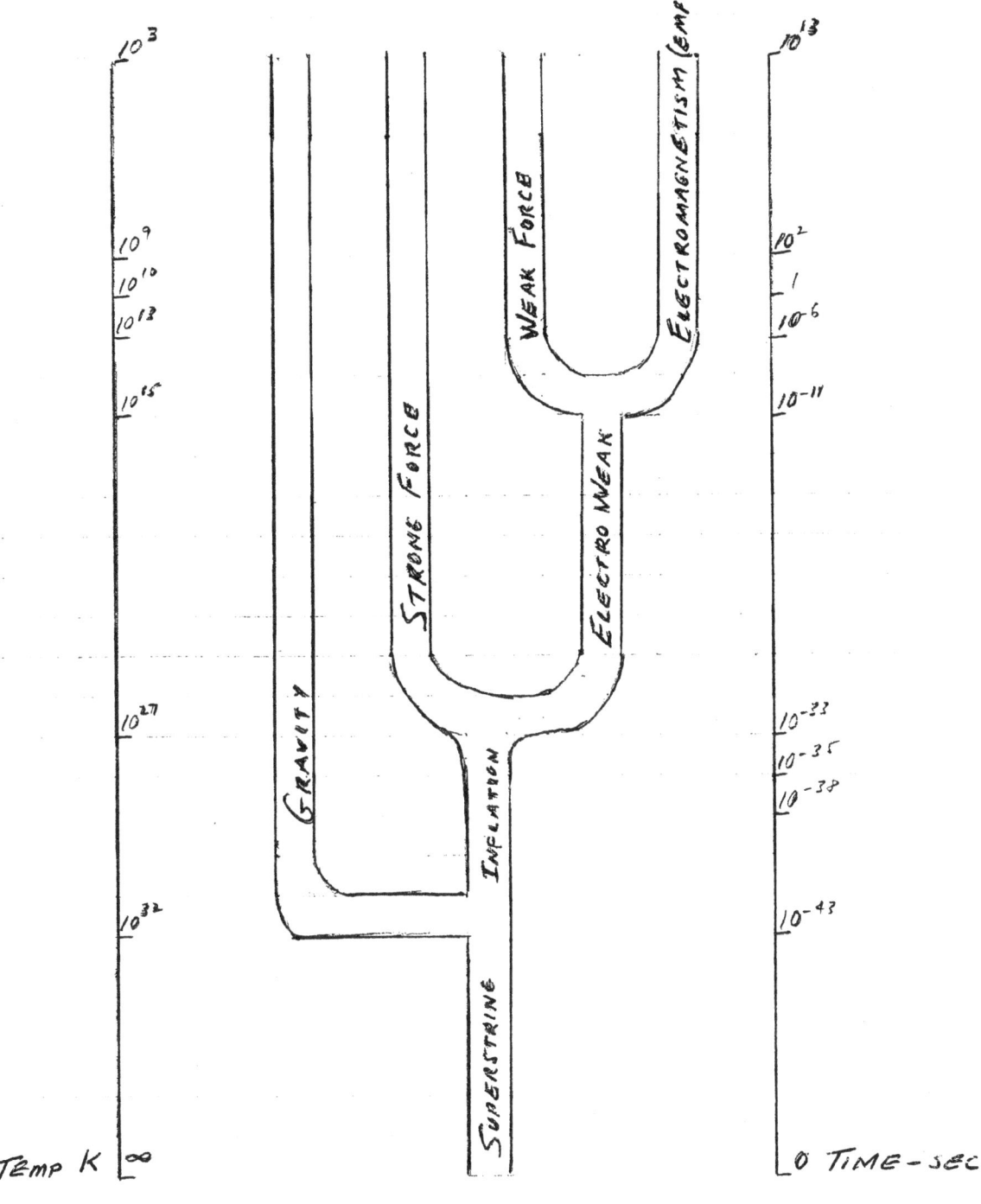

t = Planck Time = 10^{-43} seconds = the visible Universe becomes smaller than its quantum wavelength.

λ = Planck Length = 10^{-33} centimeters = the distance light has traveled in Planck Time.

$\lambda = ct = 3 \times 10^{10}$ cm/sec x 10^{-43} sec = 3×10^{-35} cm (Gamma ray $\lambda = 10^{-14}$ cm).

ALTERNATIVES TO THE BIG BANG

1. Instead of beginning as a state of infinite density, the Universe of space, time and matter comes into being with a finite density and continues in a state of expansion.

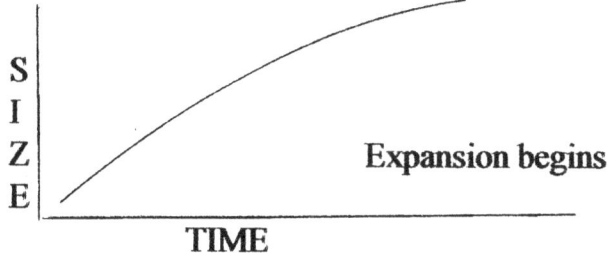

2. The Universe "bounces" into a state of expansion from a previous state of maximum but finite contraction.

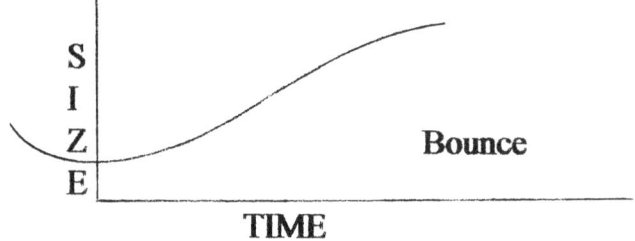

3. The Universe suddenly begins its expansion from a static state, which it has resided for past eternity.

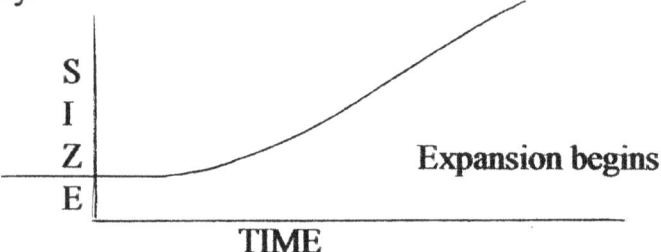

4. The Universe gets even smaller in the past without reaching a state of zero size. It had no beginning.

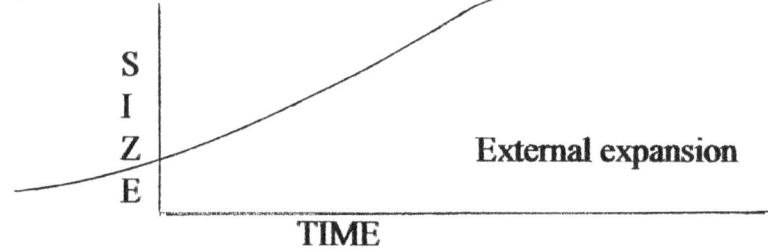

THE FORMATION OF THE ELEMENTS

Where did all the elements come from? How were they formed? Have they always existed or were they created later? We now know that hydrogen and most of the helium present are *primordial* - formed after the big bang. All other elements result from stellar nucleosynthesis - formed by nuclear fusion *in the heart of stars*.

Following the big bang, a proton captured an electron and the first hydrogen atom was assembled. At a temperature of ten million Kelvin four protons (four hydrogen nuclei) combine to form a helium nucleus. The result is a nucleus of two protons and two neutrons which capture a pair of electrons to become a helium atom. This reaction causes a release of two positrons and two neutrinos plus energy. The positrons immediately interact with nearby free electrons, producing high-energy gamma rays through matter-antimatter annihilation. The neutrinos escape carrying energy.

At a temperature of one hundred million Kelvin, helium nuclei can overcome their mutual electrical repulsion. This causes a *triple-alpha reaction*. The net result of this reaction is that three helium-4 nuclei are combined into one carbon-12 nucleus, releasing energy in the process. When this carbon-12 nucleus collides with a helium-4 nucleus at six hundred million Kelvin an oxygen-16 nucleus is formed with a release of energy.

Oxygen-16 can also be formed through another process combining the hydrogen nucleus. This is a longer process that can occur at lower temperatures:

carbon-12 + hydrogen-1 = nitrogen-13 + energy
nitrogen-13 + hydrogen-1 = carbon-13 + positron + neutrino
carbon-13 + hydrogen-1 = nitrogen-14 + energy
nitrogen-14 + hydrogen-1 = oxygen-15 + energy
oxygen-15 + hydrogen-1 = nitrogen-15 + positron + neutrino
nitrogen-15 + hydrogen-1 = carbon-12 + helium-4
carbon-12 + helium-4 = oxygen-16 + energy

The alpha particle (helium nucleus) addition is common up through iron-52. Elements heavier than iron are formed by another process.

oxygen-16 + helium-4 = neon-20 + energy
neon-20 + helium-4 = magnesium-24 + energy
magnesium-24 + helium-4 = silicon-28 + energy
silicon-28 + helium-4 = sulfur-32 + energy
sulfur-32 + helium-4 = argon-36 + energy
argon-36 + helium-4 = calcium-40 + energy

calcium-40 + helium-4 = titanium-44 + energy
titanium-44 + helium-4 = chromium-48 + energy
chromium-48 + helium-4 = iron-52 + energy
iron-52 + helium-4 = nickel-56 + energy

Elements heavier than iron are formed from neutron capture and decay. Iron-59 plus a neutron decays in one month to form cobalt-59. Cobalt-59 plus a neutron decays into nickel-60. Nature normally chooses the simplest and more direct avenues to solve its problems. Sometimes extreme temperatures are present when certain elements are available giving direct combinations. At one billion Kelvin carbon-12 can combine with another carbon-12 nucleus to form magnesium-24 with a corresponding release of energy. At 1.2 billion Kelvin two oxygen-16 nuclei can join to form sulfur-32. Extremely high temperatures are required for the fusion process - some higher than others. Element formation depends on what nuclei are available and the temperature present. It all began following the big bang and continues today in the interior of stars. The neutron decay in these cases is the conversion of the neutron into a proton, electron and neutrino.

COSMIC ABUNDANCE OF THE ELEMENTS

ELEMENT		ABUNDANCE (% OF TOTAL)
Hydrogen	H	73
Helium	He	25
Oxygen	O	0.8
Carbon	C	0.3
Neon	Ne	0.1
Nitrogen	N	0.1

Except for hydrogen and helium, every element in the Universe, including every element on Earth and in your body, has been manufactured inside a star.
We are from the stars!

GALAXIES

Harvard astronomer Harlow Shapley investigated the distribution of stars with respect to the Milky Way. His analysis determined that the center of our galaxy was about thirty thousand light years from our solar system in the direction of the constellation of Sagittarius. Further observations revealed that there were two gaps in the star counts toward the center of the galaxy - the first being one thousand light years away and the other six thousand light years distant. There is another gap when looking toward the edge of our galaxy. These gaps were correctly interpreted by the star count astronomers as having been produced by spiral arms of stars. This determined the shape of the Milky Way as a spiral galaxy consisting of three distinct spiral arms.

Analyzing the rotational speeds in different parts of the galaxy verified this finding because the inner portions were faster than the outer portions in their orbits. This action tends to expand the outer portions through time causing the outer spiral arms to increase speed to the point that the spiral gaps close, creating a disk-shaped galaxy. The shape of the galaxy can also be used to determine its age. Galaxies come in all shapes, sizes and forms.

Our galaxy has a diameter of one hundred thousand light years, is approximately two thousand light years thick with a galactic center fifteen thousand light years in diameter and contains over one hundred billion stars. A massive black hole exists at the center of our galactic center initiating the energy of our motion. We belong to a "local group" of galaxies numbering twenty or so. Ten galaxies lie within one million light years of galaxy and ten more lie one to two million light years away. The Andromeda galaxy is located in this outer group and resembles our galaxy, except it is twice as large and contains over four hundred billion stars. Clusters like this are found throughout space, some containing hundreds or thousands of galaxies. Calculations suggest that at least four hundred billion galaxies exist in one form or another.

To name a few of our close neighbors: the Virgo cluster, fifty million light years away contains some twenty-five hundred galaxies; the Pegasus cluster, one hundred ninety million light years away contains five hundred galaxies; and the Centaurus cluster five hundred million light years away contains three hundred galaxies. These galaxies are in motion within each cluster. An example is that the galaxies in the Virgo cluster are moving with a speed of about 750 kilometers per second (1,677,750 miles per hour) within the cluster. Our own Milky Way is lazily making its way through our cluster at only a million miles per hour or so. Local superclusters may have diameters of one hundred million light years, separated by vast intergalactic voids.

Galaxies grow by repeated merging of smaller objects. Galactic formation is a "bottom-up" process, opposite to the "top-down" process of star formation. The "top-down" process is one in which a large cloud fragments into smaller pieces that eventually become stars. Our observations of galaxies that are further away from us - those that are near the limit of our observational capabilities - are much dimmer and smaller than any of our neighboring galaxies. These distant galaxies appear to us as they were four, six or eight billion years ago. Therefore, it appears that the young galaxies we observe at these distances will have grown larger and brighter through the accumulation process by the present era.

Galaxies appear in many forms. According to the classification scheme developed by Edwin Hubble, galaxies may be broadly divided into three major types: elliptical, spiral and irregular. The most massive ones are the ellipticals. These are smooth, featureless, almost spherical systems with little or no gas or dust. Stars move around the center in a busy pattern with most of these stars being very old. Supermassive black holes are believed to reside in virtually every elliptical galaxy.

Spiral galaxies, such as our own Milky Way, are highly flattened and organized structures in which stars and gas move in circular or near-circular orbits around the center. These are also known as disk galaxies. The pinwheel-like spiral arms are filaments of hot young stars, gas and dust. At their center, spiral galaxies contain bulges which also contain massive black holes. Almost a third of spiral galaxies have a rectangular structure toward the center, possibly caused from instabilities in the disk. There are barred spiral galaxies - those that appear to have a heavy spoke which ranges from one edge of the galaxy to the other through the galactic center.

Irregular galaxies are those that do not fit into the elliptical or spiral classifications. Some appear to be ellipticals or spirals that have been violently distorted by a recent encounter with a neighbor. Others are isolated systems with a shapeless structure and exhibit no signs of any disturbance. The Magellanic Cloud galaxies we observe from the Southern Hemisphere are irregular galaxies.

We can assume that our galaxy originated as an irregular shaped system with gas, dust and young stars in this volume. As time passed, mergers with other small systems occurred causing this volume to increase. As stars formed during these growth stages, there was no specific direction in which they moved or no particular order in their locations as they began their star-making processes. In time, the gravitational effect of these masses caused a rotation to begin in earnest with the result that the gas, dust and stars fell into a galactic plane and formed our spinning disk. The older, heavier stars migrated to the outer portion of the disk with the less dense gas and dust being contained toward the galactic center - later to become new stars.

Due to the fact that the Milky Way galaxy is a spiral galaxy, the stars close to the center take less time to orbit the galactic center than those further out. As billions of years pass, this momentum will cause the spirals to move faster, closing the gaps as the outer stars speed up, making our galaxy near the shape of a true disk around its center.

It is hard for us to view a lot of our own galaxy. Due to our location near the outer edge, when we look toward the center, there are billions of stars blocking our view of this area. It is much easier looking toward the edge because there are several billion less stars in our way. Remember, our galaxy contains over one hundred billion stars. We can study our galaxy by observing other spiral galaxies to gather the details. Look at the Andromeda Galaxy to visualize our shape.

The consensus is that galaxies began to form about eight billion years ago. The light that we are now observing from the most distant galaxies establishes their location from us at some eight billion light years *at the time the light was emitted.* Due to the expanding Universe these galaxies are now located several billion light years further away. The light now being emitted will reach the Milky Way Galaxy some ten to twelve billion years in the future. Remember, you are looking back in time and we are viewing these galaxies as they *were* - not how they presently appear. If they are observing light from our galaxy this present moment, they are also seeing us as we *were.*

It is not possible for us to observe a galaxy forming because the process is too slow. Researchers piece the puzzle together by observing many different galaxies, each at a different phase in its evolutionary history. This type of research is a time machine as they compare the furthest galaxies to each one in sequence as they approach nearer to our own. Astronomers traditionally use the word "evolution" to refer to the life cycle of a single object, such as a star or a galaxy - the way it changes as it gets older. They are not referring to Darwinian evolution with one variety of star or galaxy being replaced by another cosmic species.

MAIN GALAXIES IN LOCAL GROUP

GALAXY	DISTANCE	DIAMETER
Milky Way		100,000 LY
Large Magellanic Cloud (LMC)	0.15×10^6 LY	31,000 LY
Small Magellanic Cloud (SMC)	0.18×10^6 LY	13,000 LY
Andromeda Group (M31)	2.1×10^6 LY	110,000 LY
(M32)	2.1×10^6 LY	2,000 LY
(M33)	2.2×10^6 LY	38,000 LY
Sculptor	0.35×10^6 LY	5,000 LY
Fornay	0.75×10^6 LY	11,000 LY
NGC 205	2.1×10^6 LY	6,000 LY
NGC 6822	1.8×10^6 LY	7,000 LY
IC 1613	2.1×10^6 LY	10,000 LY

NEAR GALACTIC GROUPS AND CLUSTERS

	DISTANCE	NUMBER GALAXIES
Sculptor Group	8.2×10^6 LY	
Ursa Major Group (M81)	8.6×10^6 LY	200
Virgo Cluster	50×10^6 LY	2,500
Fornay Cluster	54×10^6 LY	
Pegasus Cluster	130×10^6 LY	100
Perseus Cluster	160×10^6 LY	500
Centaurus Cluster	500×10^6 LY	300

OBJECTS IN THE SKY

Objects in the sky are different sizes. Listed are some general dimensions that may present some perspective to these sizes:

Red Giant	250 times larger than our Sun
Sun	100 times larger than a White Dwarf
White Dwarf	700 times larger than a Neutron Star
Neutron Star	3 times larger than a Black Hole

RED GIANT

A red giant occurs when a mature star grows old and begins to run out of fuel which sustains the fusion process to which it had become accustomed. To maintain its life, it must have more fuel - hydrogen. The outer edges of the star reach out into space for the individual atoms of hydrogen that exist there and use them for the continuation of the fusion process. It continues this growth process, its volume increasing hundreds of times, until the internal gravity of the star causes it to cease its growth. It was a star, became a red giant and now will gravitationally collapse until it becomes a white dwarf, neutron star or a black hole. Our Sun will become a red giant in about four to five billion years from now as it expands past Mercury, Venus and reaches Earth's orbit at its maximum expansion.

WHITE DWARF

A white dwarf is an approximately Earth-sized star that does not have a source of nuclear energy in its interior. It has a radius of only a few thousand miles with a density of several hundred tons per cubic inch. The star is supported by means of a form of pressure that arises when the densities at the star's interior are so high that the usual orbits of the electrons about their nucleus cannot exist and the electrons are pushed much closer to the nucleus. A white dwarf can be supported in this way as long as its mass does not exceed 1.4 solar masses - cannot weigh more than 1.4 times the mass of our Sun, a maximum of 2.786×10^{30} kilograms (3.0566×10^{27} tons) maximum.

NEUTRON STAR

A neutron star is formed as a result of a supernova explosion or the collapse of a star in the later stages of its life that is between 1.4 to 3.0 times the mass of our Sun - a solar mass (SM). The collapse of the star causes the gravitational pressure to push the electrons in orbit of the nucleus into the protons in the nucleus to form neutrons which then join the neutrons already present in the nucleus - reverse of the beta decay process.

The smallest neutron star of 1.4 solar masses (1.4 SM) would be gravitationally compressed into a sphere 17.6 miles (28.4 kilometers) in diameter. The total weight of this neutron star would be 3.0378×10^{27} tons (2.76×10^{33} grams). The largest neutron star would be gravitationally complressed into a sphere 22.8 miles (36.8 kilometers) in diameter and weigh 6.5819×10^{27} tons (5.98×10^{33} grams). Stars below 1.4 SM end their life as white dwarfs and those with 3.0 SM or greater end up as black holes.

SUPERNOVA

A supernova is an immense stellar explosion which can increase a star's intrinsic brightness by as much as a billion times. The explosion blows off a major portion of the star to form an expanding gas cloud such as the Crab Nebula. The remaining material forms a dense object such as a neutron star. Neutrinos occur when electrons and protons in the star's collapsing core merge to form neutrons. The neutrinos precede the light because they escape during the collapse, whereas the first light of an explosion is emitted only after the supernova shock has penetrated the body of the star to the surface. The supernova's neutrino luminosity is many tens of thousands of times greater than its optical energy output.

On February 23, 1987 astronomers in the Southern Hemisphere observed the closest and brightest supernova to be seen in nearly four centuries. The supernova occurred in a dwarf galaxy companion of our Milky Way Galaxy called the Large Magellanic Cloud which contains several billion stars. The Large Magellanic Cloud is a satellite galaxy held captive by the Milky Way's gravity. The Supenova was designated SN1987A (SN - supernova; 1987 - year observed; A - first in sequence).

During the first second of the explosion, the energy emitted by the dying star was greater than the energy emitted by the entire visible Universe. The collapse generated an intense burst of neutrinos - massless, chargeless particles emitted in nuclear reactions. On February 21, 1987, one hundred billion neutrinos from the exploding star passed through the bodies of every person on Earth. About twenty hours before the supernova was detected optically, a thirteen second burst of neutrinos was simultaneously recorded by underground detectors in Japan and the United States. For the first time, astronomers have received information from beyond the solar system by radiation outside the electromagnetic spectrum.

The star that exploded was twenty times as massive as the Sun (20 SM) and had lived for ten million years. The larger the star, the shorter the life due to faster fuel consumption to maintain fusion. In its youth, it was considerably hotter than the Sun and fifty thousand times brighter. Blue in color originally, it turned red, and then blue again in the last years of its life, evolving at an accelerating rate at the end. In rapid succession, nuclear fires burning at the center of the star produced carbon, oxygen, neon, magnesium, silicon, sulfur, argon, calcium, titanium, chromium, iron and other elements. In a matter of seconds the star collapsed and rebounded in the explosion observed on Earth.

PULSAR

Pulsar B1257+12 is something like an atomic nucleus a few miles in diameter spinning at ten thousand revolutions per minute. It is a supernova remnant. There was a colossal catastrophe that blew off most of the mass of that star. Going around it are at least three planets, two roughly of Earth-like mass, one roughly of lunar mass; a little closer in position than Mercury, Venus and Earth. Three planets of roughly Jovian mass have been discovered orbiting around the stars 51 Pegasi, 47 Urase Majoris and 70 Virginis. Pulsars are rapidly rotating neutron stars emitting electromagnetic radio in radio and visible light frequencies.

QUASAR

Some years ago astronomers noticed a very unusual star in the heavens. A measurement of the distance to the "star" showed that it was two billion light years from us - far beyond the boundary of our galaxy. If this were an ordinary star, it would be very faint at that great distance - too far to be seen. The fact that the star could be seen at all in spite of its great distance indicated that it must be enormously brighter than an ordinary star. When allowance was made for the star's great distance, its true brightness turned out to be equal to that of hundreds of billions of ordinary stars.

Some other examples of the strange stars were found. One was a trillion times brighter than an ordinary star. These points of light could not be ordinary stars. These objects became known as quasi-stellar objects - QUASARS. The extraordinary brightness of quasars was only one of their unusual properties. Even stranger was the fact that the tremendous flow of energy was coming from a small point in space - smaller than our solar system. It soon became apparent that quasars were located in the center of galaxies. To justify the generation of this enormous amount of energy a black hole must be assumed. Could quasars be giant black holes?

A massive black hole sits in the center of a spiral galaxy, surrounded by billions of stars. The stars circle around the black hole under the pull of its gravity. Gradually they spiral toward it. As each star draws close to the black hole, its gaseous body is torn apart by the black hole's powerful gravitational force. The atoms of gaseous matter within the disintegrating star, picking up speed under the attraction of the black hole, move faster and faster. As the atoms approach the boundary of the black hole, they collide with one another. The collisions heat the gas, and the hot gas radiates energy into space. This energy is what we see when we observe a quasar. This is the result of a collection of billions of stars into a single unit - a giant black hole. This black hole is a billion times more massive than an "ordinary" black hole.

After a time, many stars in the inner part of the galaxy are gone because they have been torn apart and consumed by the giant black hole. After a relatively short interval of time, a few hundred million years or so, very few stars are left. With its source of energy gone, the quasar fades into darkness. Where the quasar once blazed, a galaxy of ordinary appearance appears, but with a quiet black hole at its center. Our Milky Way and all the neighboring galaxies are like this - ordinary. The nature of quasars being so distant agrees with the fact they are the younger development of galactic evolution as we look back in time.

One of the oldest quasars observed lies 12.2 billion light years away with the youngest having been active a billion years ago. If people on this "oldest" galaxy we observe are observing us, they would see us in the same context as we see them. We are the same distance from them as they are from us causing each observer to see only quasars - not each other as we presently exist. Each of us is "ordinary" now. Quasars became numerous eleven billion years ago with the result that they are the well developed galaxies of today.

OTHER ACTIONS

Certain stars in given regions of the sky are moving together (same direction). These are called "local" or "galactic" clusters. The Hyades cluster in Taurus is an example of a local stellar cluster. The comparison of photographs of the stars in the

Ursa Major constellation show that all the stars in the Big Dipper are not moving together and that this constellation will lose its present shape ten thousand years from now. This, along with the effects of precession, will cause Deneb to become the pole star in seven thousand years.

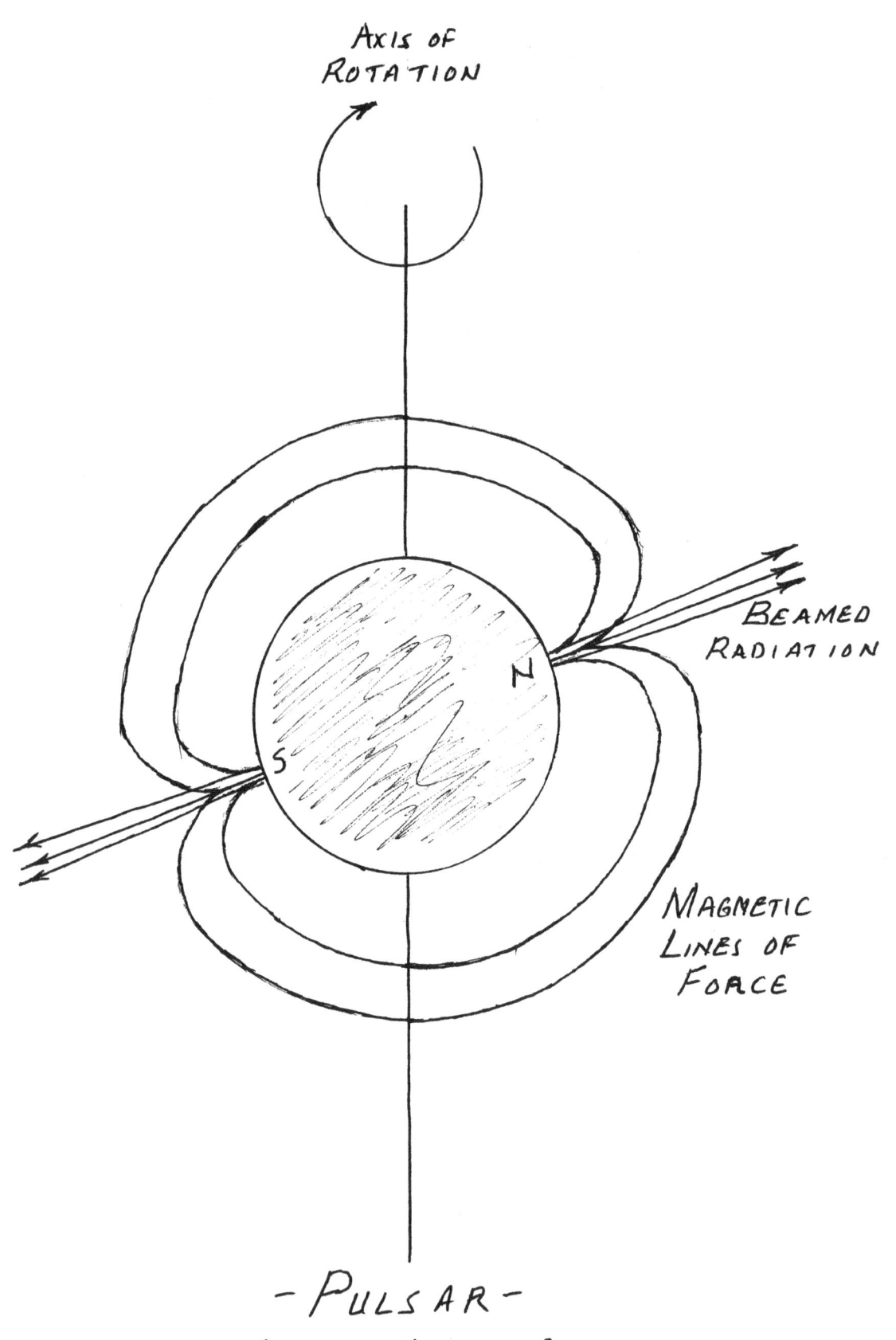

AXIS OF
ROTATION

N

BEAMED
RADIATION

MAGNETIC
LINES OF
FORCE

- PULSAR -
SPINNING NEUTRON STAR

URSA MAJOR

100,000 YEARS AGO

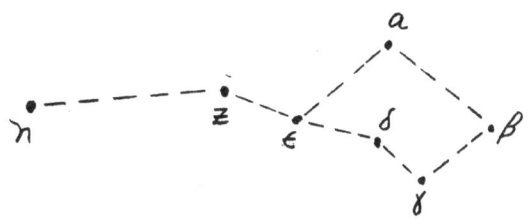

PRESENT
SHOWING RELATIVE DIRECTION OF MOVEMENT

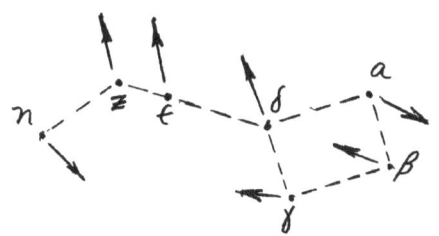

100,000 YEARS IN THE FUTURE

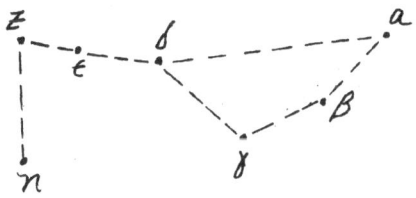

THINGS DOWN HERE

SIZE OF THE MILKY WAY GALAXY

If the solar system were the size of a dinner plate, the nearest neighboring stars would be a few kilometers (few miles) away and the Milky Way Galaxy would cover the United States of America. The solar system would be located near the "Four Corners" - the geographic point that Arizona, Colorado, New Mexico and Utah meet.

SOLAR SYSTEM

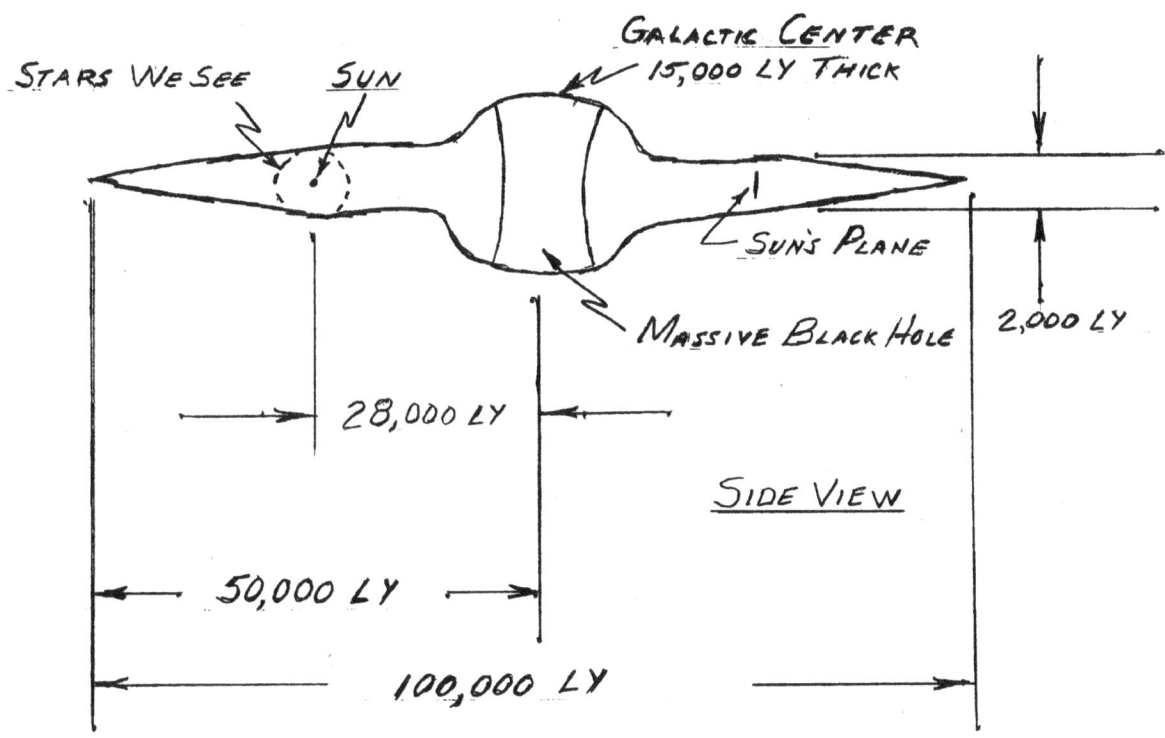

The solar system not only orbits the galactic center in a nearly circular orbit but also bobs up and down through the galactic plane in an oscillating motion. This motion carries the Sun two hundred light years above and below the plane as it makes its circuit of the galactic center. The period between successive passages of the plane is over thirty-three million years. About seven passages are made through the plane in each orbit. The cycle completes one revolution every two hundred forty million years for the outermost edges of the galaxy. The Sun completes a revolution every two hundred twenty million years at a speed somewhere between one hundred fifteen to one hundred fifty miles per second - four hundred fourteen thousand to five hundred forty thousand miles per hour.

STARS NEAREST TO THE SUN

Alpha Centauri	4.40 light years	25.87 trillion miles
Bernards Star	5.97 light years	35.10 trillion miles
Sirius	8.60 light years	50.56 trillion miles
Cygni	11.20 light years	65.84 trillion miles
Vega	27.00 light years	158.76 trillion miles

Sun's Plane - Side View

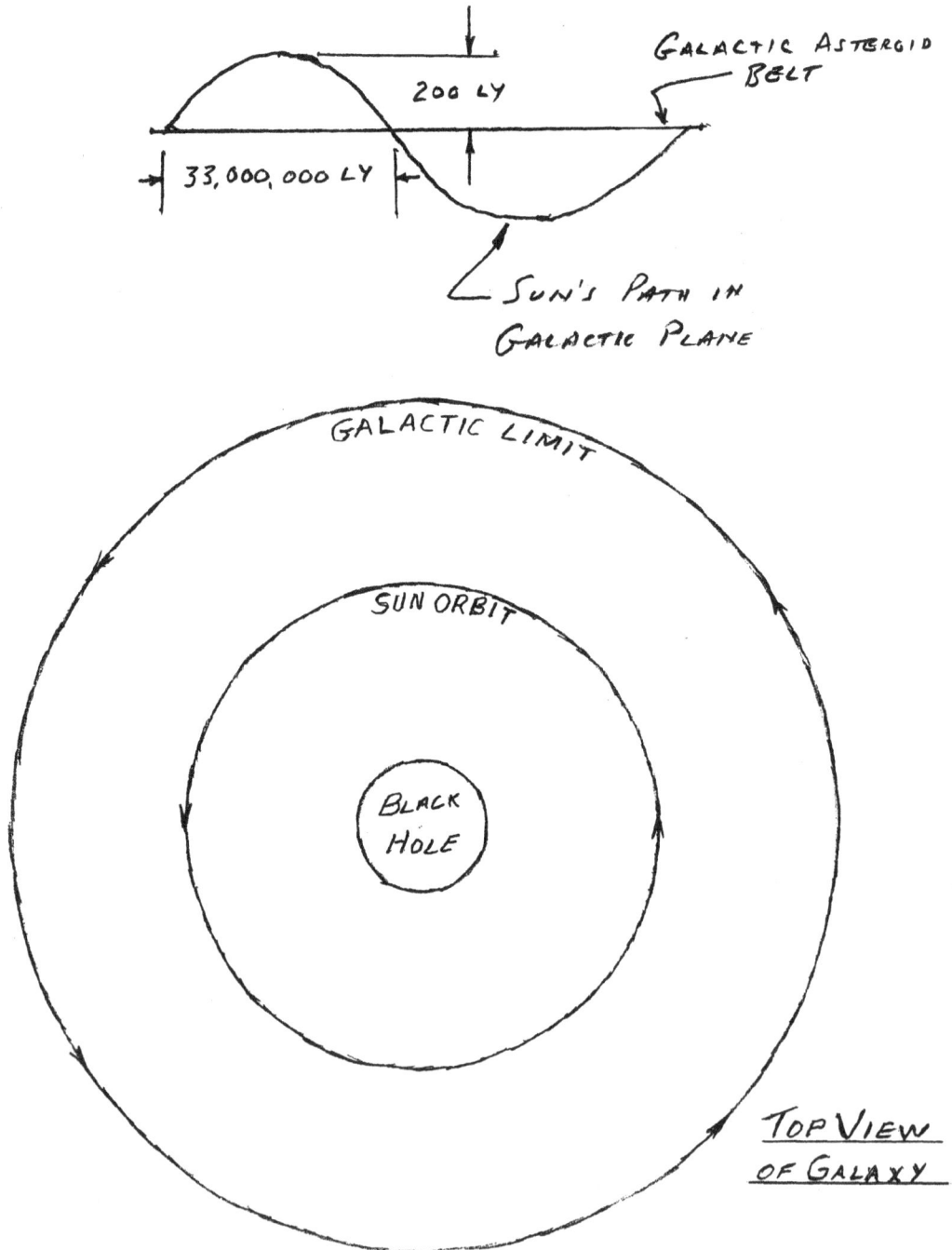

200 LY

GALACTIC ASTEROID BELT

33,000,000 LY

SUN'S PATH IN GALACTIC PLANE

GALACTIC LIMIT

SUN ORBIT

BLACK HOLE

TOP VIEW OF GALAXY

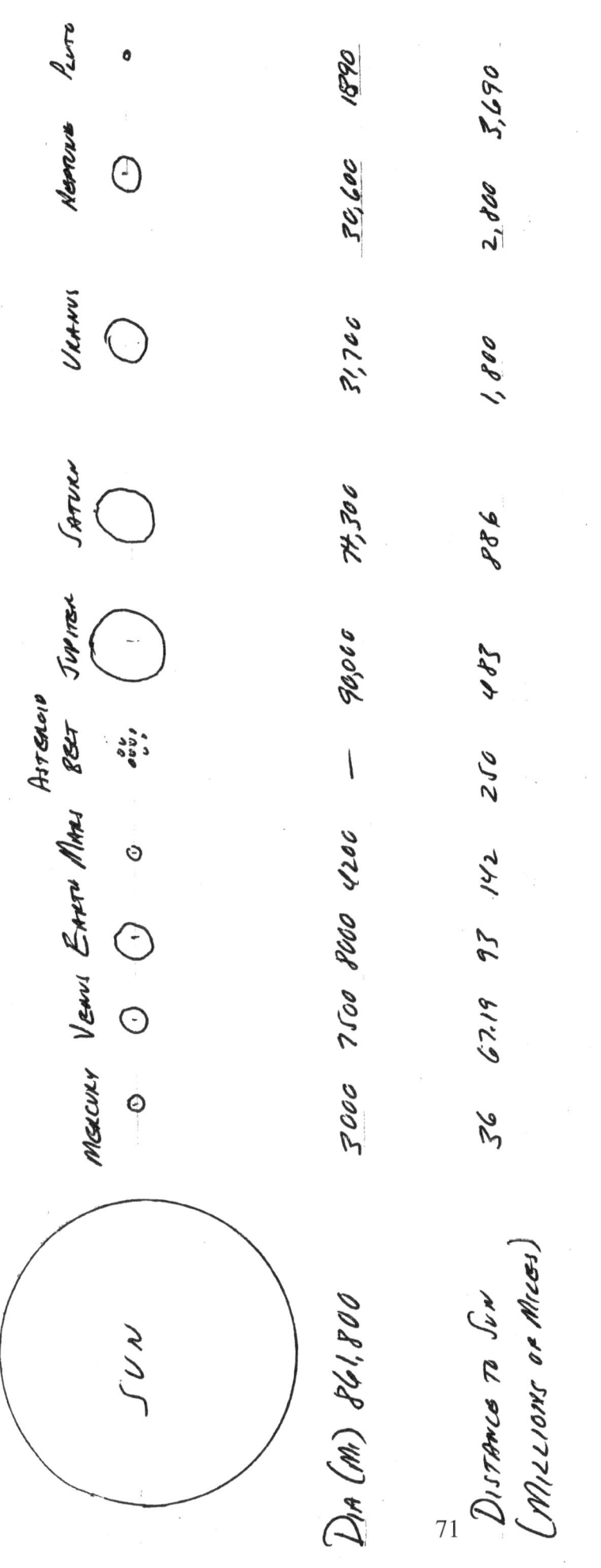

	Mercury	Venus	Earth	Mars	Asteroid Belt	Jupiter	Saturn	Uranus	Neptune	Pluto
DIA (MI)	3000	7500	8000	4200	—	90,000	74,300	31,700	30,600	1890
DISTANCE TO SUN (MILLIONS OF MILES)	36	67.19	93	142	250	483	886	1,800	2,800	3,690

DIA (MI) 861,800 (SUN)

ELECTRON SPHERES 10^4 LARGER THAN NUCLEUS

SUN AS NUCLEUS – $861,800 \times 10^4 = 8.6 \times 10^9$ mi ELECTRON FIELD

PLUTO ORBIT SAME RATIO AS ELECTRON FIELD

71

A MODEL OF THE SOLAR SYSTEM

Most of us have studied models of our solar system located in planetariums, observatories, universities, science centers and museums. These are very interesting to observe because they represent each object from what was known at the time each of the models was constructed. Every new model is more accurate than the one previously constructed due to the latest information we have received from satellites, the Hubble Space Telescope or from other means. These models represent the latest information known - we are learning more each day. These models are accurate to the nth degree *except* for one specification - the proper measurements (sizes relative to each other and the distances between each).

To properly display a model of our solar system showing the proper sizes and distances involved so that one could observe and realize the immensity involved, a space of over one and one-half miles in diameter and one-half mile high would be required. The actual dimensions of this model are shown in the table at the end of this discussion. For this narrative describing the model, the sizes and distances are accurate using common objects from everyday life to represent the model's sizes and distances, rounded off to the nearest amount. The orbit velocities are not correct because the true model movement would be too slow to observe. The velocities chosen for each planet are a true representation of their relative speeds to each other with the time for each to complete an orbit shown by how many orbits would be completed during one orbit of Pluto. To put the entire solar system into true motion, the Sun's velocity would be over one hundred fifty miles per hour - the building would have to be moving at this speed - so we will not place the Sun in motion, it will remain stationary.

The size of the model was determined by the smallest object involved - Pluto. One of the smallest objects a human eye can observe without too much trouble is a grain of sand. This was chosen to represent Pluto, a model diameter of twenty-four thousandths of an inch - halfway between one sixty-fourth and one thirty-second of an inch. All sizes and distances are generated from this determination. The many asteroids would only be dust particles because the largest asteroid is only one-third the size of Pluto.

Let us enter this world's largest building and study this model. As we enter the building and adjust to its enormity, it appears EMPTY! We are given diagrams of this exhibit and paired off into golf carts to explore this vastness. Some choose to find Pluto and work inward toward the Sun. We choose to drive directly to the center of the building - over eight-tenths of a mile away - and start our journey outward from the Sun.

We drive toward the center of the building and cross painted lines on the floor. Each line is a different color and the name of a planet is written on it showing the direction and velocity it is moving. These lines represent the path of the planet identified - follow this line and look up one-quarter of a mile and there is your planet. We arrive at the center of the building and notice that we are still in an empty building. There is a circle on the floor twelve inches in diameter with the word "SUN" enclosed. We look up one-quarter of a mile and until we finally see a BASKETBALL! If this is the largest object in this model of the solar system, we are in trouble and need to change our mode of transportation to view objects located one-quarter of a mile above the floor. We change to a hovercraft and ascend to the height of one-quarter of a mile above the floor to continue our observations. All of the planets, except Pluto, will generally be at this height, the ecliptic (plane of the planets relative to the Sun), in the building.

To anticipate your question of, "Why is your stupid model so high - every other model is located at eye level so you don't have to fly to see it!" Don't blame me - it is Pluto's fault. All the planets, except Pluto, orbit the Sun similar to Earth's orbit - basically level to Earth's orbit; some orbit a little above and others orbit a little below, but generally in the same plane (ecliptic). A model of these planets orbiting the Sun could be made for eye-level observation. Pluto's orbit is inclined a little over seventeen degrees from the level, and, being furthest from the Sun, determines the size of the building. This angle causes Pluto, our grain of sand, to orbit touching the floor at one end of the building to hitting the ceiling, one-half mile above the floor, at the other end of the building over one and one-half miles away. A GRAIN OF SAND determines the size of this volume - the size of the building!

After establishing our proper altitude, we are sitting by a basketball - our SUN! We have seen basketballs before and decide to fly over and observe Mercury. We travel forty-two feet (fourteen yards) from the basketball and wait. All of a sudden a small metal bearing, half the size of an ordinary BB, goes sailing by. We look at each other, and before we could say, "Did you see that?" it passed by again. You have to concentrate to see an object this small passing you every six seconds traveling at thirty miles per hour. This one-half sized BB completes 1,024 orbits by the time the grain of sand, Pluto, traveling at three miles per hour, completes *one* orbit of the basketball. Now we are beginning to realize why we thought the building was empty when we first entered - space is made up of a great quantity of NOTHING! Sitting in our hovercraft, we can look back and recognize the basketball, but when we look out through the rest of the building it still looks EMPTY! Oops, we almost got in Mercury's way as it sped by.

We leave Mercury behind as we travel to view Venus twelve yards away (thirty-six feet from Mercury's orbit). We refer to our diagram of the exhibit to determine what we are to observe. Venus is seventy-eight feet (twenty-six yards) from the basketball. We are looking for a BB that will pass us every sixteen and one-half seconds traveling at nineteen and one-half miles per hour. This BB - Venus - will complete 363 orbits during Pluto's one orbit. Looking back, we can still recognize the basketball. Now for EARTH!

We move until we are one hundred eighty feet (thirty-six yards) from the basketball and look for another BB. This BB, Earth, flies by us at eighteen and one-half miles per hour every twenty-five seconds. There is a grain of sand, our Moon, four and one-half inches from the BB. Our Moon and Pluto are almost the same size. This grain of sand is moving around the BB. We have seen our first satellite of a satellite. This combination completes 247 rotations to each of Pluto. We can look back and recognize the basketball.

We move out almost twenty yards until we are one hundred sixty-five feet (fifty-five yards) from the basketball - our Sun. Mars is only a third larger than Mercury causing us to search for another small bearing about half the size of a BB. Here it comes at fifteen miles per hour, passing us every forty-five seconds. If we lingered here, Mars would pass us 131 times by the time Pluto, our grain of sand, completes one orbit. We can still see the basketball. We move out until we are one hundred yards, the length of a football field, from the basketball. We are in a lot of dust - we have entered the asteroid belt. This dust is moving past us at eleven miles per hour.

We now move out another two hundred sixty feet to observe Jupiter. We are now over one-tenth of a mile from the basketball. We can barely see the basketball and request that a bright light be spotted on it so that we can locate it. We eventually see a billiard ball, Jupiter, pass us traveling at eight miles per hour. We have to wait four minutes and forty-two seconds for it to pass again. We have finally arrived at something large enough for us to see without squinting. This billiard ball would pass us only twenty-one times during Pluto's singular orbit. We have noticed that the further from the basketball we move, the slower the objects become as they pass our hovercraft. There is also a longer time between each pass.

Our next stop is Saturn, one thousand feet (almost two-tenths of a mile) from the basketball. It only comes by every fifteen minutes at five miles per hour, so we go looking for it - a ping-pong ball. It would complete its path eight times as our grain of sand, Pluto, completed one. We can find the basketball's location only due to the brightness of the light - we can see the light but not the basketball.

Now we double our distance from Saturn to reach Uranus - almost twenty-one hundred feet (four-tenths of a mile) from the basketball. We have to search for it because it travels at four and one-third miles per hour. We find a marble that completes it's orbit every thirty-five minutes, completing it three times to each of Pluto's. We can barely see the spotlight that is shining on the basketball.

We continue until we are six-tenths of a mile from the light of our basketball. We have to search for another marble - Neptune - traveling at only three and one-third miles per hour. It takes one hour and seven minutes for Neptune to complete one orbit. We have quite a search on our hands to locate this marble and finally find it as it makes one and one-half orbits to Pluto's one. We are almost to the edge of the building - only two-tenths of a mile to go.

Here is where the trouble begins. We have to find a grain of sand, eight-tenths of a mile from the basketball, traveling at three miles per hour requiring one hour and forty-five minutes to complete one orbit. It is not going in a circle in the path we have taken parallel to the floor of the building. It travels from the floor at one end of the building to the top of the building, one-half mile high, at the opposite end of the building over one and one-half miles away! It is almost a five mile trip searching this orbit before the grain of sand is spotted. We are at the edge of the building as we observe Pluto and look back toward where the basketball is supposed to be located - we can barely see a point of light. We have traveled a long way to see the planets as they are represented in the solar system. Comets come from a long way outside this building.

As we disembark our hovercraft and prepare to leave the building where a truly representative model of the solar system (one you can barely see) is assembled, we look back toward the center of the building and it appears to be nothing more than just another empty building, but we know better. Space is made up of a lot of *nothing*!

A MODEL OF THE SOLAR SYSTEM

OBJECT	SIZE	SIZE	DISTANCE	DISTANCE	SPEED	DIST/ORBI	TIME/ORBIT	ORBITS/PLUTO
SUN	12.000 IN.	BASKETBALL						
MERCURY	0.048 IN.	1/2 DIA. BB	42 FT.	14 YD.	30 MPH	0.05 MI.	6 SEC.	1,024 TIMES
VENUS	0.108 IN.	BB	78 FT.	26 YD.	19 1/2 MPH	0.09 MI.	16 1/2 SEC.	363 TIMES
EARTH	0.108 IN.	BB	108 FT.	36 YD.	18 1/2 MPH	0.13 MI.	25 SEC.	247 TIMES
MARS	0.060 IN.	1/2 DIA. BB	165 FT.	55 YD.	15 MPH	0.20 MI.	45 SEC.	131 TIMES
ASTEROIDS		DUST	300 FT.	100 YD.		0.36 MI.		
JUPITER	1.560 IN.	BILLIARD BALL	560 FT.	0.1 MI.	8 MPH	0.60 MI.	4M 42 SEC	21 TIMES
SATURN	1.080 IN.	PING-PONG BALL	1000 FT.	0.2 MI.	5 MPH	1.50 MI.	15 MIN.	8 TIMES
URANUS	0.480 IN.	MARBLE	2090 FT.	0.4 MI.	4 1/3 MPH	2.50 MI.	35 MIN.	3 TIMES
NEPTUNE	0.480 IN.	MARBLE	3250 FT.	0.6 MI.	3 1/3 MPH	3.80 MI.	1 HR 7 MIN.	1-1/2 TIMES
PLUTO	0.024 IN.	GRAIN OF SAND	4300 FT.	0.8 MI.	3 MPH	5.00 MI.	1 HR 45 MIN	1 TIME

NOTE: ACTUAL SPEEDS FOR THIS MODEL WOULD BE MUCH SLOWER.

PLUTO'S 17.2 DEGREE INCLINATION - DIAMETER OF ORBIT - 1.60 MI.
BASE OF ORBIT   - 1.53 MI.
HEIGHT OF ORBIT   - 0.47 MI.

PLUTO'S ORBIT WOULD BE 1/4 MILE ABOVE CENTER AND 1/4 MILE BELOW CENTER

SOLAR SYSTEM FORMATION

Similar to the forces that formed our galaxy, the origin of the solar system involves the combination of rotation and gravitation. The action begins within the galaxy as an irregular cloud made of gas and dust. This cloud exhibits slow rotation. The random molecular motions are later replaced by gravity, the mutual attraction of the atoms and molecules in the cloud for one another. The cloud begins to contract with the outer edges being drawn inward. The density of the cloud increases as a fixed amount of matter squeezes itself into progressively smaller volumes. As it contracts, the cloud spins faster. As the gas, dust and condensations that make up the cloud spin faster around their common axis of rotation, the centripetal force increases. This will slow down and eventually stop the contraction in the plane of rotation. The matter in this equatorial plane continues to fall in. The initially irregular cloud now becomes a flattened disk. The further the disk collapses, the more rapidly it rotates and the denser it becomes at the center. The collapse slows when the disk is spinning so fast that matter is ejected from the periphery. The center of the disk could be the beginning of a star.

A star is formed when a large amount of gas, mostly hydrogen, starts to collapse in on itself due to its gravitational attraction. As it contracts, the atoms of gas collide with each other more and more frequently and at greater and greater speeds, heating the gas. Eventually, the gas will be so hot that when the hydrogen atoms collide they no longer bounce off each other but instead unite to form helium. The heat released in this reaction, which is like a controlled hydrogen bomb explosion, is what makes the star shine. This additional heat also increases the pressure of the gas until it is sufficient to balance the gravitational attraction causing the gas to stop contracting. The time for a star to contract from an interstellar gas cloud to a main sequence star has been calculated to be one hundred thousand years for a twenty solar mass star, one million years for a one solar mass star and one hundred million years for a one-tenth solar mass star.

This very same cloud action within galaxies is happening continually throughout the Universe. Most of these actions result in binary stars - twin, or double, stars - due to the rotation and gravitation over enormous periods of time. A few may evolve into some type of planetary systems. Finding a system similar to ours would be difficult due to the unique factors of the distance and period that Earth orbits relative to its heat source, the Sun.

The central condensation formed a star. Our Sun is a second generation star. The original star formed during this same sort of action in the formation of the Milky Way galaxy. It condensed, formed our giant star and then, a hundred million

77

years or so later, blew itself apart as a supernova. These normal birth pains were just part of a natural result of cause and effect. This cloud (supernova) became our Sun and its planets - our solar system. The interstellar matter compressed to the density that initiated thermonuclear reactions which caused the Sun to begin shining as a star. Other smaller nearby condensations formed the planets; each sweeping out a wide strip of nearby fragments as each grew in size. The result was the regular spacing of the newly formed planets we have today. Our Sun is a star, and a very average star at that, but with one unique factor: It is very close to us, some three hundred thousand times closer than our nearest neighbor, Alpha Centauri. This glowing ball of gas is held together by its own gravity and is powered by nuclear fusion at the surface of its core.

The Sun rotates, but not as a solid body. Being a gas, it spins differently - faster at the equator and slower at the poles similar to the gaseous planets Jupiter and Saturn. The solar photosphere rotates once every twenty-seven days at the equator but only once every thirty-one days at the poles. The rotation direction is similar to Earth and most of the other planets - west to east. The diameter of the Sun is 861,800 miles at the equator.

The composition of the Sun is primarily hydrogen (the fuel used in the fusion process) at 91.2 percent abundance in atoms or 71.0 percent of the total mass. Helium provides almost all of the remainder with 8.7 percent of the atoms, or 27.1 percent of the total mass. Carbon, nitrogen, oxygen, neon, magnesium, silicon, sulfur and iron constitute the remaining one-tenth of a percent of atoms or nine-tenths of a percent of the total mass. The total mass is 1.99×10^{30} kilograms (2.189×10^{27} tons). The Sun contains 99.865 percent of the mass of the entire solar system.

Electromagnetic radiation and fast moving particles are emitted from the Sun. The radiation, our sunlight, moves away from the photosphere at the speed of light, reaching Earth, ninety-three million miles away, in eight minutes and twenty seconds. The particles, called plasma, travel more slowly at nine hundred thousand to one million six hundred twenty thousand miles per hour reaching Earth in a few days. This constant stream of solar particles is our solar wind. Plasma consists of separated hydrogen atoms parts - individual protons and electrons.

Another observation of the energy generated by the Sun is to place a plate that is one foot square above the atmosphere of the Earth. We receive the full impact of the Sun's energy at this location without any interference from our atmosphere. The amount of energy falling on the one square foot detector would correspond to a flow of a little more than one hundred watts, enough to power an ordinary light bulb. From this we compute that the total energy falling on all the square feet of the surface of Earth would amount to about one hundred seventy trillion kilowatts. Enough falls in one-half hour to equal all energy generated by human energy in a year. Because the Sun is a sphere, we receive approximately four ten-billionths of its generated energy. The Sun is radiating energy into space at the rate of 10^{24} kilowatts.

There is plenty of hydrogen, fused to form helium with the release of energy, in the Sun to serve as fuel. It ignited some 4.6 billion years ago, and at the rate of consuming six hundred million tons per second of hydrogen, will continue to burn for another five billion years or so. This is converted into five hundred ninety-six million tons per second of helium. The four million tons per second difference is the conversion to energy which we observe as sunlight. Seventy billion neutrinos produced by nuclear reactions inside the Sun pass through every square centimeter of Earth every second.

A young, sun-like star with orbiting blobs of dust and rock has been discovered and may be forming planets. This would give astronomers their first chance to observe the evolution of a planetary system like our own. The star, named KH 15D, is twenty-four hundred light years from Earth. KH 15D is only about three million years old and has evidence that a single gaseous planet similar to Jupiter has formed and is circling the star closer that the orbit of Mercury around our own Sun. If further study verifies this as a planet, then planet formation can start very early after a star is born. The orbiting blobs could be other planets beginning to form. Almost one hundred planets have been found in other areas and nearly all of them are large gaseous bodies the size of Jupiter or larger.

The processes that are going on in this inner disk region could be analogous to what was going on in the formation of Earth. This could present information on our origins by helping us to understand how Earth and the planets in the solar system originated.

THE SUN'S LIFE CYCLE

The Sun is a typical fairly low-mass star, steadily burning through its hydrogen fuel and turning its interior into helium. The helium mostly resides in a central core that is inert as far as nuclear reactions are concerned: the fusion takes place at the surface of the core. Therefore, the core itself is unable to contribute to the crucial heat generation needed to hold the Sun up in the face of crushing gravitational forces. To prevent collapse, the Sun expands its nuclear activity outward in search of fresh hydrogen. Meanwhile, the helium core is gradually reduced. As time passes, the Sun's appearance will gradually change as a result of the internal modifications. It will grow in size, but its surface will cool somewhat, giving it a reddish color. This trend will continue until the Sun turns into a red giant star, over two hundred fifty times as large as it is today. The red giant phase marks the beginning of the end for a low-mass star.

Even though a red giant is relatively cool, its large size gives it a huge radiating surface, which means a greater overall luminosity. The Sun's planets will face a hard time, some four billion years from now, as the increased heat engulfs them. The Earth will become uninhabitable long before the atmosphere will be stripped as the oceans boil away. As the Sun's diameter increases searching for more hydrogen to fuse, Mercury, Venus and then Earth will be enveloped in its fiery envelope. Our planet will be reduced to a cinder, still maintaining its orbit even after incineration; the density of the Sun's red-hot gasses will be so low that conditions will approximate a vacuum, exerting little drag on Earth's orbit.

Following a period of complicated activity, low and medium mass stars finally yield to gravity and shrink. The shrinkage is relentless, and continues until the star is compressed to the size of a small planet, becoming a white dwarf. Because white dwarfs are so small, they are very dim even though their surface temperatures can be much greater than that of our Sun. None of these white dwarfs now in existence are visible from Earth without the aid of a telescope.

It is our Sun's destiny to become a white dwarf in the far future. When the Sun reaches that phase, it will continue to remain hot for many billions of years; its huge bulk will be so compacted that it will trap its internal heat more efficiently than the best known insulator. Because the internal nuclear furnace will have shut down for good, there will be no reserves of fuel to replenish the slow leakage of heat radiation into the cool depths of space. Very, very slowly, the dwarf remnant of what was once our mighty Sun will cool and dim until it begins its final change of form as it gradually becomes solid. Eventually it will fade out completely, merging quietly into the blackness of space accompanied by its system - all its planets that will only observe nighttime for eternity in their modified orbits.

KEPLER'S 1ST LAW

THE ORBITAL PATHS OF THE PLANETS ARE ELLIPTICAL WITH THE SUN AT ONE FOCUS

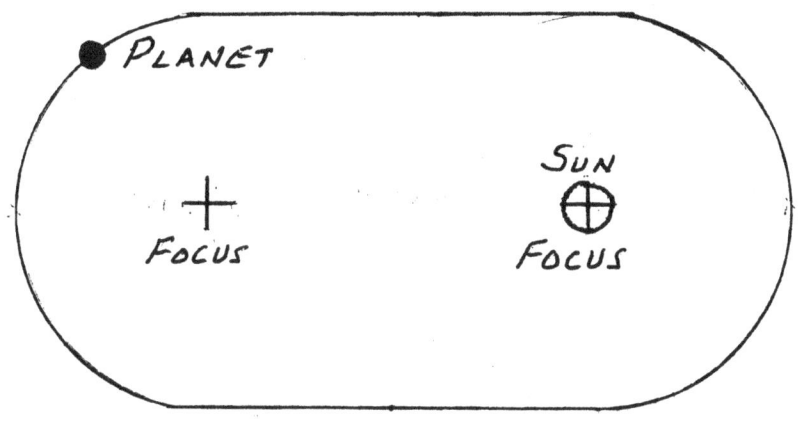

KEPLER'S 2ND LAW

AN IMAGINARY LINE CONNECTING THE SUN TO ANY PLANET SWEEPS OUT EQUAL AREAS OF THE ELLIPSE IN EQUAL INTERVALS OF TIME.

$$\text{AREA}: AB_{SUN} = CD_{SUN} = EF_{SUN}$$
$$\text{TIME } AB = \text{TIME } CD = \text{TIME } EF$$

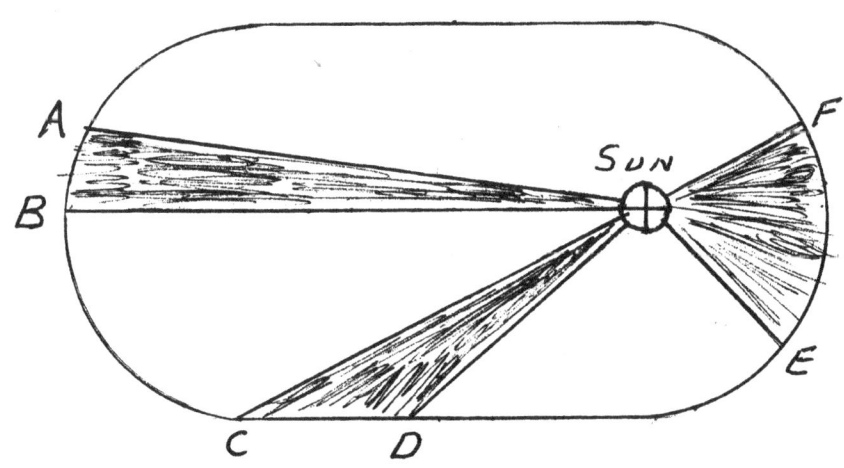

KEPLER'S 3RD LAW:

THE SQUARE OF THE PLANET'S ORBITAL PERIOD IS PROPORTIONAL TO THE CUBE OR IT'S SEMI-MAJOR AXIS. $T^2 = a^3$

$$\frac{T^2}{a^3} = 1$$

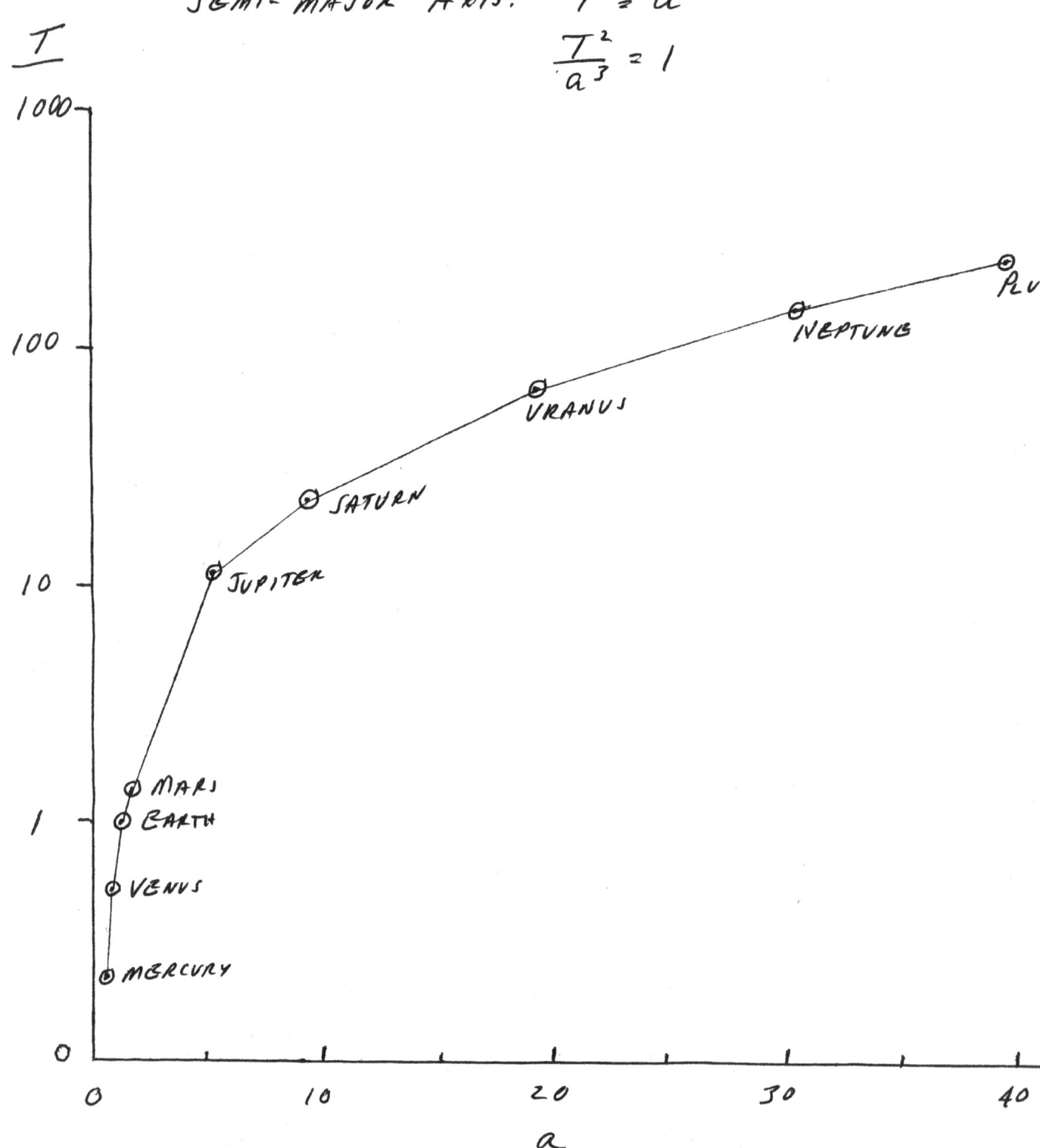

THE CAUSE OF ELLIPTICAL ORBITS

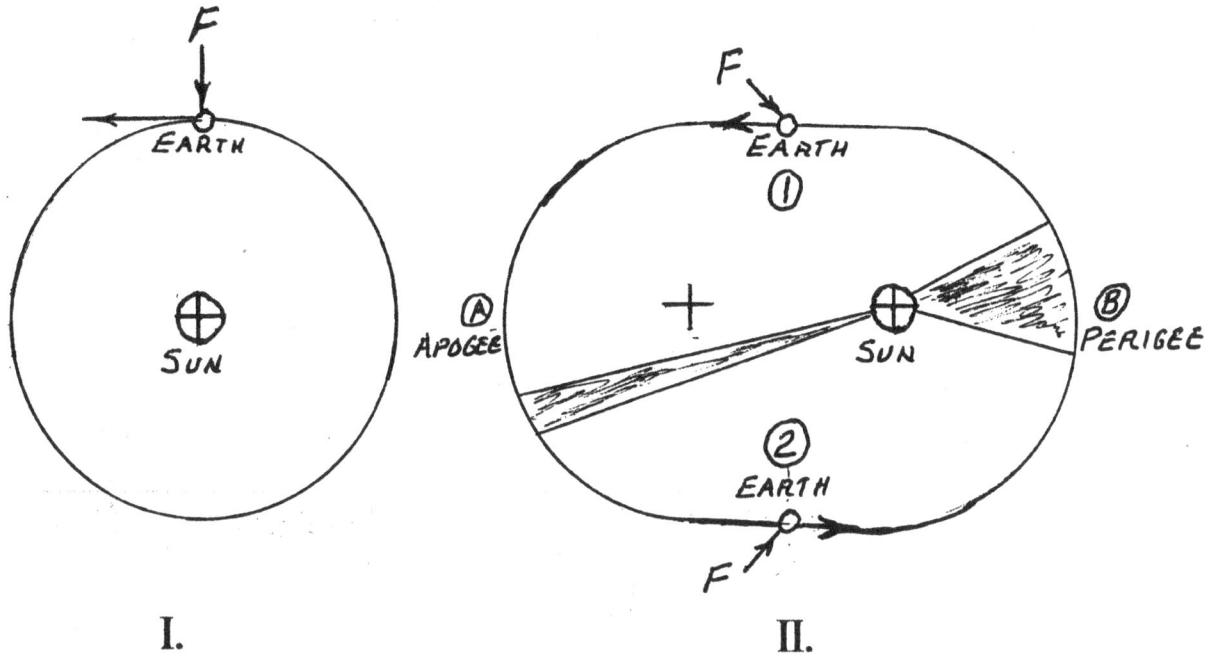

I.   II.

I.   Circular orbit about a stationary object. Satellite has centripetal force equally distributed following the laws of gravitation.
Sun velocity = 0
Earth velocity = 66,588 miles per hour.

II.   Place the object in motion and the satellite appears to assume an elliptical orbit with the Sun at one focus (Kepler's 1ˢᵗ law). The actual distance between the two focus points in Earth's orbit is only three million miles - perigee being almost ninety-two million miles with the apogee being almost ninety-five million miles. The actual path of the satellite around the Sun is sinusoidal through space as the Sun moves along its path. The velocity of the satellite in the opposite direction to movement (1) creates momentum carrying the satellite past the center of the circular path as it swings around the back of the Sun (A) at its slowest pace. Momentum is gained as the satellite speed is added to the Sun's speed, speeding up for the second half of its orbit (2). The satellite will pass closer to (B) due to the Sun's direction of movement being toward the satellite's path. Momentum is increased and velocity is increased - equal areas swept out in equal time (Kepler's 2ⁿᵈ law).
Earth's forward velocity (1) = 433,412 miles per hour.
Earth's forward velocity (2) = 566,588 miles per hour.

WHY CERTAIN PLANETS HAVE RINGS

Our Moon orbits Earth due to the influence of gravity. Each object has an effect on the other. The Moon causes our tides to occur - the force being greatest on the side nearest the Moon and smallest on the opposite side. The Moon is held together by internal forces, its own gravity. We see our tides vary due to the influence of the Moon but cannot see the tidal effects that Earth has on the surface of the Moon. If the Moon were to move closer to Earth the tidal force on it would increase. The closer the Moon came to the Earth, it would bulge toward the Earth. Instead of being a sphere, its shape would appear in the shape of a doorknob, getting flatter the nearer it orbited Earth. As the Moon gets closer, it would reach a point where the tidal forces tending to stretch it out become greater than the internal forces holding it together. At that point, our Moon would be torn apart by Earth's gravity. The pieces of the Moon would then begin their individual orbits around Earth, eventually spreading around Earth in the form of a ring. What once was our Moon would now only be a ring.

For any given planet and any given moon, this critical distance, inside of which the moon is destroyed, is known as the *tidal stability limit,* or the *Roche limit.* The French astronomer Edouard Roche first analyzed the problem of this type of tidal fracturing of a solid body in 1850. He proved mathematically that there is a distance from the center of each massive sphere at which a body approaching the sphere would be torn apart by the sphere's gravitational tidal action. If the average density of the moon is the same as that of the parent planet, then the Roche limit is just 2.4 times the radius of the planet. If our Moon were to close to within ten thousand miles of the center of Earth, we would have a ring orbiting only six thousand miles above the surface. The moons of Earth and Mars are too far from their parent planets to have this potential future.

The Jovian planets - Jupiter, Saturn, Uranus and Neptune - formed and had many moons in their systems. As the moons encountered their respective Roche limit, they were converted into the rings we see today on each. This action would occur in any galaxy anywhere in the Universe involving planets and moons that meet this criteria. Sufficiently small moons can survive even within the Roche limit because they are held together mostly by electromagnetic forces, not by gravity.

Saturn's rings have a thickness of only tens of meters as they stretch across several hundred thousand kilometers - proportionally as thick as a sheet of tissue paper spread over a football field. Uranus' rings are slightly out of round due to their self-gravity resisting the tendency to smear into a circular band. The faintest known rings have optical depths between 10^{-8} and 10^{-6} - particles are spread far apart. Jupiter's rings and Saturn's outermost rings are this faint. Because these particles do not collide very often, they seldom form a flattened disk. These particles are comparable in size to smoke particles. Neptune's densest ring is not a smooth band; it contains discontinuous arcs that together encompass less than a tenth of the circumference. These structures should spread fully around the planet in about a year. Images from the Hubble Space Telescope and ground based observatories find that the positions of the arcs have shifted little in the past fifteen years.

Several lines of evidence now suggest that most rings are young. First, tiny grains must lead short lives. Second, some ring moons lie very close to the rings. Third, icy ring particles should be darkened by cometary debris, yet they are generally bright. Fourth, satellites just beyond Saturn's rings have remarkably low densities, as though they are rubble piles. Finally, some moons are embedded within rings. Unless replenished, faint rings should disappear within just a few thousand years. Variations in the composition, history and size of the planets and satellites would naturally account for the great diversity of rings.

MERCURY

Mercury, the nearest planet to the Sun, is visible to the naked eye only when the Sun's light is blotted out - just before dawn or just after sunset. Mercury is named after the Roman god who was the fleet-footed messenger for the other gods. The ancients originally believed this companion to the Sun was two different objects - the connection between the planet's morning and evening appearances took some time to establish. The early Greek astronomers had two separate names for Mercury - Hermes in the evening and Apollo in the morning. Mercury is the Roman name for the Greek god Hermes. Later Greek astronomers became aware that the two "planets" were really different alignments of a single body.

Mercury rotates west to east on its axis at about six miles per hour and consists of rock and metal and has an appreciable magnetic field indicating the presence of a large metallic core, which also accounts for the planet's high density. The surface of Mercury appears similar to the Earth's Moon due to the presence of craters - impacts that occurred in the past. Erosion of these craters, if any, is negligible showing the pattern of planet construction. Mercury does not have an atmosphere.

Diameter	4,878 km	3,030 miles
Mass	3.3×10^{23} kg	363×10^{18} tons
Density	5.44 gm/cm^3	0.198 lb/in^3
Surface gravity	3.7 m/sec^2	12 ft/sec^2
Escape velocity	4.25 km/sec	9,503 mi/hr
Surface temperature	-173 C to +427 C	-279.4 F to +800.6 F
Satellites	None	
Axis (distance to Sun)	57.9 million km	36 million miles
Mean orbital velocity	47.87 km/sec	107,037 mi/hr
Sidereal year	87.969 days	
Sidereal day	58.646 days	
"Noon to noon" day (solar day)	176 days	
Orbital distance	364 million km	226 million miles
Ecliptic inclination	7 degrees	
Axis tilt	2 degrees	
Spin-orbit resonance	3:2 - 3 rotations for every 2 revolutions	

$3 \times 58.65 = \underline{175.95}$ (3 rotations) = 3 times sideral day

$2 \times 88 = \underline{176}$ (2 revolutions) = 2 times sidereal year

VENUS

Venus is named in honor of the Roman goddess of love and beauty. In its bulk properties Venus is almost a carbon copy of Earth. These two planets are similar in size, density and chemical composition. They orbit at comparable distances from the Sun. At formation they must have been almost identical, yet they are now about as different as two terrestrial planets could be. While Earth is an energetic world, rich with life, Venus is an uninhabitable inferno, with a dense, hot atmosphere of carbon dioxide, lacking any trace of oxygen or water. Our only common factor is that both consist of rock and metal.

Being the second planet from the Sun, Venus orbits within Earth's orbit and is visible above the horizon for at most three hours before the Sun rises or after it sets. This is the reason it is called the "morning star" or the "evening star," depending on where it happens to be in its orbit. This planet appears brighter than any star. Venus' spin is from east to west - retrograde (opposite) to the rotation of Earth and the other planets - at only four miles per hour.

The atmosphere consists of 96.5 percent carbon dioxide with almost all of the remaining 3.5 percent being nitrogen. The top layers of clouds are composed of sulfuric acid, created by the reactions between sulfur dioxide and water. Sulfur dioxide is an excellent absorber of ultraviolet radiation. Particles of sulfur are suspended in the atmosphere in and near the cloud layers and account for Venus' characteristic yellowish color. The reason Venus' atmosphere is so hot is because its atmosphere is much thicker and denser than Earth's, a much smaller fraction of the infrared radiation leaving the planet's surface actually escapes into space. The result is a much stronger greenhouse effect than on Earth and a correspondingly hotter planet. The outgoing infrared radiation is not absorbed at a single point in the atmosphere; instead, absorption occurs at all atmospheric levels. Venus, being thirty percent closer to the Sun, receives almost twice the amount of sunlight than Earth. This evaporated what water was available and created great amounts of carbon dioxide.

The planet's surface appears to be both smooth, resembling rolling plains with modest highlands and lowlands, and rough and rocky in other areas. Rift valleys appear to be present with other variable features. Only two or three continental-sized features decorate the landscape, and these contain mountains comparable in height to those on Earth. The highest peaks rise some 14 kilometers (8.7 miles) above the level of the deepest surface depression. On Earth, the distance from the top of Mount Everest to the deepest section in the ocean, Challenger Deep at the bottom of the Marianas Trench on the eastern edge of the Philippine plate, is about 20 kilometers (12.5 miles).

Diameter	12,102.6 km	7,516 miles
Mass	4.88×10^{24} kg	$5,368 \times 10^{18}$ tons
Density	5.27 gm/cm³	0.192 lb/in³
Surface gravity	8.88 m/sec²	28.86 ft/sec²
Escape velocity	10.4 km/sec	23,254 mi/hr
Surface temperature	750 K (477 C)	890.6 F
Satellites	None	
Axis (distance to Sun)	108.2 million km	67.19 million miles
Mean orbital velocity	35.01 km/sec	78,282 mi/hr
Sidereal year	224.701 days	
Sidereal day	243 days	
"Noon to noon" day (solar day)	116 days	
Orbital distance	680 million km	422 million miles
Ecliptic inclination	3.4 degrees	
Axis tilt	3.0 degrees	

MARS

Mars orbit is more elliptical than Earth's, so its distance from the Sun varies more. Mars rotates at almost the same rate as Earth and its rotation axis is inclined to the ecliptic at almost the same angle as Earth's axis. Due to its axial tilt, Mars has daily and seasonal cycles much like those on Earth, but they are more complex than those on Earth because of Mars' eccentric orbit. Mars was named after the god of war in Roman mythology. He was known as the father of Romulus and Remus, the legendary founders of Rome. Mars was once the god of agriculture but he must have been promoted and the job then went to Saturn at the time.

The surface of Mars has a wide range of geological features - huge volcanoes, deep canyons, vast dune fields and many other geological wonders. There is as marked difference of the terrain between the northern and southern hemispheres. The northern hemisphere is made up largely of rolling volcanic plains, similar to the lunar surface. These extensive northern lava plains were formed by eruptions involving enormous volumes of lava. The plains are covered with blocks of volcanic rock, as well as with boulders blasted out of impact areas by meteoroids because the atmosphere is too thin to offer much resistance to incoming debris. The southern hemisphere consists of heavily cratered highlands lying several kilometers above the level of the lowland north. Most of the dark regions visible from Earth are mountainous regions in the south.

The major geological feature on the planet is the Tharsis bulge. It is a region, roughly the size of North America, on the Martian equator that rises some ten kilometers higher than the rest of the Martian surface. To the east of Tharsis lies Cyryse Planitia and to its west lies a region known as Isidis Planitia. These features are wide depressions, hundreds of kilometers across and up to three kilometers deep.

Liquid water once existed in great quantity on the surface of Mars. The runoff channels are found in the southern highlands. They are extensive systems - sometimes hundreds of kilometers in total length - of interconnecting, twisting channels. These runoff channels reveal that some four billion years ago when the atmosphere was thicker, a surface of warm and liquid water was widespread. Italian astronomer, Giovanni Schiaparelli called these *channels* "canali" which was later mistranslated to "canals." The outflow channels are probably remnants of catastrophic flooding that occurred long ago. They appear only in equatorial regions and generally do not form the extensive interconnected networks that characterize the runoff channels. They are probably the paths taken by huge volumes of water draining from the southern highlands into the northern plains.

The Martian atmosphere is quite thin and composed of: 95.3 percent carbon dioxide, 2.7 percent nitrogen, 1.6 percent argon, 0.13 percent oxygen, 0.07 percent carbon monoxide and about 0.03 percent water vapor. The level of water vapor is quite variable. The atmosphere is pressure is only about 1/150th the pressure of Earth's atmosphere at sea level. Mars consists of rock and metal and rotates west to east at 550 miles per hour.

Unlike Earth's moon, Mars two moons are tiny compared with their parent planet and orbit very close to it. These two moons - Phobos and Deimos - are only a few kilometers across. These moons did not form with Mars but instead are asteroids that were slowed and captured by the outer fringes of the early Martian atmosphere. Both are quite irregularly shaped and heavily cratered. The larger of the two is Phobos which is about thirty-eight kilometers long and twenty kilometers wide and dominated by an enormous ten kilometer diameter crater named Stickney. The smaller Deimos is only sixteen kilometers long by ten kilometers wide with its largest crater being 2.3 kilometers in diameter. Phobos and Deimos move in circular, equatorial orbits as they rotate synchronously - keeping the same face permanently turned toward the planet. Phobos lies 9,378 kilometers from the center of Mars and has an orbital period of 7 hours 59 minutes. Demos lies further out at 23,459 kilometers and orbits in 30 hours 18 minutes.

Diameter	6,794 km at equator	4,219 miles
	6,751 km at poles	4,192 miles
Mass	6.42×10^{23} kg	706×10^{18} tons
Density	3.933 gm/cm^3	0.143 lb/in^3
Surface gravity	372.52 cm/sec^2	12.11 ft/sec^2
Escape velocity	5.024 km/sec	11,234 mi/hr
Surface temperature	-143 C to +27 C	-225.4 F to +80.6 F
Satellites	Two	
Axis (distance to Sun)	228 million km	142 million miles
Mean orbital velocity	24.13 km/sec	53,979 mi/hr
Sidereal year	1.88 year (687 days)	
Sidereal day	24 hr 37 min 23 sec	
Orbital distance	1.436 billion km	892 million miles
Ecliptic inclination	1.9 degrees	
Axis tilt	25 degrees average *	

* Oscillates between 15 and 35 degrees every two million years.

ASTEROID BELT

The region between the orbits of Mars and Jupiter is the location of most of the asteroids of the solar system. An asteroid is one of thousands of very small members of the solar system that orbit two hundred fifty million miles from the Sun. Asteroids travel in a moderately circular orbit in a belt roughly ninety million miles wide and shaped like a doughnut. Astronomers have learned that the asteroid belt is really two belts side by side. The inner one is made mostly of rocky lumps and the outer belt consists of mostly carbon-rich asteroids. The distance of this orbit around the Sun is over one and one-half billion miles, each asteroid traveling around forty thousand miles per hour.

Asteroids consist of rock and metal that were not accumulated into a planet and orbit in this plane in a generally even consistency. Over five hundred have a diameter of over fifty kilometers (thirty-one miles). Ceres is one thousand twenty kilometers (six hundred thirty-two miles), Vesta is five hundred fifty kilometers (three hundred forty-one miles) and Pallar is five hundred forty kilometers (three hundred thirty-five miles) in diameter. Chiron, an asteroid that orbits mainly between the orbits of Saturn and Uranus, is a moderately large body that was discovered in 1977. It is believed that the carbon containing asteroids consist of very primitive material, representative of the earliest stages of the solar system, and have not experienced significant heating or chemical evolution since the first formed some 4.6 billion years ago.

When the solar system was young, before the earth-like planets had formed, small bits of iron and rock circled the Sun everywhere, held captive by its gravity. In the inner part of the solar system, these bits of iron and rocky material collected to form Earth and the inner planets. Farther out beyond the orbit of Mars, Jupiter disrupted the process of making a planet. Fragments of rock and iron *started* to collect into larger pieces to form a full sized planet in the orbit we now call the asteroid belt. Before this accumulation could begin, Jupiter's gravitational pull disturbed their paths. Because Jupiter is so massive, the force of its gravity is greater than that of any other body in the solar system except the Sun. As a result, the asteroids, pushed and pulled by Jupiter's gravity, crashed into one another violently as they traveled around the Sun instead of colliding gently and coming together under the force of gravity. This is the primary reason the pieces of matter in the asteroid belt never accumulated into one large planetary body.

Jupiter's gravity is still pulling asteroids our of their orbits today, often setting one asteroid on a collision course with another. When a collision occurs, the shattered fragments of the two asteroids leave the belt, usually traveling in many different directions and different speeds. Some fragments are hurled further out into the solar system. Other asteroid fragments are knocked in toward the Sun by the collision. A few of these have orbits that cross the orbit of Earth. The Apollo objects are asteroids that have ventured close to Jupiter's gravity. This converts the orbits into highly elongated ones that may even intersect Earth's orbit. These asteroids vary in size. Small ones that enter Earth's atmosphere are vaporized during their fiery entry and are seen as shooting stars. Slightly larger ones may survive and hit the ground as a meteorite. Much larger ones that appear every few million years disrupt areas large enough at times to cause mass extinctions in large areas or sometimes involving the entire surface of Earth.

JUPITER

Jupiter, the King of the Roman gods, is the largest planet in the solar system. Its mass is more than twice the mass of all the other planets combined. It is composed of 86.1 percent hydrogen and 13.8 percent helium along with other gases. It rotates west to east at 28,400 miles per hour producing a pronounced equatorial bulge, the amount inferring the presence of a large rocky core in its interior. Having no solid surface, the rotation rate varies from place to place in the atmosphere. Jupiter's clouds are arranged in several layers and are the product of complex and continuous chemical processes occurring in the planet's turbulent atmosphere. The various visible clouds lie at different levels with the white ammonia clouds lying over the more brightly colored layers. The colors we see are the result of chemical reactions, fueled by the planet's interior heat, solar ultraviolet radiation, auroral phenomena and lightning, all occurring at various depths below the cloud tops as seen through "holes" in the overlaying clouds. The main weather pattern on Jupiter is the *Great Red Spot*, an Earth-sized hurricane that has been raging for at least three centuries. Other, smaller weather systems, the *white* and *brown ovals*, are also observed. They can persist for decades. Jupiter radiates about twice as much energy into space as it receives from the Sun. The source of this energy is most likely the heat left over from the planet's formation some 4.6 billion years ago. Jupiter's atmosphere becomes hotter and denser with depth, eventually becoming liquid. Interior pressures are so high that the hydrogen is "metallic" in nature near the center. The planet has a large "terrestrial" core ten to twenty times the mass of Earth.

The magnetosphere of Jupiter is about a million times more voluminous than Earth's magnetosphere, and the planet has a long magnetic "tail" extending away from the Sun to at least the distance of Saturn's orbit. Energetic particles spin around magnetic field lines, accelerating Jupiter's magnetic field, producing intense radio frequency radiation.

Twenty-eight moons have been discovered to date on Jupiter. The outermost eight moons resemble asteroids and the four outer ones have retrograde orbits. The innermost of the *Galilean Satellites*, named after Galileo Galilei who discovered them, is the moon Io. The others are Europa, Ganymede and Callisto.

The Jupiter space probe, Project Galileo, announced in 1977, built by 1983 and launched in 1989 was one of the most ambitious interplanetary missions at that time. In addition to the many facts found about Jupiter, the mission also explored a few of Jupiter's many moons. It arrived in the area in 1995 to begin its mission of explorations.

The moon Io is the most volcanic body in the solar system. Rocky Io is tortured by Jupiter's mighty gravitational pull. With a solid surface that rises and falls in three hundred foot tides, Io changes its face daily with new eruptions of gas and superheated lava, some of which reach temperatures as high as 3,140 F. To picture Europa, consider Earth's moon and cover it with a hundred mile thick shell of water, some of it frozen. Ganymede, the solar system's largest moon, is bigger than Mercury. It pulsates with its own magnetic field and may have once erupted with volcanoes of ice or water. Callisto, Jupiter's most distant moon, has something eating away at its surface, but even more mysterious is what may lie beneath its crater face - an ocean of salty water.

Diameter	142,984 km	88,676 miles
Mass	1.9×10^{27} kg	2.1×10^{24} tons
Density	1.32 gm/cm^3	0.048 lb/in^3
Surface gravity	26.06 m/sec^2	85.47 ft/sec^2
Escape velocity	61 km/sec	136,396 mi/hr
Surface temperature	-148 C to -73 C	-153.8 F to -99.4 F
Satellites	28	
Axis (distance to Sun)	778 million km	483 million miles
Mean orbital velocity	13.06 km/sec	29,202 mi/hr
Sidereal year	11.862 years	
Sidereal day	9 hrs 55 min 30 sec	
Orbital distance	4.83 billion km	3 billion miles
Ecliptic inclination	1.3 degrees	
Axis tilt	3.0 degrees	

JUPITER'S LARGEST MOONS

MOON	DISTANCE FROM JUPITER		DIAMETER	
	KM	MI	KM	MI
Io	422,000	262,200	3,632	2,257
Europa	671,000	416,900	3,126	1,942
Ganymede*	1,020,000	664,900	5,276	3,278
Callisto	1,833,000	1,170,000	4,820	2,995

* Solar system's largest moon.

SATURN

Saturn was the outermost planet known to ancient astronomers and orbits the Sun at almost twice the distance of Jupiter. Named after the father of Jupiter in Greek and Roman mythology, also known as the god of agriculture. Saturn, as does Jupiter, rotates very rapidly from west to east at a speed of 22,800 miles per hour with an atmosphere consisting of 92.4 percent molecular hydrogen, 7.4 percent helium, 0.2 percent methane and 0.02 percent ammonia. Weather systems are seen on Saturn, as on Jupiter, although they are less distinct. Saturn has a weaker gravity and a more extended atmosphere than Jupiter with the planet's overall butterscotch color being due to the cloud chemistry similar to that occurring in Jupiter's atmosphere. Saturn, like Jupiter, has bands, ovals and turbulent flow patterns powered by convection motion in the interior. It is the least dense planet and the only one that is lighter than water.

Saturn, as with Jupiter, emits far more radiation into space than it receives from the Sun. Saturn's conducting interior and rapid rotation produce a strong magnetic field and an extensive magnetosphere large enough to contain the planet's ring system and innermost sixteen moons. From Earth, the main visible feature of Saturn's rings are the A, B and C Rings and the Cassini and Encke Divisions. The Cassini Division is a dark region between the A and B rings. The Encke Division lies near the outer edge of the A ring. The rings are made up of trillions of icy particles ranging in size from dust grains to boulders, all orbiting Saturn like so many tiny moons. The total mass is comparable to that of a small moon. Both divisions are dark because they are almost empty of ring particles because it is believed that planetary rings have a lifetime of only a few tens of millions of years. The fact that rings are seen around the four Jovian planets means that they must constantly be reformed or replenished, perhaps by material chipped off moons by meteoritic impact or by the tidal destruction of entire moons. Each of Saturn's rings had started out as a moon within Saturn's Roche limit but has been fragmented into a ring by Saturn's tidal action.

The medium-sized moons of Saturn are made up predominately of rock and water ice. They show a wide variety of surface terrains and are heavily cratered, tidally locked by the planet's gravity into synchronous orbits. The innermost mid-size moon Mimas exerts influence over the structure of the rings. The moon Iapetus has a marked contrast between its leading and trailing faces, while Enceladus has a highly reflective appearance, possibly the result of water "volcanoes" on its surface. Saturn's small moons exhibit a wide variety of complex motions. Several moons "share" orbits. The moon Hyperion undergoes chaotic rotation - constantly tumbling in an unpredictable way as it orbits the planet.

Saturn's large moon Titan is the second largest moon in the solar system, also larger than the planets Mercury and Pluto. Its thick atmosphere obscures the moon's surface and may be the site of a complex cloud and surface chemistry. The existence of Titan's atmosphere is a direct consequence of the cold conditions that prevailed at the time of the moon's formation. Titan is the only moon in the solar system to possess a substantial atmosphere composed of compounds of nitrogen, carbon and hydrogen which is believed to resemble Earth's atmosphere during its infancy. It is also the only known body in the solar system besides Earth whose surface is covered by a liquid. Its oceans are composed of liquid methane at temperatures of -175 degrees Celsius and its continents are made of ice.

Diameter	120,660 km	74,930 miles
Mass	5.684×10^{26} kg	625.24×10^{21} tons
Density	0.71 gm/cm^3	0.026 lb/in^3
Surface gravity	12.9 m/sec^2	41.9 ft/sec^2
Escape velocity	39.4 km/sec	88,098 mi/hr
Surface temperature	-178.6 C	-388.88 F
Satellites	32	
Axis (distance to Sun)	1.427 billion miles	886 million miles
Mean orbital velocity	9.7 km/sec	21,689 mi/hr
Sidereal year	29.46 years	
Sidereal day	10 hrs 39 min 24 sec	
Orbital distance	8.97 billion km	5.57 billion miles
Ecliptic inclination	2.5 degrees	
Axis tilt	26.7 degrees	

SATURN'S LARGEST MOONS

MOON	DISTANCE	ORBIT PERIOD	DIAMETER	
Tethys	295,000 km	1.89 days	1,050 km	652 miles
Dione	377,000 km	2.74 days	1,120 km	696 miles
Rhea	527,000 km	4.52 days	1,530 km	950 miles
Titan	1,220,000 km	16.0 days	5,150 km	3,198 miles
Iapetus	3,560,000 km	79.3 days	1,440 km	894 miles

URANUS

The three outermost planets were unknown to the ancients. British astronomer William Herschel discovered the planet Uranus in 1781. The tradition of using names from Greco-Roman mythology was continued and the planet was named Uranus, the father of Saturn. Uranus rotates east to west at 5,800 miles per hour around a large rocky core. The axis is tilted ninety-eight degrees from the perpendicular - tipped on its side pointed toward the Sun. For half a year the "North Pole" points toward the Sun and the "South Pole" faces the Sun the other half of the year. This produces some extreme seasonal effects and it is not known how the tilt became so extreme. The atmosphere consists of 84 percent molecular hydrogen, 14 percent helium and 3 percent methane. Uranus has atmospheric clouds and flow patterns that move around the planet with the rotation with wind speeds of two hundred to five hundred miles per hour.

Uranus has five major moons and at least sixteen smaller moons. The five largest moons are probably composed of ice and rock. The two outermost moons, Titania and Oberon, are heavily cratered and show little indication of any geological activity. Ariel does appear to have experienced some activity in the past. It shows signs of resurfacing in places and exhibits surface cracks. Miranda displays a wide range of surfaces, including ridges, valleys, large oval faults and many other difficult geological features. These were possibly caused by catastrophic disturbances that happened several times.

The ring systems surrounding this planet were discovered in 1977. The main rings are named Alpha, Beta, Gamma, Delta and Epsilon and range from forty-four thousand to fifty-one thousand kilometers from the planet's center. These rings are dark and widely spaced. There are a total of eleven rings.

Diameter	52,290 km at equator	32,472 miles
	51,036 km at poles	31,694 miles
Mass	8.7×10^{25} kg	95.7×10^{21} tons
Density	1.2 gm/cm^3	0.044 lb/in^3
Surface gravity	8.3 m/sec^2	26.98 ft/sec^2
Escape velocity	20.833 km/sec	46,603 mi/hr
Surface temperature	-215 C	-743.88 F
Satellites	21	
Axis (distance to Sun)	2.88 billion km	1.79 billion miles
Mean orbital velocity	6.8 km/sec	15,205 mi/hr
Sidereal year	84.01 years	

Sidereal day	11 to 24 hours	
Orbital distance	18.19 billion km	11.3 billion miles
Ecliptic inclination	0.8 degrees	
Axis tilt	98.0 degrees	

URANUS' LARGEST MOONS

MOON	DISTANCE	ORBIT PERIOD	DIAMETER	
Miranda	130,000 km	1.41 days	485 km	301 miles
Ariel	191,000 km	2.52 days	1,160 km	720 miles
Umbriel	266,000 km	4.14 days	1,190 km	739 miles
Titani	436,000 km	8.71 days	1,610 km	1,000 miles
Oberon	583,000 km	13.5 days	1,550 km	963 miles

NEPTUNE

Eighteenth century astronomers quickly discovered a small discrepancy between Uranus' predicted position and the observed position. Half a century later the discrepancy had grown to a quarter of an arcminute, much too large to have been an observational error. The logical conclusion was that an unknown body must be exerting a gravitational force on Uranus. Astronomers realized that there had to be another planet in the solar system disturbing Uranus' motion. In the 1840's, two mathematicians independently solved the difficult problems of determining this new planet's mass and orbit. In 1846 a German astronomer, Johann Galle, found the new planet within one or two degrees of the predicted position on his first attempt. Neptune was named after the god of the sea in Roman mythology.

The atmosphere is similar to Uranus. It consists of 84 percent molecular hydrogen, 14 percent helium and 2 percent methane. Neptune rotates west to east at 5,750 miles per hour. Although it lies at a greater distance from the Sun, Neptune's upper atmosphere is actually slightly warmer than Uranus'. Neptune has an internal energy source causing it to radiate 2.7 times more heat than it receives from the Sun. This source could be heat left over from the planet's formation and this combination of extra heat and less haze may be responsible for the greater visibility of Neptune's atmospheric features, as its cloud layers lie at higher levels in the atmosphere than do Uranus'. Neptune displays several storm systems similar in appearance to those seen on Jupiter. The largest such storm is known as the Great Dark Spot.

Neptune has a rocky core about the size of Earth. It has eight known moons with the inner moon, Triton, discovered in 1846 and the outer moon, Nereid, discovered in 1949. Voyager 2 discovered the additional smaller moons. Four dark rings surround Neptune, three are quite narrow and one is quite broad and diffuse.

Diameter	49,528 km	30,678 miles
Mass	1.03×10^{26} kg	113.3×10^{21} tons
Density	1.66 gm/cm^3	0.06 lb/in^3
Surface gravity	13.8 m/sec^2	44.85 ft/sec ^{2}
Escape velocity	18.46 km/sec	41,300 mi/hr
Surface temperature	-213 C	-179.2 F
Satellites		8
Axis (distance to Sun)	4.5 billion km	2.8 billion miles
Mean orbital velocity	5.4 km/sec	12,075 mi/hr

Sidereal year	164.8 years	
Sidereal day	18 hrs	
Orbital distance	28 billion km	17.6 billion miles
Ecliptic inclination	1.8 degrees	
Axis tilt	28.7 degrees	

NEPTUNE'S MOONS

MOON	DISTANCE	ORBIT PERIOD	DIAMETER	
Proteus	118,000 km	1.12 days	415 km	258 miles
Triton	354,000 km	5.88 days*	2,760 km	1,714 miles
Nereid	5,520,000 km	360.2 days	200 km	124 miles
		*Retrograde		

PLUTO

By the end of the nineteenth century observations of the orbits of Uranus and Neptune suggested that Neptune's influence was not sufficient to account for all of the irregularities in Uranus' motion. It was noticed that some other unknown body could also affect Neptune's orbit. Percival Lowell calculated where the supposed ninth planet should be. He died in 1916 and in 1930 American astronomer Clyde Tombaugh found the missing planet only six degrees from Lowell's predicted position. The new planet was named Pluto for the Roman god of the dead who presided over eternal darkness. Its discovery was announced on March 13, 1930 on Percival Lowell's birthday.

Pluto consists of rock and metal and rotates west to east at thirteen miles per hour. Its surface composition is nitrogen, carbon monoxide, methane and water ice. It has one moon, Charon, named after the mythical boatman who ferried the dead across the river Styx into Hades, Pluto's domain. The surface composition of Charon is water ice. The moon's orbital period is 6.39 days, weighs 1.5×10^{22} kg and has a diameter of 1,200 kilometers orbiting at an average distance of 19,600 kilometers from Pluto. Because Pluto is neither terrestrial nor Jovian in its makeup and because of its similarity to the ice moons of the outer planets, it is obvious that Pluto is not a "true" planet, it is probably an escaped planetary moon or a large ice chunk of debris left over from the formation of the solar system. Pluto's orbit, at times, can pass inside the orbit of Neptune, and at these times, Neptune's orbit is the furthest from the Sun.

Due to the fact that Charon is half the size of Pluto, astronomers consider them a double planet. The prevailing theory of how the Pluto-Charon system formed is that Pluto collided with another large body in the distant past and that much of the debris from this impact went into orbit around Pluto and eventually coalesced to form Charon. Because it appears that a similar collision led to the creation of Earth's moon, the study of Pluto and Charon is expected to shed some light on that subject as well.

Diameter	2,246 km	1,395 miles
Mass	1.46×10^{22} kg	16×10^{18} tons
Density	2.29 gm/cm^3	0.083 lb/in^3
Surface temperature	-223 C	-369.4 F
Satellites	One	

Axis - min. (distance to Sun)	4.4 billion km	2.75 billion miles
max.	7.3 billion km	4.58 billion miles
mean	5.94 billion km	3.69 billion miles
Mean orbital velocity	4.74 km/sec	10,599 mi/hr
Sidereal year	247.69 years	
Sidereal day	6.4 days	
Orbital distance	37.40 billion km	23.25 billion miles
Orbit diameter	11.87 billion km	7.38 billion miles
base	11.38 billion km	7.07 billion miles
height	3.52 billion km	2.19 billion miles
Ecliptic inclination	17.2 degrees	

KUIPER BELT

The Kuiper Belt is located outside Pluto's orbit. This belt is similar to the Asteroid Belt except that it has far more mass, far more objects (especially of large sizes) and a greater supply of ancient, icy and organic material left over from the birth of the solar system. Extrapolating from the small fraction of the sky that has been surveyed so far, investigators estimate that the Kuiper Belt contains approximately one hundred thousand objects larger than one hundred kilometers across. Pluto and Charon could have been part of this belt that was to have been formed into another gaseous planet that some disturbance may have prevented.

The size, shape, mass and general nature of the Kuiper Belt appear to be much like the debris belts seen around other nearby stars, such as Vega and Fomalhaut. Computer models used to simulate the formation of this belt almost five billion years ago as the planetary system was coalescing from a whirling disk of gas and dust show that it was approximately one hundred times as massive as it is today. There was once enough solid material in this belt to have formed another planet the size of Uranus or Neptune. Computer simulations reveal that larger planets like Neptune would have naturally grown from debris fields like this belt in a very short time had nothing disturbed the region. Undoubtedly something disturbed this belt in the past causing it to lose much of its mass and prevent its formation into a massive object.

SOLAR SYSTEM'S TEN LARGEST MOONS

PLANET	NAME OF MOON	DIAMETER
Jupiter	Ganymede	3,278 miles
Saturn	Titan	3,198 miles
Jupiter	Callisto	2,995 miles
Jupiter	Io	2,257 miles
Earth	- - -	2,159 miles
Jupiter	Europa	1,942 miles
Neptune	Triton	1,714 miles
Uranus	Titania	1,000 miles
Uranus	Oberon	963 miles
Saturn	Rhea	950 miles

MASS OF THE SOLAR SYSTEM

OBJECT	MASS	DENSITY
Sun	2.189×10^{27} tons	0.050 lb/in^3
Mercury	363×10^{18} tons	0.198 lb/in^3
Venus	$5,368 \times 10^{18}$ tons	0.192 lb/in^3
Earth	$6,578 \times 10^{18}$ tons	0.200 lb/in^3
Mars	706×10^{18} tons	0.143 lb/in^3
Jupiter	2.1×10^{24} tons	0.048 lb/in^3
Saturn	625.24×10^{21} tons	0.026 lb/in^3
Uranus	95.7×10^{21} tons	0.044 lb/in^3
Neptune	113.3×10^{21} tons	0.060 lb/in^3
Pluto	16×10^{18} tons	0.083 lb/in^3

Total Planets 2.947271×10^{24} tons

Asteroids, moons $6,575 \times 10^{18}$ tons (est.)

Total System $2.191953846 \times 10^{27}$ tons

Mass of Sun equals 99.865 percent of the solar system's total mass.

It would require over one thousand Jupiters to equal the mass of the Sun.

Excluding Jupiter, all planets, moons and asteroids of the solar system equal only 40.34 percent of Jupiter's mass - a total of 0.847271×10^{24} tons.

COMETS

Comets were formed ultimately out of interstellar grains within the solar nebula just before the planets and moons formed some 4.6 billion years ago. Many comets collided with each other forming larger bodies and then hitting the new planets helping them to grow. Their collisions with Earth produced most of the water for the oceans after millions of years of this action. Our planet also seems to have been formed from such objects with some ice and rock. Many other comets were gravitationally ejected from the solar system but maintain orbits at various enormous distances around the Sun. These distances are directly proportional to the length of time they may pass in their orbit around the Sun - ten years, one hundred years, one thousand years, ten thousand years or longer. Like everything else of which we have evidence, comets are born, live for a time and then die as they break apart into small rock and dust particles. These may remain in space until captured by some gravitational attraction and attached to it or burned up (separation of atoms) as they enter some atmosphere.

It is the nucleus, the densest and heaviest part of the comet, which moves along the orbit. The nucleus and coma together form that part of the comet commonly known as the head. Measurements performed on a dozen comets have shown the radius of the head may range from one hundred meters to about fifty kilometers (thirty-one miles). The tail may stretch two hundred fifty million kilometers (one hundred fifty-five million miles). The contents of the head are usually a "dirty snowball" which does not change its orbit due to gravitation, or a solid rock covered with ice that can be gravitationally influenced. The path near the outer planets usually determines which.

Naked-eye comets are visible for between a week and six months. Comets do not streak across the sky; they rise and set with the stars. The comet lives in the nearly perfect vacuum of interplanetary space and the tails of some naked-eye comets have stretched from horizon to horizon. During this orbit which has been in existence for millions of years, atoms of hydrogen, nitrogen, carbon and oxygen are collected and become molecules of ice - water vapor (H_2O), ammonia (NH_3) or methane (CH_4). Due to the low temperature of space, ice is formed from these gases. The coma and tail seen is nothing more than the outer layer of these molecules being warmed off into space by the solar wind. We see them because sunlight reflects off these molecules at night like the Moon. Some meteors may be worn out cometary nuclei because their ice coating has finally worn away. This can change their orbit and in time Earth's gravity causes those approaching our orbit to enter our atmosphere and burn up.

The lifetime of a comet is limited and the comets that are now being observed cannot have been long-time members of the inner solar system. The Oort Cloud is located in the far reaches of the solar system and supplies the comets that orbit our Sun. This cloud could contain the remnants of the primordial cloud that collapsed and resulted in our solar system. The Oort Cloud is named after Jan Hedrik Oort, the Dutch astronomer who concluded that there is a vast reservoir of comets trapped in sort of deep-freeze storage located at the fringes of the solar system. The comets in storage, whatever their origin, share the Sun's motion very exactly and definitely are members of the solar system though they are perturbed by the nearby stars.

If Halley's comet were to impact Earth:

   Halley's comet has a mass of a trillion tons.

   Comet impact energy = Mass x V^2

   Energy = one trillion tons x $(30 \text{ km/sec})^2$ = 400 billion kilotons of TNT
      Twenty billion times more powerful than the atom bomb
      dropped on Hiroshima.

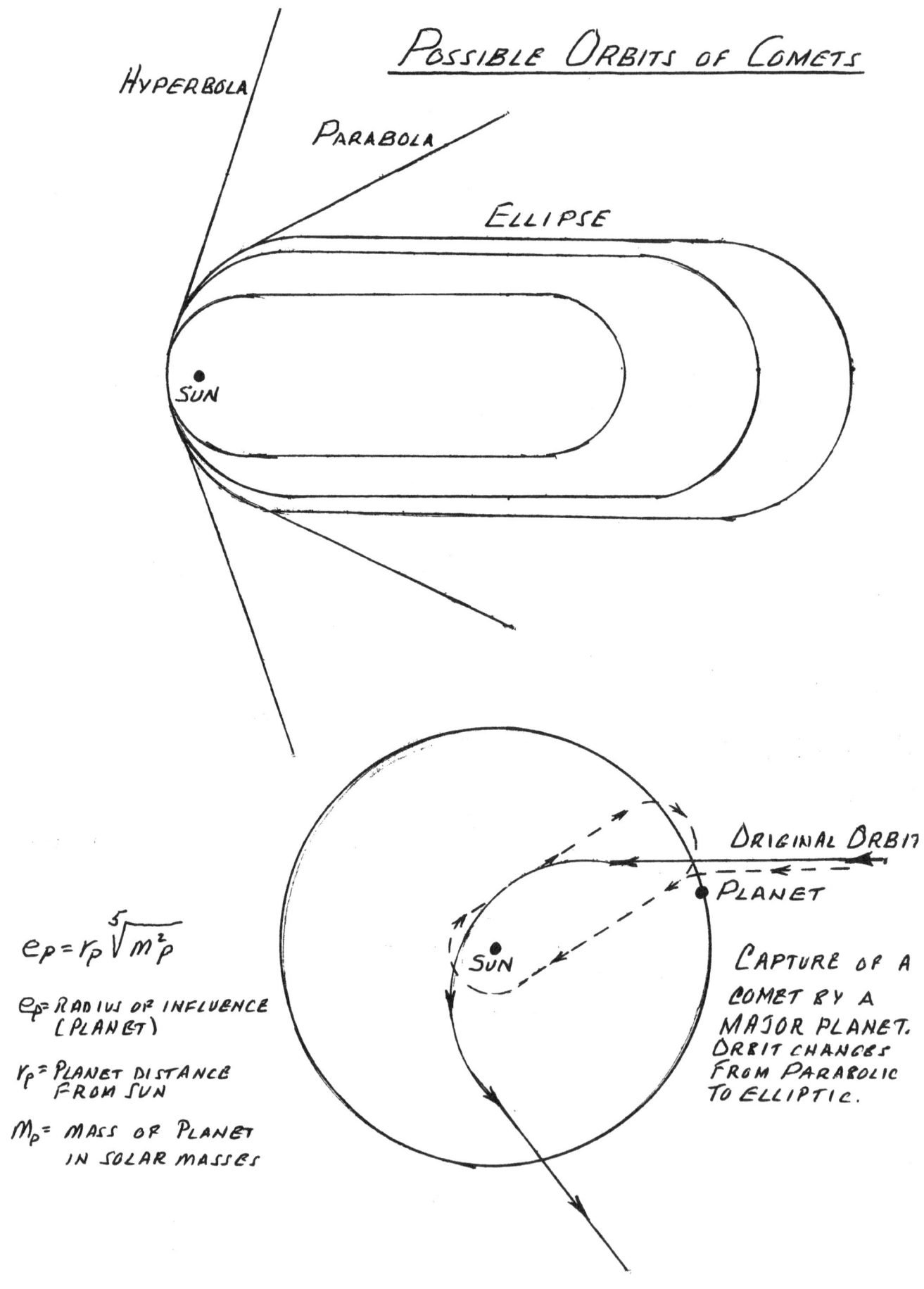

POSSIBLE ORBITS OF COMETS

HYPERBOLA

PARABOLA

ELLIPSE

SUN

$$e_P = r_P \sqrt[5]{m_P^2}$$

e_P= RADIUS OF INFLUENCE
(PLANET)

r_P= PLANET DISTANCE
FROM SUN

m_P= MASS OF PLANET
IN SOLAR MASSES

ORIGINAL ORBIT

PLANET

SUN

CAPTURE OF A
COMET BY A
MAJOR PLANET.
ORBIT CHANGES
FROM PARABOLIC
TO ELLIPTIC.

UNIDENTIFIED FLYING OBJECTS
U F O's

On February 9, 1913 at 9:05 PM, a number of astronomers observed some strange celestial objects crossing the sky over Toronto, Canada. First they spotted a bright red object approaching from the northwest moving parallel to the horizon. More objects in small groups of two, three and four appeared from the same direction and followed the first along the same path. They all had small bright tails that made them look like rockets. At the end of this occurrence the astronomers heard a booming sound, as of distant thunder. The period of all they observed did not last more than three and one-half minutes. The swarm was observed not only in Toronto but also in several other places along its trajectory on an arc reaching down to Cape St. Rocco in Brazil. These objects hit the ground soon after passing Brazil. These objects were called Cyralliads for lack of knowing their origin. It is most likely cometary debris that, if it appeared today, would be promptly called UFO's.

This is the reason that the first word of this description is *UNIDENTIFIED*, because it is not known what the object may have been at the time. Many phenomena occur in nature that are unexplainable at the time, but research usually finds a logical explanation. Many have been found to be gas pockets, reflections from dust particles in space that have erratic paths and even such mundane objects as weather balloons. It would be safe to assume that these are NOT space invaders investigating us for some dire experiment or capture.

There have been and will continue to be objects observed in our skies or in space that appear in shapes or actions unfamiliar to us. Who really knows what these objects are until they are IDENTIFIED - this is the time they cease being UFO's. Venus has the distinction of being identified as an UFO more than anything else.

LAW OF UNIVERSAL GRAVITATION

$$F_g = G(M_1 M_2 / d^2)$$

F_g = Resultant force of attraction due to gravity.
G = Universal gravitational constant
    = 6.673×10^{-8} dyne cm^2/gm^2 (dyne = cm-gm/sec^2)
    = 6.673×10^{-11} newton m^2/kg^2 (newton = m-kg/sec^2)
M_1 = Mass of object (1).
M_2 = Mass of object (2).
d = Distance between masses.

Isaac Newton's *Principa*: (1) A particle outside a uniform spherical shell is drawn toward the shell as if all the mass of the shell were concentrated at its center. (2) A particle inside the shell is drawn equally in all directions and experiences no force at all. At the center of the Earth, or any other massive object, weight equals zero. If you would like to lose "weight" you must descend in a mine-shaft toward the center of Earth until you are satisfied.

Because protons and neutrons have mass, all matter must have mass. Contained in the phenomenon of mass is another phenomenon - gravity. The denser the object, the greater an object of less density is attracted to it. You weigh almost six times as much on Earth as you would on the Moon because of this. The Moon and the Earth are attracted to each other, but the Moon is more attracted to the Earth because of Earth's greater mass. If the Moon were half as far away as it is presently, the force of attraction to the Earth would be four times greater and the tides would be much higher. If the Moon were twice as far away as it is presently, the force of attraction to the Earth would become one-quarter of present values and the tides would be decreased proportionally. The law of gravity states that every object in the Universe is exerting a gravitational force on every other object. From the practical point of view, the largest mass in the area exerts controlling forces.

An object released near the Earth's surface will fall sixteen feet in the first second. An object thrown horizontally will also fall sixteen feet. Even though it is moving horizontally, it still falls the same sixteen feet in the first second. An object like a bullet shot horizontally will go a long way in one second - perhaps two thousand feet or further - but it will still fall sixteen feet in first second. What happens if we shoot the bullet faster and faster? Do not forget that the Earth is a sphere - its surface is curved. If we shot the bullet fast enough, as it falls its sixteen feet each second, it would be at just the same height above the ground as it was before. It still falls at the rate of sixteen feet per second, but the Earth curves away, so it "falls" around the Earth. The Earth falls away from the bullet at the same rate as the bullet falls, placing the bullet in an "orbit." If the speed of the bullet were 4.9 miles per second (17,640 miles per hour) and moving just above Earth's atmosphere where there is no air resistance, it would become a satellite and would orbit indefinitely. The Moon is a satellite of Earth and if its orbital speed were modified it would move into a different orbit. If its orbital speed were suddenly reduced to zero, the Moon would fall immediately to Earth the same as a rock thrown into the air.

The laws of gravity also require that certain speeds in space are required to maintain these orbits. The farther satellites are from Earth, less velocity is required to maintain these orbits. At 300 miles above Earth a velocity of 17,034 miles per hour is required. At the height of our GPS (Global Positioning Satellites) located 11,000 miles above Earth, their velocity to maintain orbit is only 9,092 miles per hour. The geosynchronous (geocentric) satellites located 22,245 miles above Earth travel at only 6,870 miles per hour - maintaining a position above a fixed point on Earth. The formulas for determining this data are:

$$g_1 d_1^2 = g_2 d_2^2$$
$$g_1 = V^2/r$$

g_1 = Value of the satellite's gravity.
g_2 = Value of Earth's gravity - 32 ft/sec^2.
d_1 = Satellite's distance above the center of Earth.
d_2 = Earth surface distance from the center of Earth.
V = Velocity to maintain orbit.
r = The same as d_1.

A satellite has no weight, it has mass and inertia. Its centripetal force equals its attraction to the gravity of Earth explaining why there is only one height to place satellites above Earth that does not appear to move, but continually remain in a fixed position relative to a spot on Earth. This is the intersection of *geocentric velocity* and *velocity to maintain orbit* lines.

We know the effect the Moon has on the tides of our oceans. The effect of gravitational forces is greatest on the closest object. For this reason, Venus has about 58 percent of Jupiter's gravitational effect on Earth even though it has only 0.25 percent of Jupiter's mass. Venus' closest distance to Earth is only 26 million miles while Jupiter is 390 million miles away at its closest approach to Earth. Some people worry about the lining up of all the planets in the same direction as Earth causing catastrophes. These people may be assured that nothing such as this will ever happen. Using the laws of gravitation, the combined force of all the planets are only equal to 2.2 percent of the Moon's influence on Earth. The Sun's relative force is considered infinite due to our orbit of this body. The planetary alignment will have no effect that anyone will notice. Nearby objects have more effect gravitationally than these planets. A person weighing two hundred pounds standing next to an aircraft carrier or a large building would be attracted by these objects over one thousand times greater than the force of the Moon. One of our nearest stars, Alpha Centauri is located 4.4 light years away has no effect on anything on Earth at ALL! So much for astrology.

Using the law of universal gravitation:

Between Earth and:

Sun	$16,700,000 \times 10^{26}$ G
Mercury	74×10^{26} G
Venus	530×10^{26} G
Moon	$90,000 \times 10^{26}$ G
Mars	20×10^{26} G
Jupiter	920×10^{26} G
Saturn	500×10^{26} G
Uranus	2.1×10^{26} G
Neptune	0.8×10^{26} G
Pluto	Negligible at 10^{26} G

Between a 200 pound man and:

Moon	140,000,000 G
Building 50 feet distant weighing 140,000 tons	140,000,000 G
Aircraft carrier 0.1 mile distant	1,000,000 G
Another 200 pound man 50 feet distant	100 G
Alpha Centauri 4.4 light years distant	0.33 G

Two men located fifty feet apart exert three hundred times more force between them than either experience from the star Alpha Centauri.

VELOCITY REQUIRED TO MAINTAIN SATELLITE IN ORBIT

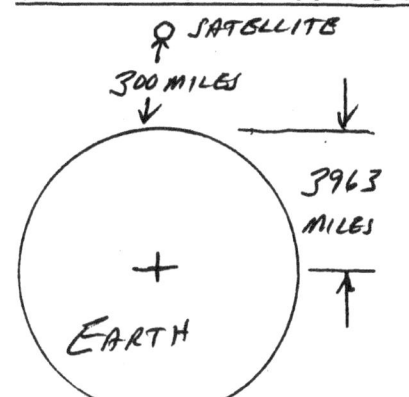

SATELLITE VALUE OF GRAVITY $-g_1$

EARTH " " " $-g_2 = 32.09\ \mathrm{FT/SEC^2}$

SATELLITE DISTANCE FROM

EARTH CENTER $-d_1 = 4263\ \mathrm{MI}$

EARTH SURFACE DISTANCE FROM

EARTH CENTER $-d_2 = 3963\ \mathrm{MI}$

NEWTON'S UNIVERSAL LAW OF GRAVITATION

$$g_1 d_1^2 = g_2 d_2^2$$

$$g_1 = \frac{g_2 d_2^2}{d_1^2} = \frac{32.09(3963)^2}{(4263)^2} = 27.73\ \mathrm{FT/SEC^2}$$

$$g_1 = \frac{V^2}{r}\quad V = \sqrt{g_1 r} = \sqrt{(27.73)(4263)} = 24,983\ \mathrm{FT/SEC}$$
$$(r = d_1) \qquad\qquad\qquad = 17,034\ \mathrm{MI/HR}$$

__GPS__ $g_1 = 2.25\ \mathrm{FT/SEC^2}$
(11,000 MI) $d_1 = 14,963\ \mathrm{MI}$
$V = 13,335\ \mathrm{FT/SEC}$
$= 9,092\ \mathrm{MI/HR}$

__GEOCENTRIC__ $g_1 = 0.73\ \mathrm{FT/SEC^2}$
(22,245 MI) $d_1 = 26,208\ \mathrm{MI}$
$V = 10,076\ \mathrm{FT/SEC}$
$= 6,870\ \mathrm{MI/HR}$

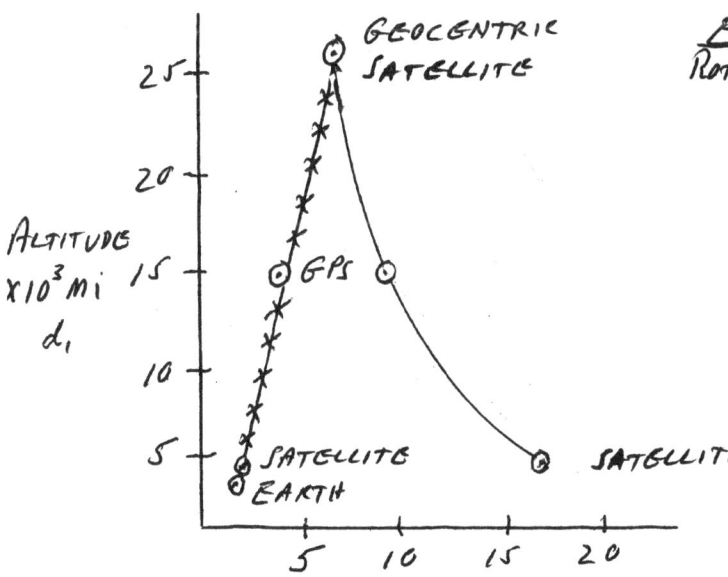

$$\frac{\text{EARTH } R}{\text{ROTATION } V} = \frac{3963}{1037}$$

300 MI = 1115 MI/HR
GPS = 3915 MI/HR
GEOCENTRIC = 6858 MI/HR

×—×—×—×—×—×—× GEOCENTRIC VELOCITY

———————————— VELOCITY TO MAINTAIN ORBIT

113

CELESTIAL MECHANICS

THE STUDY OF MOTIONS OF GRAVITATIONALLY INTERACTING OBJECTS SUCH AS PLANETS AND STARS

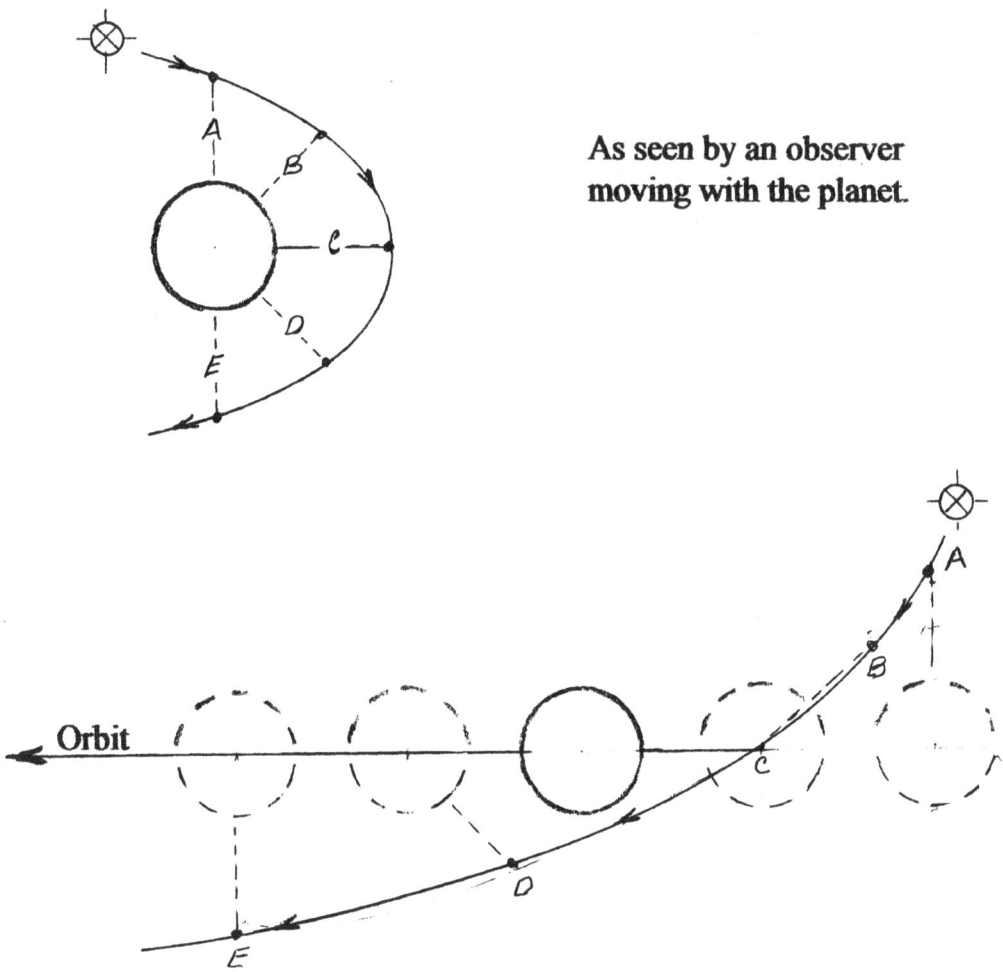

As seen by an observer
moving with the planet.

Orbit

As seen by a stationary observer
situated elsewhere.

The gravitational slingshot of a spacecraft close to a planet is used to speed it up by the energy received from being pulled toward the planet. This method is used to speed up a spacecraft on its journey to the outer planets when the inner planets are available en route to its destination.

Celestial Mechanics!!! Just hearing these words causes people to drop to their knees in fear. What is it about these words that make things seem ominous? The only reason that comes to mind to answer this question is that people do not have the vaguest idea of the definition - what these words mean. Celestial mechanics is simply the study of motions of gravitationally interacting objects such as stars, planets, asteroids and even satellites and spacecraft. When we pick up a rock and drop it, the rock falls to the ground - the gravity described by Isaac Newton. Newton's laws of gravity also explain why the planets orbit the Sun and objects in our solar system act as they do. Einstein's general theory of relativity does the same thing from another point of view. Both are correct and both are explained in this book.

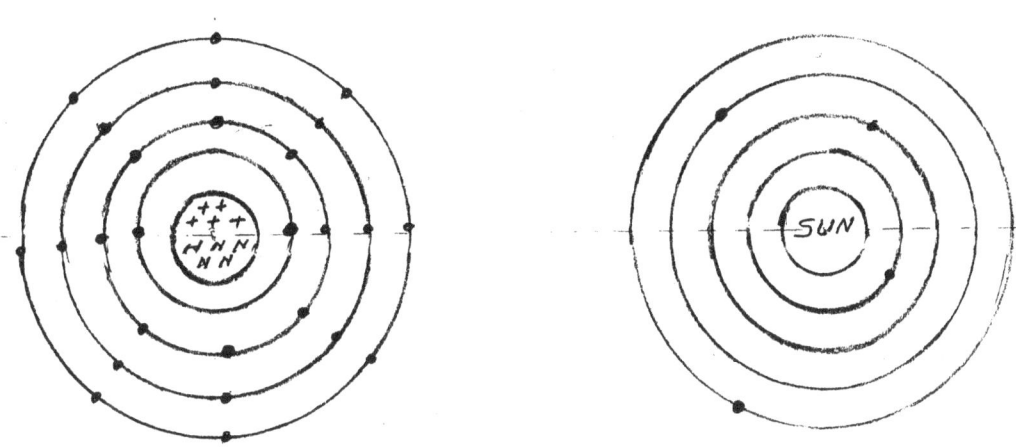

When we look at a classical picture of the solar system, we are looking at a picture that appears to be the same as the picture of an atom - a circle in the middle of a lot of larger circles going around it. As a matter of fact, and strictly a coincidence, the relative size of the Sun in respect to the orbit of Pluto is almost the same as the size of the nucleus of an atom is to its outer ring, or shell, of electrons. Other than size, there is another major difference between the electrons in *orbit* around the nucleus of an atom and the *path* of the planets as they travel through space with the Sun. The electrons moving around the nucleus of an atom are moving so fast that the actual movement creates what would appear as a sphere - a hollow ball. When I say fast, I mean FAST! These electrons orbit the nucleus creating this sphere at 1,000,000,000,000,000 times each second! One million billion times each second. The planets move much slower.

The planets move so slowly that when we look at them at night they appear stationary. We know the Sun is moving around the Milky Way at approximately 540,000 miles per hour. Mercury is located some 36 million miles from the Sun traveling at 107,037 miles per hour and requires 88 days to complete its orbit - one year on Mercury but only one-half of a day. A "noon to noon" occurs every 176 days which is TWO years on Mercury because it rotates so slowly on its axis. As we continue further out in the solar system, Venus is 67 million miles from the Sun traveling at 78,282 miles per hour and takes 225 days to complete its year - an orbit. This is almost two "noon to noon" days per year - its rotation is also very slow compared to the rest of the planets. We know Earth is 93 million miles from the Sun traveling at 66,588 miles per hour taking a little over 365 days per orbit. Our Moon orbits us at 2,286 miles per hour 238,172 miles away about every 28 days.

Mars is located 142 million miles from the Sun traveling at 53,979 miles per hour and its orbit is 1.88 years, which is 687 days. The asteroid belt is 250 million miles from the Sun. Notice how the speed required to orbit the Sun is becoming less and less the further each planet is from the Sun. Jupiter is 483 million miles from the Sun traveling at a speed of only 29,202 miles per hour taking almost 12 years to complete an orbit. Saturn is almost twice as far from the Sun as Jupiter - 886 million miles traveling at 21,689 miles per hour and requiring 29-1/2 years to complete each orbit. Notice that the time complete each orbit increases the further each planet is from the Sun. This is natural because of the slower velocity of each as the path to travel increases - slower velocities and longer paths. Uranus is almost a billion miles further out than Saturn at 1.8 billion miles traveling at 15,205 miles per hour with an orbit time of 84 years. Neptune's numbers are 2.8 billion miles, 12,075 miles per hour and almost 165 years to complete one orbit. Last is Pluto and its moon, both probably misplaced asteroids or moons, averaging 3.7 billion miles from the Sun going at only 10,600 miles per hour and taking almost 248 years to complete an orbit. All these distances, velocities and times are based on the Earth frame of reference. These speeds are decidedly slower than the orbiting electrons.

Electrons *orbit* the nucleus. If the nucleus is stationary, the electron shells also appear stationary. Move the nucleus and the *orbiting* electrons go right along with it maintaining their proper positions. Electrons *orbit* the nucleus. We also, mistakenly, use the term *orbit* when referring to planets and the Sun. Planets *appear* to *orbit* the Sun when shown in the classical manner. Planets do not *orbit* a stationary Sun - they travel in a *path* that can be described as eventually going around the Sun - in space-time. The Sun itself is moving very fast in the Milky Way galaxy carrying us with it, not in an *orbit*, but causing the planets to trace out a sinusoidal *path* as we move through space.

We will use the example of Earth's *path* through space for one year. We see that Earth's location in each season is:

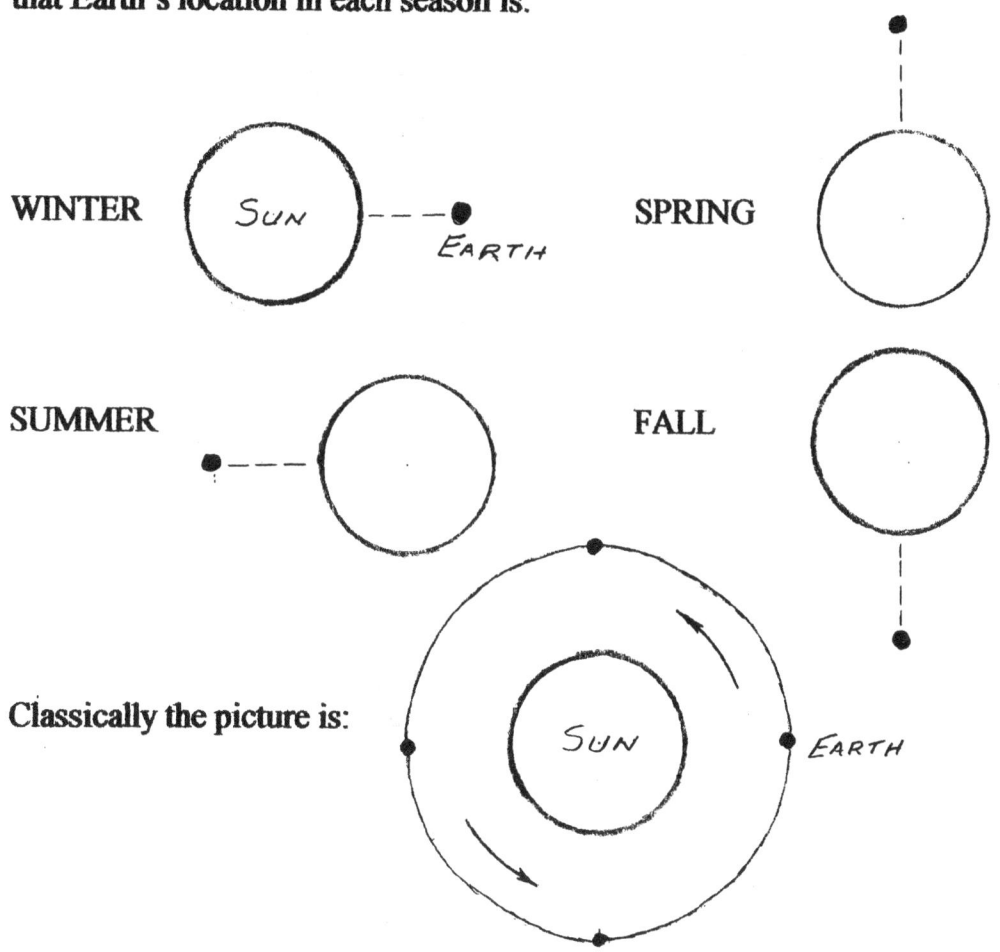

WINTER   SPRING

SUMMER   FALL

Classically the picture is:

Since the Sun is moving at a speed of almost 540 thousand miles per hour through space, we have to move along with it. Our true *path* through space is:

- TOP VIEW -

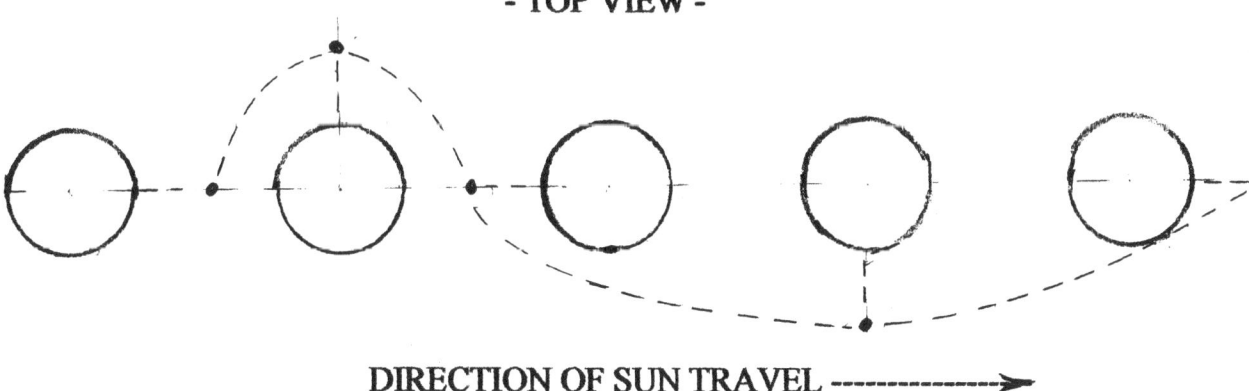

DIRECTION OF SUN TRAVEL ------------>
BASIC SPACE-TIME DIAGRAM

Now we can understand why we on Earth are not just going around in circles - not an *orbit*. Our *path*, as also the other planets and any objects in the solar system is sinusoidal - similar to the sine wave we see on an oscilloscope when viewing various frequencies. Add Mars to the picture and the space-time diagram becomes a little more complicated. It takes almost two years for Earth and Mars to be at their nearest point to each other. You do not launch satellites to Mars any time you choose, you launch when these planets relative location to each other will result in the least travel time both going and returning. This is very important if manned exploration flights are planned. Notice the *paths* that each Earth and Mars travel. To add a little spice to our picture refer to the Viking I satellite trip to Mars that lasted ten months. When you know what you are doing, results are easy to attain. If you are bored some rainy afternoon and are looking for something to do, design a space-time diagram for our solar system for a period of ten years. You will then realize that we do not *orbit*, we travel a *path*.

The distance Earth travels each year:

Sun travel in galaxy at 500,000 MPH	4,383,000,000 miles
Earth path of Sun	584,336,234 miles
Total	4,977,336,234 miles

Almost FIVE BILLION MILES each year.

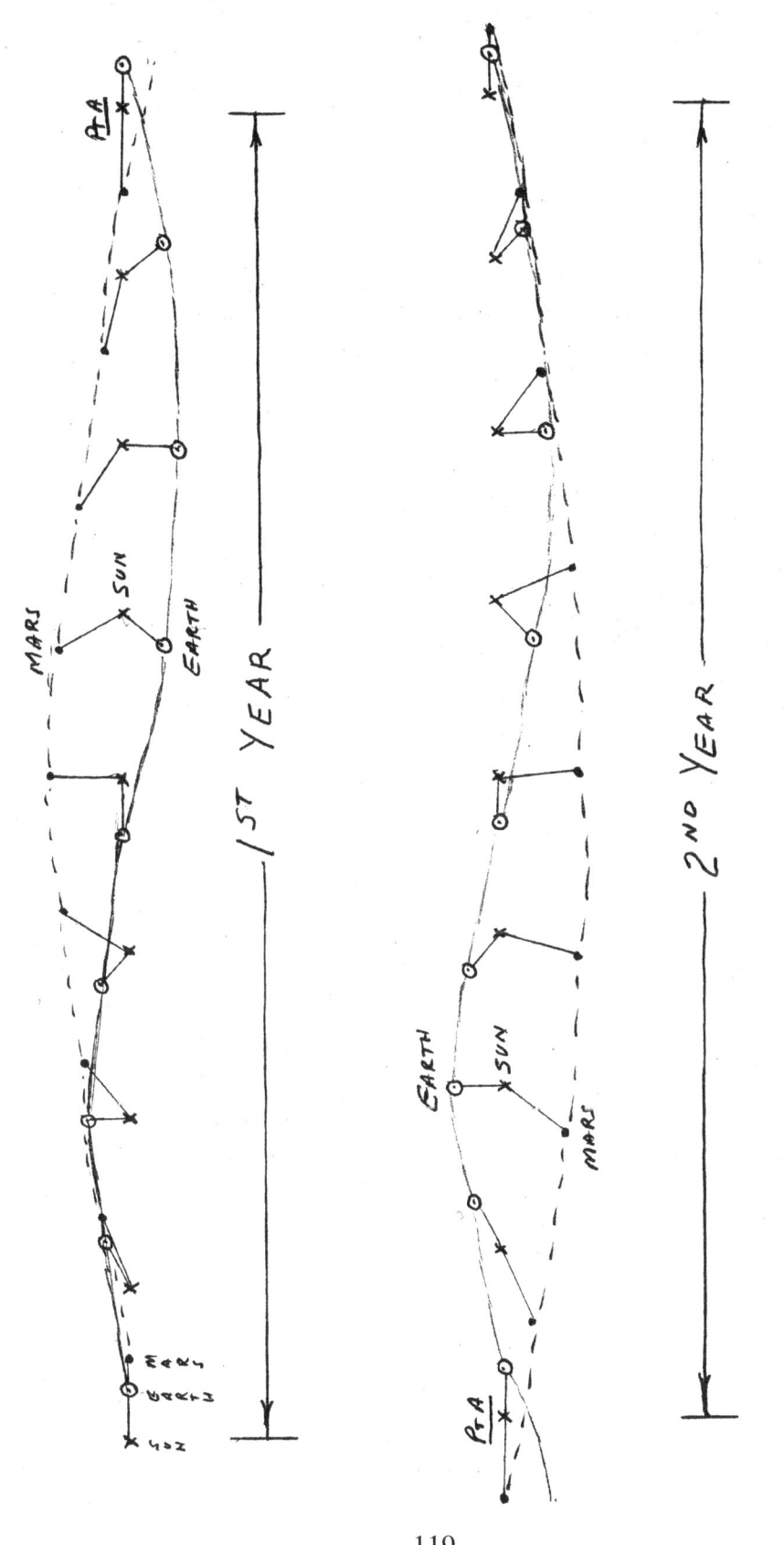

1ST YEAR

2ND YEAR

EARTH-MARS SPACE-TIME DIAGRAM

119

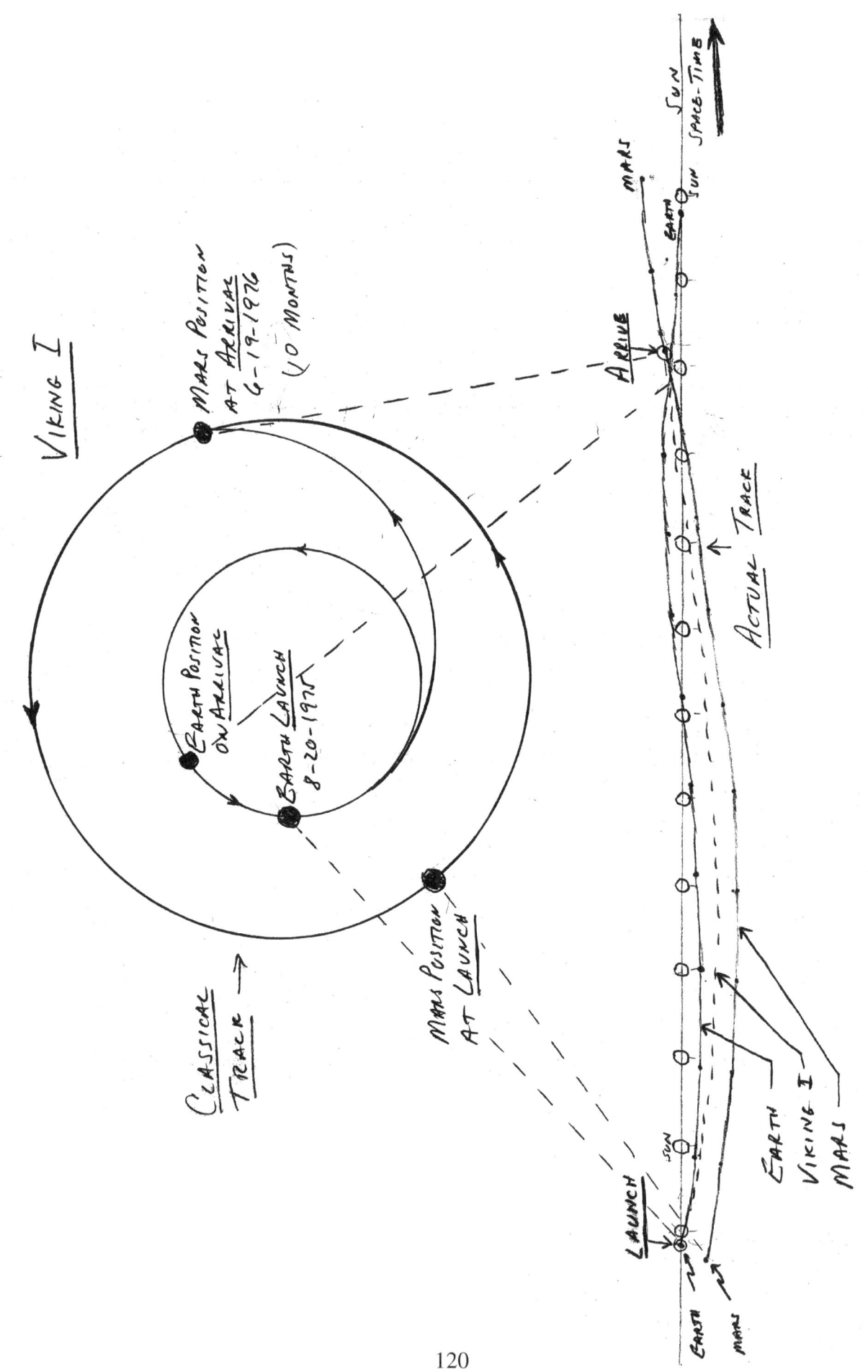

VIKING I

CLASSICAL TRACK →

MARS POSITION AT ARRIVAL
6-19-1976
(10 MONTHS)

EARTH POSITION ON ARRIVAL

EARTH LAUNCH
8-20-1975

MARS POSITION AT LAUNCH

ARRIVE

ACTUAL TRACE

MARS

EARTH

SUN SPACE-TIME

SUN

EARTH
VIKING I
MARS

LAUNCH

SUN

EARTH

MARS

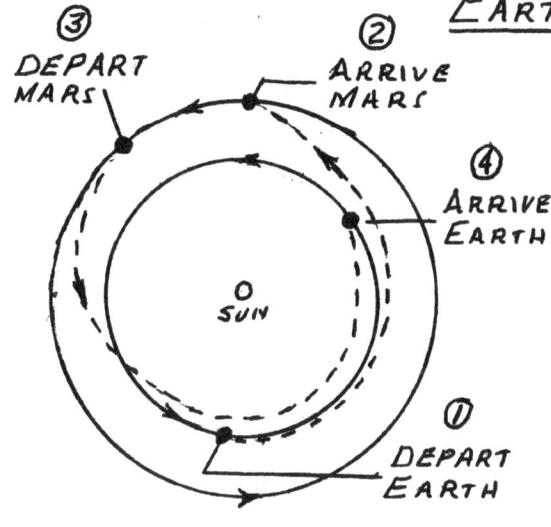

③ DEPART MARS

② ARRIVE MARS

④ ARRIVE EARTH

SUN

① DEPART EARTH

SHORTEST RETURN JOURNEY
OPPOSITION CLASS MISSION
DURATION 519 DAYS WITH A
STAY ON MARS OF 20 DAYS.

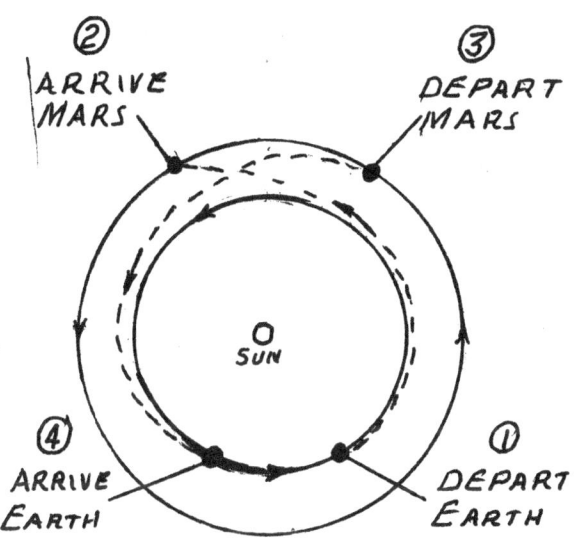

② ARRIVE MARS

③ DEPART MARS

SUN

④ ARRIVE EARTH

① DEPART EARTH

CONJUNCTION CLASS MISSION
DURATION 693 DAYS WITH A
STAY ON MARS OF 60 DAYS.

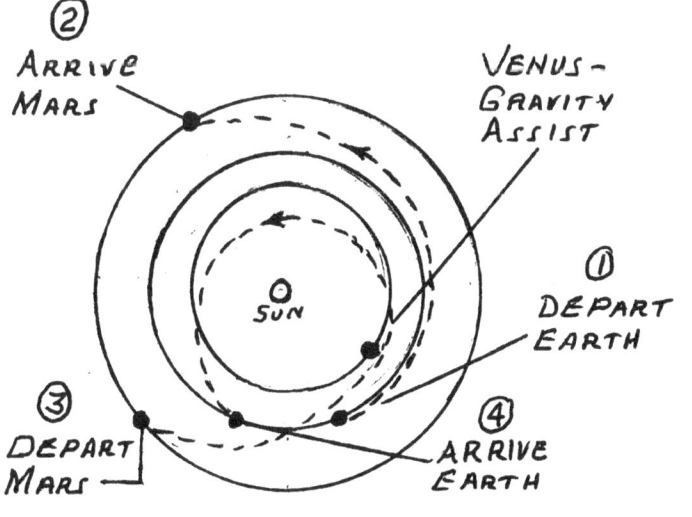

② ARRIVE MARS

VENUS-GRAVITY ASSIST

SUN

① DEPART EARTH

③ DEPART MARS

④ ARRIVE EARTH

VENUS SWINGBY MISSION
DURATION 1009 DAYS WITH A
STAY ON MARS OF 560 DAYS

OTHER PLANETARY SYSTEMS

Since there are hundreds of billions of stars in hundreds of billions of galaxies, the odds are very favorable in finding life in existing planetary objects in both the Milky Way galaxy and the rest of the Universe. We do not expect them to resemble our solar system in any way - no two of our planets are even alike. We know that most binary star systems are that - binary (double star) systems. The ones that did not become binary systems more than likely developed in the manner of our solar system - a star with a lot of debris that may take some shape in time. The shapes could be rock and metal like our inner planets or gaseous similar to the outer planets. We would not be able to see any of the smaller planets because the reflection from them would not be enough to traverse the many light years required to reach us. Those that are Jupiter's size or larger may be seen because their reflection would be so much brighter due to the size of the surface area available for reflection and the gaseous nature of the planets.

One must remember that there are many other planet-sized objects in space that are near the size of Earth. These could be the white, brown or dark dwarf stars that have collapsed into their present state emitting no light. These should not be orbiting anything, unless this is a binary system in which both stars are dwarfed, and would appear to be traveling alone. More and more discoveries of other planetary systems are becoming more frequent as our technology improves in being able to recognize and find these objects. Many that have been found orbit too close to their star to be able to have any hope of life developing on them - they are too hot! Others are the Jovian type which cannot support life (unless a moon or so on them is able to provide this service), but the searching will continue.

What are the odds of life existing elsewhere? Very good! If only one billionth of the stars have planets and only one billionth of a percent of these contain life; how many planets would have life? What if the percentage was reduced to one millionth?

Number of galaxies in the Universe	400 billion
	400×10^9
	4×10^{11}
Average number of stars in a galaxy	200 billion
	200×10^9
	2×10^{11}

One billionth $= 10^{-9}$
One billionth of a percent $= 10^{-11}$
One millionth $= 10^{-6}$
One millionth of a percent $= 10^{-8}$

Number of stars in the Universe $= (4 \times 10^{11})(2 \times 10^{11}) = 8 \times 10^{22}$
Number of stars with planets $\quad = (8 \times 10^{22})(10^{-9}) = 8 \times 10^{13}$

Number of planets with life @ 10^{-11} percent $= (8 \times 10^{13})(10^{-11}) = 800$
Number of planets with life @ 10^{-8} percent $= (8 \times 10^{13})(10^{-8}) = 800,000$

The calculations show that at least eight hundred potential planetary systems are active if only one billionth of a percent is used. If the amount were dropped to only one millionth of a percent, eight hundred thousand potential systems appear. So the odds are that we are not alone.

DIRT AND STUFF

History is subject to geology.
> Will and Ariel Durant

Nature has no goal in view, and final causes are only human imaginings.
> Baruch Spinoza

The Earth-Moon system could be described as a double planet due to their similar spherical shapes and relative sizes.

EARTH FORMATION

Earth was formed with only so many atoms. These same atoms exist today in many places. They can not be created or destroyed, they can only be altered and can change places with other atoms or join them. The Earth was formed from the swirling fragments of the solar nebula in a process of growing larger through the accumulation of smaller bodies - absorbing them. High-velocity impacts generated heat and eventually the upper layers of the planet melted to form a global ocean of liquid rock. At some point during the period of growth, Earth was struck by another growing world about the size of Mars - one with a mass of about 10 percent of that of Earth. The smaller, Mars-sized planet was completely destroyed and even the larger Earth was shattered to its core. Some of the ejected material continued to orbit the Earth as a giant ring, which cooled and collapsed to form our Moon. If the Moon-forming impact had been just a little larger, the Earth would have broken up.

Venus spins in the opposite direction from its orbital motion about the Sun, probably as the result of a late collision that struck it a glancing blow and reversed its direction of rotation. Notice the difference of Earth's rotation of one thousand miles per hour at the equator versus the four miles per hour rotation of Venus in the opposite direction. The small planet Mercury appears to be the metal-rich remnant of a larger planet, stripped of most of its rocky mantle in another giant collision. It is largely a matter of luck that the final product of this chaos was the four inner planets we have today: Mercury, Venus, Earth and Mars, plus the Moon. Mercury and the Moon have similar surface appearances revealed by the many craters.

As the rain of cometary materials continued, the Earth built up a thick atmosphere of carbon dioxide and other compounds and developed shallow oceans of liquid water rich in dissolved organic materials. Such an environment is exactly what was required for the origin of life. The first self-replicating molecules must have formed in these early seas. Following the end of the heavy bombardment, the Earth has continued to experience occasional impacts. Some of these, like the Cretaceous impact of sixty-five million years ago, severely disrupted the environment and redirected the course of biological evolution. Impacts, taking place at the right level, appear to have played a crucial role in the history of life. Too many impacts and the planet would have been sterilized. Too few, and evolution may stagnate in a static, mild environment. Just the right number of impacts, and we have the development of *Homo sapiens*.

As the Earth finally settled down into its present general shape, many factors were set into motion. The first noticeable object that continued to follow Earth was our Moon. The Moon was orbiting Earth under the same principles that caused

Earth to orbit the Sun. This Earth-Moon combination causes our orbit around the Sun to follow the center-of-gravity of the system, the barycenter, rather than the center-of-gravity of the Earth. The next noticeable fact to be observed after one complete orbit of the Sun by Earth was that the weather was always changing - seasons occurred. Along with these change in seasons, the Sun appeared to move higher in the sky and then lower in the sky as these seasons changed. This movement caused the daylight hours to be longer when the Sun was higher in the sky and shorter when the Sun was lower in the sky. This fact resulted in realizing that Earth was not rotating on an axis that was perpendicular to its orbital plane around the Sun. The axis of the Earth was tilted! Today we know the exact amounts of Earth's tilt, rotational speed, orbital speed and precession. We also know all the factors of movement associated with the Moon. Our continents and oceans appear to be stationary, but we know that the actions of plate tectonics are continuous.

To simplify and explain some of these actions we will refer to the diagram:

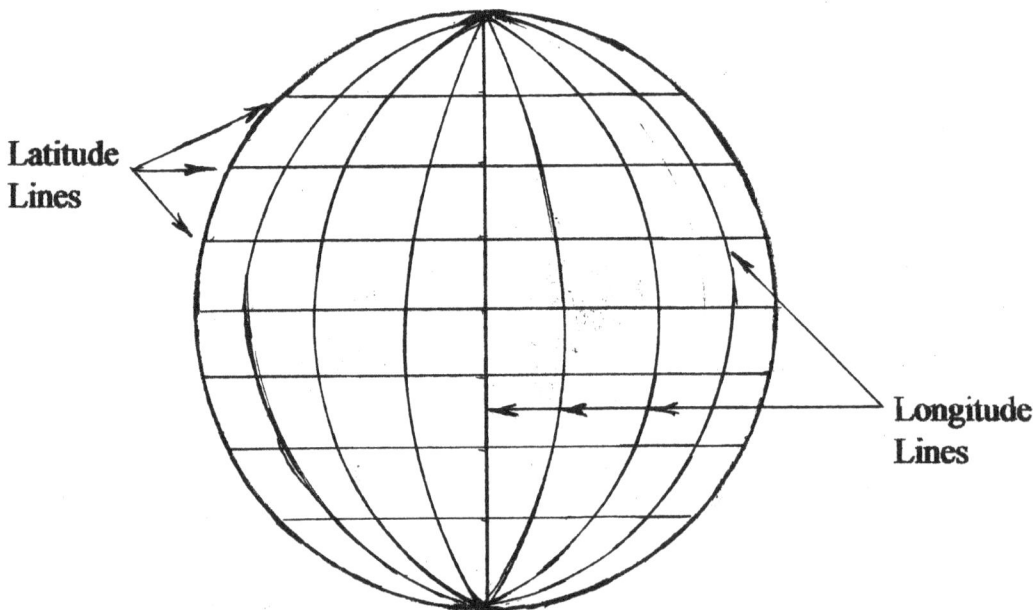

Earth is divided into latitude and longitude lines. The latitude lines are drawn parallel to the equator (0^0) northward to the North Pole ($+90^0$) and southward to the South Pole (-90^0). The longitude lines are drawn from pole to pole. Greenwich, England is located at zero degrees longitude with the progression to 180 degrees West (toward the United States and the Pacific Ocean) and 180 degrees East (toward Europe and Asia). These longitude lines meet in the Pacific Ocean west of Midway Island at the International Date Line.

For convenience, the number of hours in a day on Earth was decided to be twenty-four. Since the distance around Earth is 360 degrees, one hour passes as the Earth rotates fifteen degrees. We know that the Earth rotates at 1,037 miles per hour at the equator and approaches zero at the poles. Therefore, the speed of rotation decreases as one moves northward or southward from the equator. To understand why this occurs is simply explained using the longitude lines. The distance between these longitude lines is greatest at the equator and decreases as one proceeds north or south from the equator. At every fifteen degree mark, one hour passes on the longitude line. If you wanted to jump up in the air and watch the Earth spin beneath you, you would have to be in a plane traveling west at 1,037 miles per hour located above the equator. If you placed the Sun at the proper position at this speed, you could travel forever with the Sun not appearing to move at all! Look down at the Earth you are passing over - it moves under you quickly. Since you are positioned in a relatively constant position to the Sun, when you look down at the Earth, you are seeing it as it moves under you. The Earth is spinning under you at this speed at this location on the equator. If another plane had taken off at the same time and in the same direction as you at thirty degrees north (or south) latitude, its speed would only be 898 miles per hour to observe the Earth spinning beneath it. It would reach the next fifteen degree longitude mark at the same time as you. A plane leaving forty-five degrees north (or south) latitude at the same time would have to travel just under the speed of sound at 733 miles per hour to proceed to the next longitude line marker of fifteen degrees. At sixty degrees north (or south) latitude your speed west would only have to be 518 miles per hour - all the planes would be in a line above the curvature of Earth headed west. Each plane would pass each longitude line at exactly the same time. At ninety degrees north (or south) you would be at either the North or South Geographic Pole. Here you could just sit in a chair, looking south (or north) at a line of planes as you pivoted on your spot, always looking at the Sun. If all five observers (equator, 30^0, 45^0, 60^0 and the pole) maintained these position for four weeks, they would observe a complete orbit of the Earth by the Moon. Since the same surface of the Moon faces Earth, its rotation on its axis is only ten miles per hour.

Light from a single source (the Sun) hitting a sphere can only illuminate half of it at a time. Tilt this sphere and only half of the sphere continues to be illuminated. From the diagrams we see that half of the Earth is always in daylight, the other half being dark. In the WINTER (Winter Solstice diagram), notice that the point nearest the Sun becomes the Tropic of Capricorn because this point is perpendicular to the incoming rays of the Sun. The South Pole and areas around it are in daylight while areas around the North Pole are dark - these areas are known as the Arctic and Antarctic Circles. These areas vary over long periods of time as the tilt of the axis of Earth varies between 21.39 and 24.36 degrees. This creates a boundary of limits for the circles to vary between 68.61 and 65.64 degrees latitude. It is presently at 23.45 degrees making the latitude of 66.55 degrees north (or south) areas either six months of darkness or six months of light: $23.45^0 + 66.45^0 = 90^0$.

During the SPRING (Vernal Equinox diagram), Earth's orbit has placed the equator at the point perpendicular to the incoming rays of the Sun. The Vernal Equinox is the term used when the Sun appears to cross the equator heading north on March 21 each year - *Vernal*: Greek for *spring*; *Equinox*: time of year when daylight and nighttime hours are *equal*. Further orbit places the Tropic of Cancer perpendicular to the incoming rays of the Sun - SUMMER (Summer Solstice diagram). The announcement of FALL (Autumnal Equinox diagram) is when the equator again becomes perpendicular to the incoming rays of the Sun - the Autumnal Equinox (I'll let you figure out where the name, *Autumnal*, originated). As we know, this process is continuous. If the axis of the Earth were perpendicular to our plane of orbit of the Sun (the ecliptic) there would be no seasons! How would the weather react to these non-changing conditions? What would be the conditions for life on Earth?

To observe any of our sister planets, we must look directly *into* the Sun's plane during the day or directly away from this plane at night. The angle in which we search is continually changing due to our rotation, orbit and tilt. If you were located at forty-five degrees north (near Salem, Oregon) on June 21 you would have to look up 68.45 degrees to look into the Sun at noon. At night, you would only look up 21.55 degrees to search for the planets: $21.55^0 + 68.45^0 = 90^0$. At the Tropic of Cancer, you would have to look directly overhead (straight up - 90^0) to see the Sun at noon on this date.

We observe the rotation of the Moon about the Earth each month as it changes in location as it passes through the New Moon stage (the time we cannot see it because it is between us and the Sun) to the First Quarter stage, to the Full Moon and then to the Last, or Third, Quarter stage. It then returns to the New Moon stage and repeats the process. We can "see" and "feel" the Moon's motion through these observations. We accept the fact that Earth orbits the Sun due to the change in seasons. Is there a system we can use to notice this movement of a daily or weekly basis? You bet there is, but you will have to have clear nights to make your observations.

To begin your observations, select a position that gives you a clear view of the sky and is remote from other sources of light pollution. Look south (if you are in the northern hemisphere) and pick a bright star above the horizon that you can easily identify each time you search for it. Now choose a prominent landmark (a tree, a tall pole, peak of a hill, etc.) this star will pass over. When the star you have chosen is above this landmark, note the time of the observation. Return to your observation position an hour later and note the location of your star. It has moved - it is to the right of your last observation due to the rotation of the Earth. We see this every day as the Sun appears to move across the sky as we rotate under it. We have just made a nighttime verification that the Earth also rotates at night. This is fine but this one observation does not verify that we orbit the Sun.

Return to your observation point several nights later and note the time your star is above your landmark. You will notice that this star appears earlier and earlier over this landmark and knowing that its position is relatively stable to us, then it is Earth that has moved. Continue your observations and the star will continue to appear earlier over your landmark each night until it eventully has disappeared from your view, passing over the western horizon. You will have to pick several stars over the course of a year because of this condition of each disappearing as time goes by. A circle of 360 degrees and our year of 365 days requires Earth to move approximately one degree a day to complete this circle (orbit). You are observing sidereal time each night you make your observations due to Earth's orbit.

As part of each observation seeing the stars pass from east to west during the year until you return to your original star, also look north each time you make an observation. Find the Little Dipper and identify Polaris - the North Star - at the end of the handle. Notice the direction the dipper is pointed each time you make an observation. You will determine that it also moves a little each different time you look at it. It is pivoting on Polaris and when its circle is complete and has returned to its original position, a year has passed. These two methods of observation have verified that Earth orbits the Sun.

The length of Earth's day is increasing 0.003 to 0.002 seconds per century - one second each 33,333 to 50,000 years - due to the friction between the Earth and its tides. Over 525 million years ago there were 425 days in each year, slowing down by sixty days to the 365 days we presently experience. The orbit around the Sun has not changed, the rotation of the Earth has slowed. Calculations show that this slowing will continue until the rotation of Earth matches the orbit of our Moon. This will position the Moon in a stationary position over only one spot on Earth - the Moon will not rise or set.

EARTH

Diameter	12,756 km @ equator	7,922 miles
	12,714 km @ poles	7,896 miles
Mass	5.98×10^{24} kg	$6,578 \times 10^{18}$ tons
Density	5.52 gm/cm^3	0.2 lb/in^3
Surface gravity	9.807 m/sec^2	31.87 ft/sec^2
Escape Velocity	11.2 km/sec	25,043 mi/hr
Surface temperature	13 C	55.4 F
Surface area	5.1×10^{14} m^2	197,160,336 square miles
Volume	1.08×10^{21} m^2	260 billion cubic miles
Satellites	One	
Axis (distance to Sun – mean)	150,600,000 km	93,363,460 miles
Apogee (furthest)	153,098,000 km	94,920,760 miles
Perigee (nearest)	148,068,000 km	91,802,160 miles
Mean orbital velocity	29.78 km/sec	66,588 mi/hr
Sidereal year	365.212 days	
Sidereal day	23 hrs 56 min 4 sec	
Orbital distance	940 million km	584 million miles
Ecliptic inclination	0 degrees	
Axis tilt	23.45 degrees (23° 27')	
Rotation velocity @ equator	1,670 km/hr	1,037 mi/hr
Geosynchronous orbit	35,881 km	22,300 miles

MOON

Diameter	3,476 km	2,159 miles
Mass	7.35×10^{22} kg	80.85×10^{18} tons
Density	3.34 gm/cm^3	0.12 lb/in^3
Surface gravity	1.62 m/sec^2	5.265 ft/sec^2
Escape velocity	2.38 km/sec	5,322 mi/hr
Surface temperature	-171 C to +111 C	-276 F to +231.8 F
Polar temperature	-153 C	-243 F
Surface area	37.9×10^6 m^2	14,643,846 square miles
Axis (distance to Earth)	384,400 km	238,712 miles
Apogee (furthest)	406,600 km	252,092 miles

Perigee (nearest)	363,300 km	220,906 miles
Mean orbital velocity	2.808 km/sec	1,740 mi/hr
Sidereal month	27.322 days = 27 days 8 hours	
	Rises 52 minutes later each day - 656 hours	
Synodic month	29 days 12 hours - full moon to full moon - 708 hours	
Orbital distance	2.41 million km	1.5 million miles
Rotation velocity	16 km/hr	10 mi/hr
Atmosphere	None	

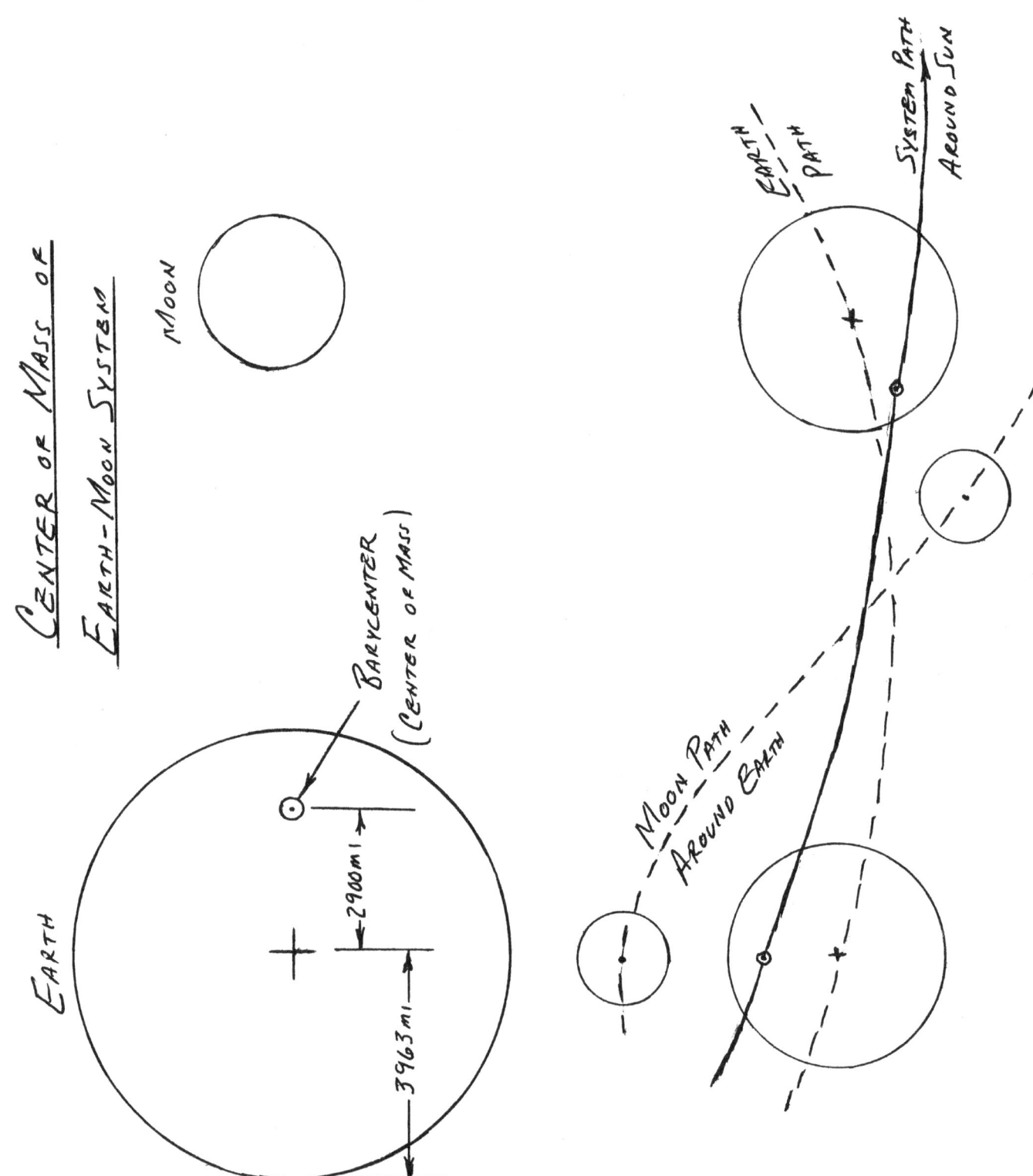

CENTER OF MASS OF
EARTH-MOON SYSTEM

MOON

EARTH

BARYCENTER
(CENTER OF MASS)

2900mi

3963mi

EARTH PATH

SYSTEM PATH AROUND SUN

MOON PATH AROUND EARTH

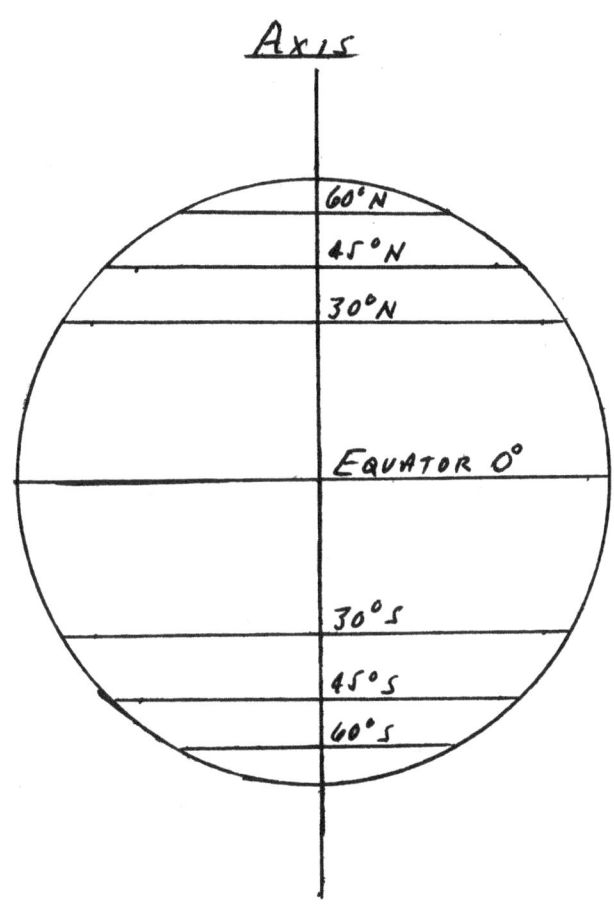

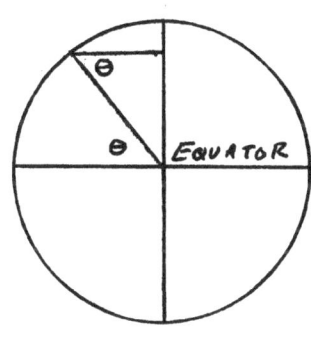

$V = 1037 \cos \theta$ 　　　　$V = Velocity$

V AT $0°$ (EQUATOR)	1037 MILES PER HOUR
V AT $\pm 30°$ LATITUDE	898 " " "
V AT $\pm 45°$ "	733 " " "
V AT $\pm 60°$ "	518 " " "
V AT $\pm 90°$ (POLES)	0 " " "

133

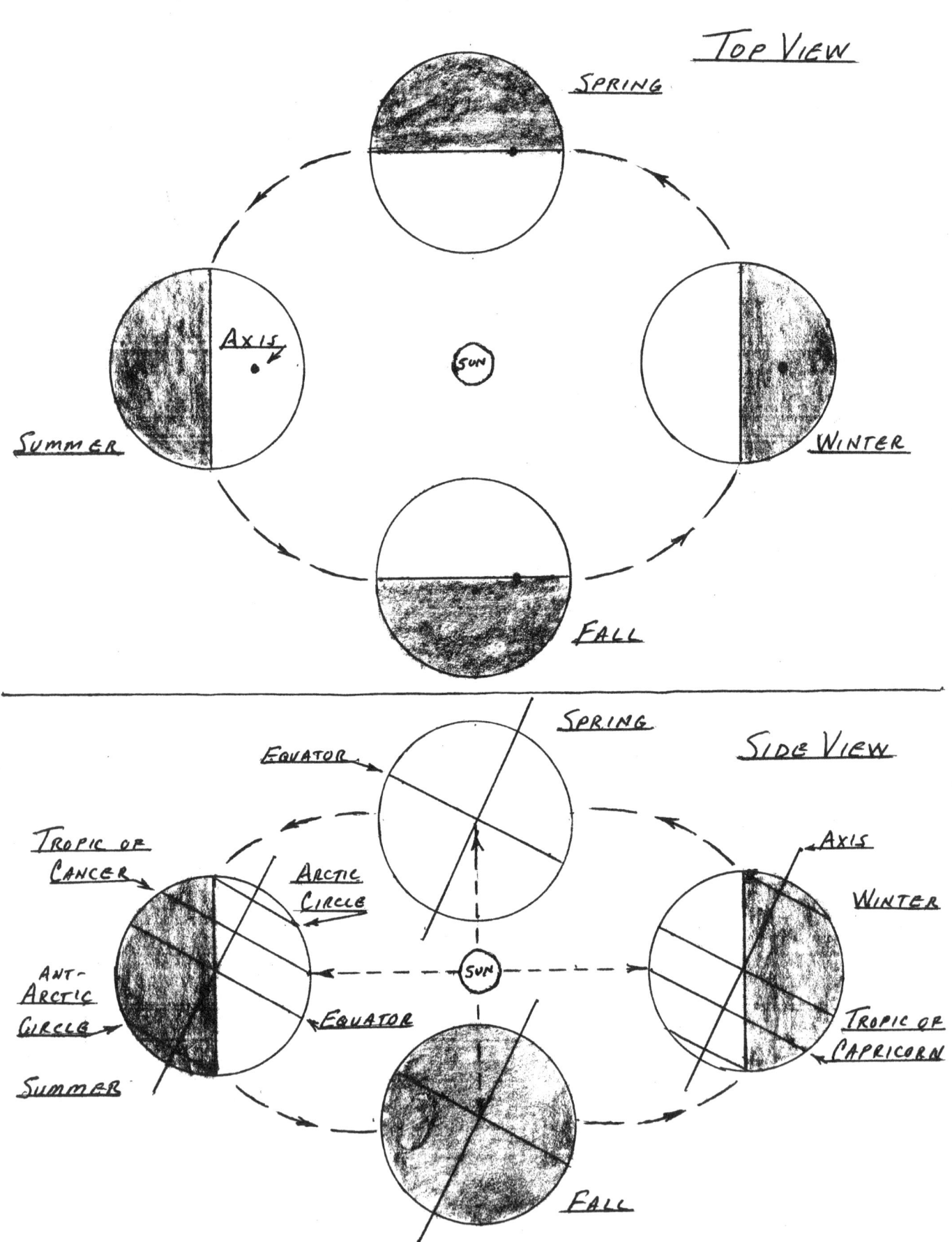

TOP VIEW

SPRING

AXIS

SUN

SUMMER

WINTER

FALL

SIDE VIEW

SPRING

EQUATOR

TROPIC OF CANCER

ARCTIC CIRCLE

AXIS

WINTER

ANT-ARCTIC CIRCLE

EQUATOR

SUN

TROPIC OF CAPRICORN

SUMMER

FALL

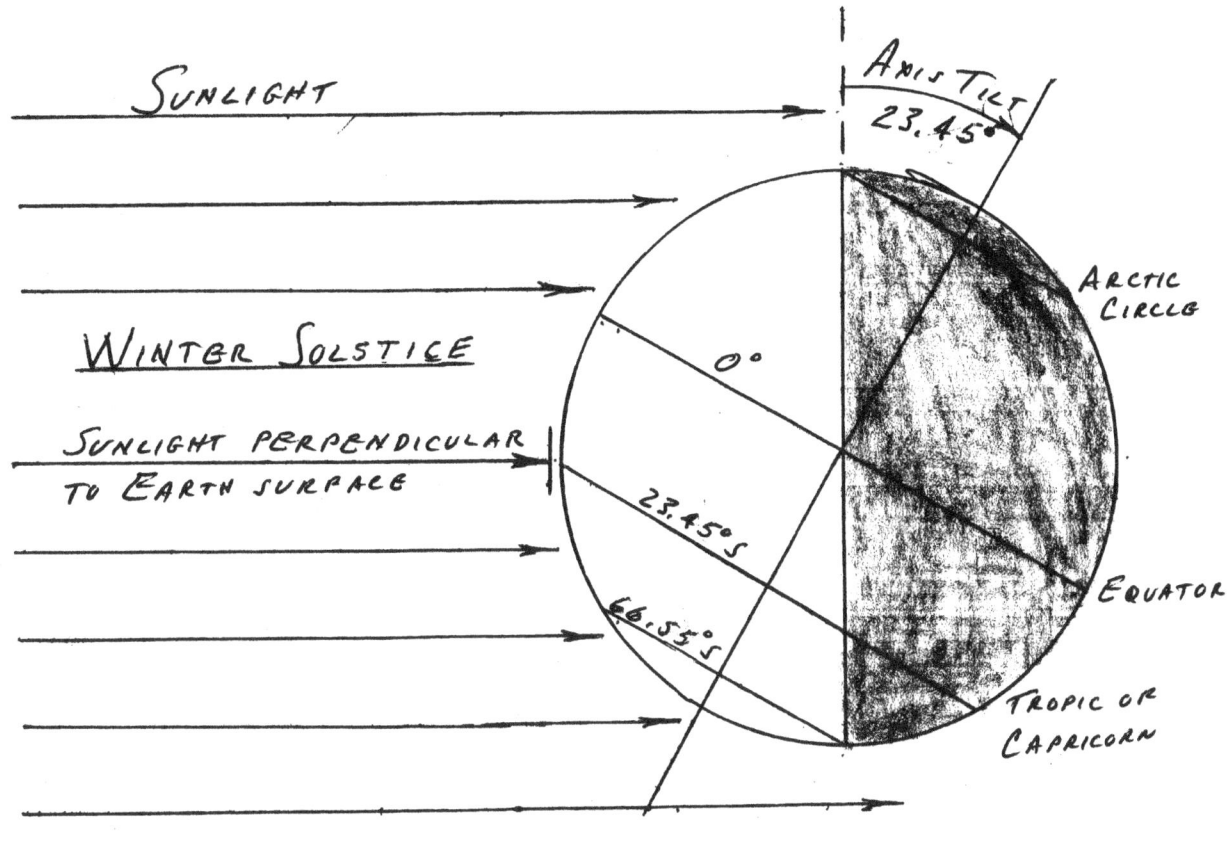

SUNLIGHT

AXIS TILT
23.45°

ARCTIC
CIRCLE

WINTER SOLSTICE

0°

SUNLIGHT PERPENDICULAR
TO EARTH SURFACE

23.45°S

EQUATOR

66.55°S

TROPIC OF
CAPRICORN

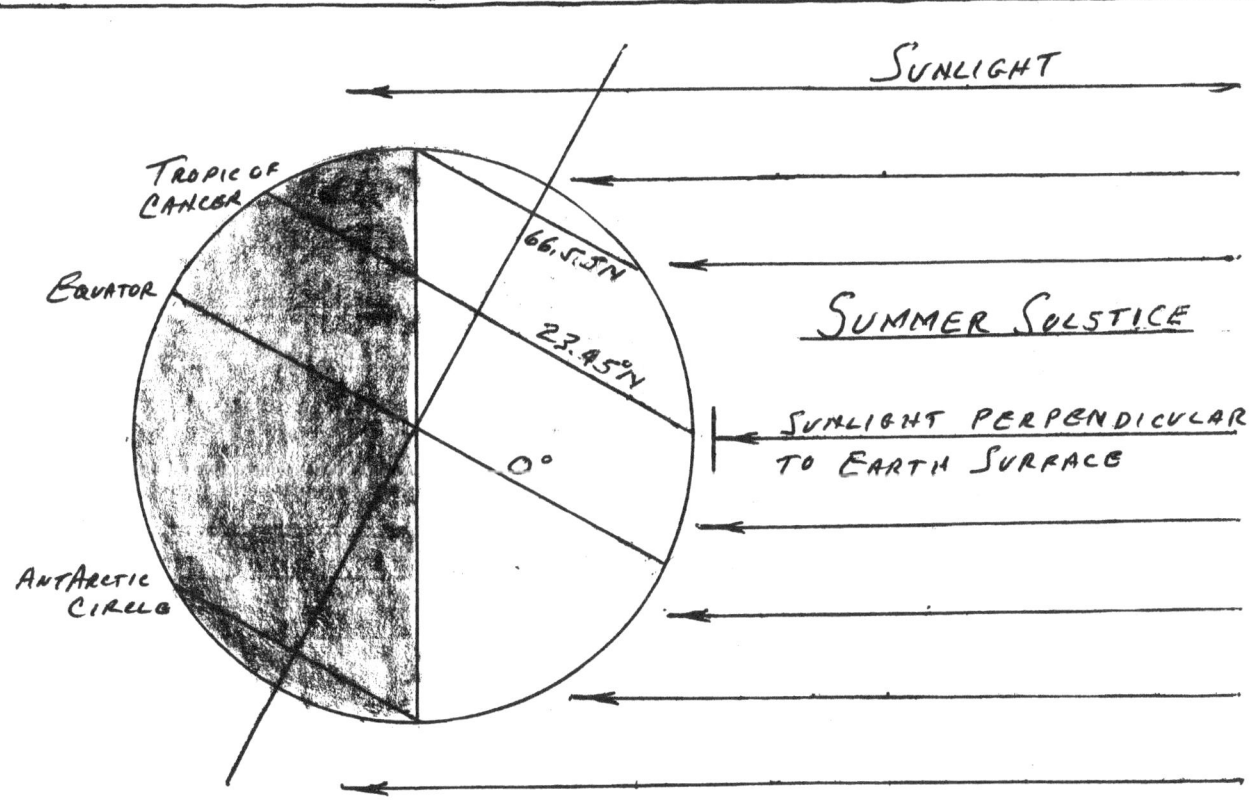

SUNLIGHT

TROPIC OF
CANCER

66.55N

EQUATOR

23.45°N

SUMMER SOLSTICE

0°

SUNLIGHT PERPENDICULAR
TO EARTH SURFACE

ANTARCTIC
CIRCLE

135

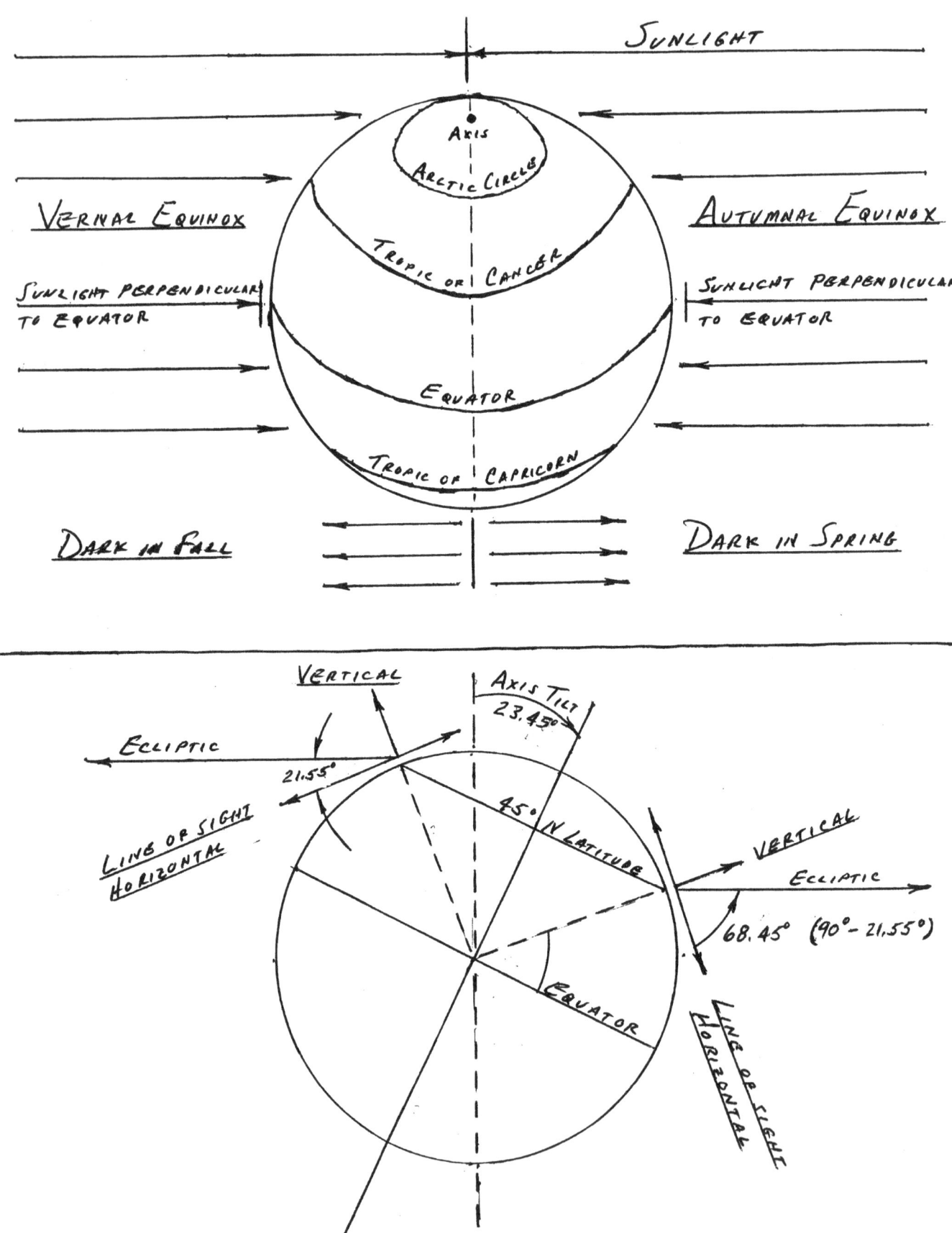

SUNLIGHT

AXIS
ARCTIC CIRCLE

VERNAL EQUINOX

TROPIC OF CANCER

AUTUMNAL EQUINOX

SUNLIGHT PERPENDICULAR TO EQUATOR

SUNLIGHT PERPENDICULAR TO EQUATOR

EQUATOR

TROPIC OF CAPRICORN

DARK IN FALL

DARK IN SPRING

VERTICAL

AXIS TILT
23.45°

ECLIPTIC

21.55°

45° N LATITUDE

VERTICAL

LINE OF SIGHT
HORIZONTAL

ECLIPTIC

68.45° (90° - 21.55°)

EQUATOR

LINE OF SIGHT
HORIZONTAL

ANGLES TO ECLIPTIC

DATE	1200	2400	DAYS
12-21	24	66	0
1	31	59	
2	38	52	
3-21	45	45	91
4	52	38	
5	59	31	
6-21	66	24	182
7	59	31	
8	52	38	
9-21	45	45	274
10	38	52	
11	31	59	
12-21	24	66	365

12-21 to 6-21 $24^0 + 0.23^0$ x DAYS = ECLIPTIC AT 1200

6-21 to 12-21 $66^0 - 0.23^0$ x DAYS = ECLIPTIC AT 1200

2400 to 1200   ECLIPTIC + $3\text{-}1/2^0$ x HOURS = ANGLE
FOR 6-21 @ 0800 $24^0 + 3\text{-}1/2^0$ x 8 = $24^0 + 28^0 = 52^0$

1200 to 2400   ECLIPTIC - $3\text{-}1/2^0$ x HOURS = ANGLE
FOR 6-21 @ 2000 $66^0 - 3\text{-}1/2^0$ x 8 = $66^0 - 28^0 = 38^0$

THE CHART IS FOR $47\text{-}1/2^0$ NORTH - SEATTLE, BUDAPEST

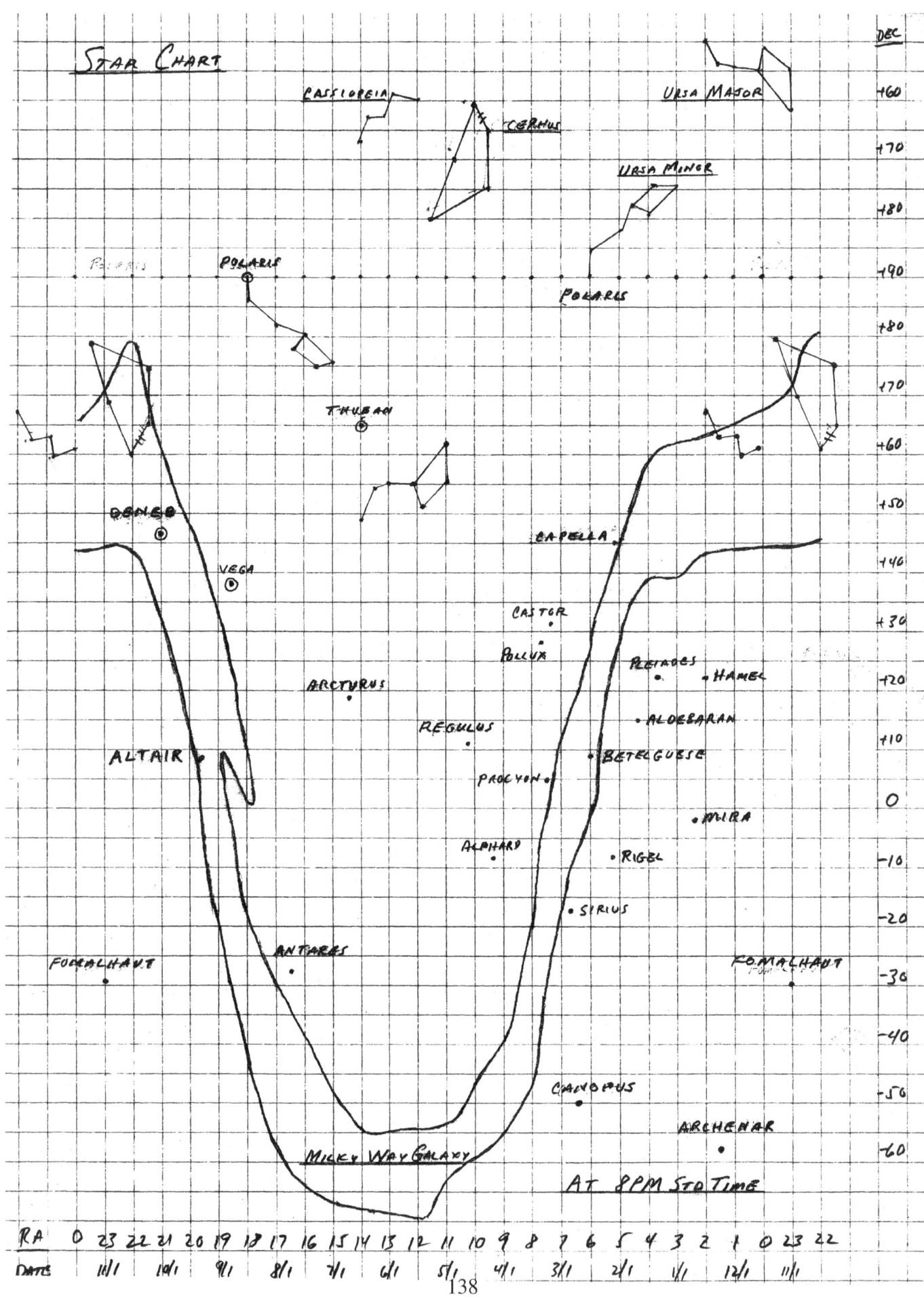

EARTH ROTATION

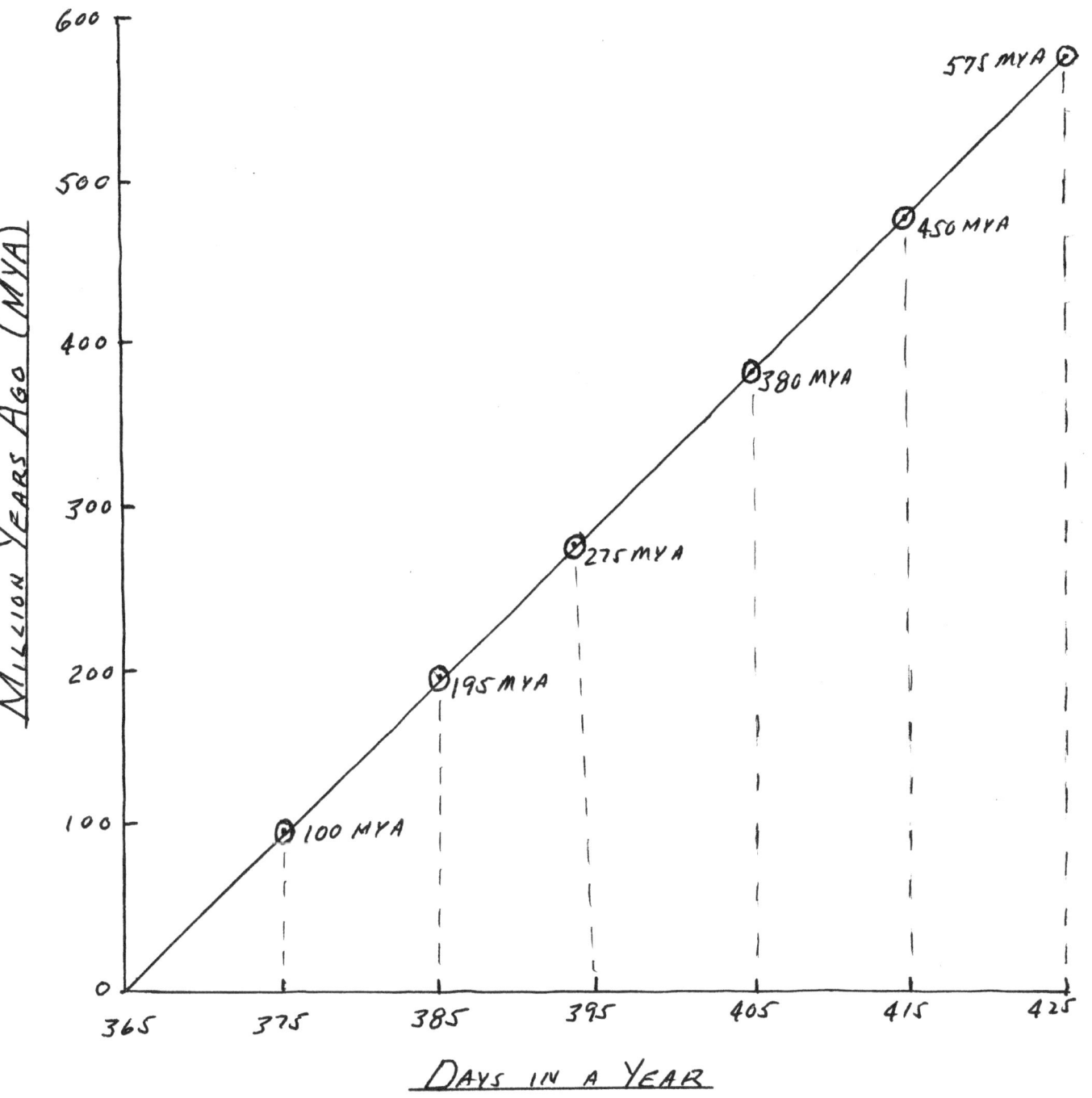

PRECESSION

Earth has many motions - it spins on its axis, it travels around the Sun, it moves with the Sun through the Milky Way galaxy, which is also moving with the other galaxies in our galactic group. We also have another movement - precession. Each of us has spun a top at one time or another. Remember when the top started, it was spinning so fast that it appeared to be standing perfectly still on its point. As it slowed, the top part of the top started going around in a circular motion while the point stayed in the same location. If you drew a line from the bottom point of the top through and out of the top of its center the circular shape of a cone would appear. This motion is called precession. This motion has occurred since the Earth was formed over four and one-half billion years ago. Imagine standing over this top as it slowly completed each cycle of the circle. If you and three of your friends stand with the tops of your heads butted together looking down on the top as it spins, you will notice that it will point to one of your friends, then to each of the others as it spins - it is *precessing* around the circle. You and your three friends have just acted as four stars looking down on Earth as time passed by - each called the polar star as the precession continued in its movement through its cycle.

When you look at a globe you notice that the shaft through it is not vertical, but is at an angle of 23.45 degrees (23^0 27'). Hold the bottom of the globe and turn the top of the globe in a circular motion. This is what the top was representing. The Sun and Moon's gravitational forces cause this action. During a complete cycle of precession, taking about 25,735 years, Earth's axis traces out the cone we discovered using the top and the globe. Superimposed on the precession of the equinoxes are two other motions of smaller amplitude. One of them is a minor regular oscillation of the Earth's axis called the *nutation*, with a period of 18.6 years. This motion amounts to about nine seconds (9") of arc and is caused by a periodic variation in the angle of the Moon's path with respect to the Earth's equator. The other motion, called the *Chandler wobble*, causes the Earth to move about its axis with a variable period of about 440 days.

There are four stars (you and your friend's heads) that appear directly above the projection of our cone as it precesses with Polaris, or the North Star, being the present pole star. Thuban was the pole star in 3000 BC, being the pole star in the era of the ancient Sumerians, Egyptians and others. Deneb will be the next pole star in about seven thousand years followed by Vega five thousand years later and then Thuban again eight thousand years after that around the year 22,000. Polaris will return as the North Star sometime in the years around 27,000. When we view the stars from our front porch, we now know why things are slowly changing.

Considering the 360 degree cycle of precession, the arc of the circle consisting of seven thousand years for the precession to move from Polaris to Deneb will be 100.8 degrees. For the five thousand years for the move from Deneb to Vega, the arc will be 72 degrees; the eight thousand years for the move from Vega to Thuban, the arc will be 115.2 degrees; for Thuban to Polaris again will take five thousand years and will cover the remaining 72 degrees to complete the cycle. The move is not noticed by life on Earth because the arc of movement is only 1.44 degrees every thousand years. $100.8^0 + 72^0 + 115.2^0 + 72^0 = 360^0$ - *so there!*

Because of Earth's precession, the length of time from one vernal equinox to the next, one tropical year, is not the same as the time required for Earth to complete one orbit - one sidereal year, which is the time required for the constellations to complete one cycle around the sky and return to their starting points, as seen from a given point on Earth. The vernal equinox occurs as Earth's equator is perpendicular to the line joining the Sun and Earth, and the Sun is crossing the equator appearing to be moving from south to north. In the absence of precession, this would occur exactly once per orbit, and the tropical and sidereal year would be identical. However, because of the slow precessional shift in the Earth's rotation axis, the instant when the equator is next perpendicular to the line to the Sun occurs slightly sooner than we would expect otherwise. Consequently, the vernal equinox drifts slowly around the zodiac over the course of the precession cycle. This is the cause of the *twenty minute discrepancy* between the tropical and sidereal year.

The tropical year is the year that our calendars measure. If our timekeeping were tied to the sidereal year, the seasons would slowly march around the calendar as the Earth precessed. Thirteen thousand years from now, summer in the Northern Hemisphere would be at its height in late February. By using the tropical year instead, we ensure that July and August will always be summer months. However, in thirteen thousand years time, Orion will be a summer constellation. Orion is presently a winter constellation.

The *sidereal* year is the time it takes for a complete revolution through the constellations. A *sidereal* day consists of 23 hours 56 minutes and 4.09 seconds with a *sidereal* year being 365 days 6 hours 9 minutes and 9.54 seconds. The *tropical* year is the time the Sun passes the vernal equinox consistently each year. A *tropical* year is 365.2422 days, or 365 days 5 hours 48 minutes 45.6 seconds (decreasing one-half a second each century). From the length of each year being shown, the *sidereal* year is 20 minutes 23.94 seconds longer than the *tropical* year. This is equal to 0.34 hours, or 0.01417 day, longer for the *sidereal* year making a gain of one day each 70.57 years. A gain of one-half year (182.625 days) at this rate would occur in 12,887.85 years, so if the *sidereal* year were used for the calendar, December 21, 14888 would have the same weather that occurred on June 21, 2000. Using the *tropical* year for the calendar maintains the current weather patterns over the centuries.

There are many more complexities to Earth's motion. The cone traced out by Earth's precession is not as clean as we initially drew. Because of the combined gravitational influence of the Moon and the planet Jupiter, the tilt of Earth's axis with respect to the ecliptic varies back and forth between 21.39 and 24.36 degrees. This variation is very slow, with one cycle completed every forty-one thousand years. Along with the other planets that make up the solar system, Earth is just going along for the ride, tracing out a sinusoidal path as it travels through space toward Vega at twenty kilometers per second (44,740 miles per hour). Earth is also losing angular momentum due to its tides. It is being transferred to the angular momentum of revolution as Earth and Moon turn about a common center of gravity - the barycenter. The Moon must then move further from Earth at 3.15 centimeters per year (twenty miles per million years) as Earth's rotation momentum decreases.

PRECESSION

SPRING

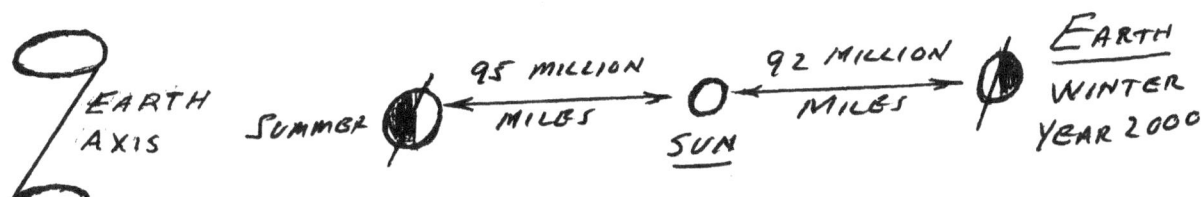

EARTH
AXIS

SUMMER
95 MILLION MILES

SUN
92 MILLION MILES

EARTH
WINTER
YEAR 2000

FALL

SEASONS AT MOST NORMAL TEMPERATURES

FALL

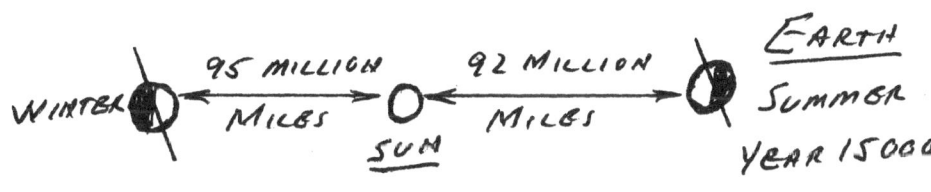

EARTH
AXIS

WINTER
95 MILLION MILES

SUN
92 MILLION MILES

EARTH
SUMMER
YEAR 15000

SPRING

SEASONS AT MOST EXTREME TEMPERATURES

WINTER - COLDEST   SUMMER - HOTTEST

CRUSTAL PLATES

The bulk of Earth's crust is composed of oxygen, silica and aluminum which form the granitic rocks that constitute most of the continents. The crust and upper brittle mantle constitute the lithosphere, which averages about sixty miles thick. The lithosphere floats on the semi-molten outer layer of the mantle called the asthenosphere, which ranges in depth from about seventy to one hundred fifty miles. This gives the outer layer of Earth a structure somewhat like a jelly sandwich, which is important for the operation of plate tectonics. Otherwise, the crust would consist of jumbled slabs of rock and Earth would be a rocky, dead planet. Life exists here because of plate tectonics.

Earth's crust is relatively thin compared to the Moon and the other terrestrial planets. Mercury, Venus and Mars have thick buoyant, nonsubducting crusts because they are either too cold or too hot and have been tectonically inactive for over two billion years. A thick buoyant crust on Earth would have remelted because of the high concentration of radioactive elements and great pressures induced by the weight of the overlaying rocks. Such a thick crust would have acted like an insulating blanket to hold the constantly generated heat from the Earth's interior, raising the internal temperature high enough to melt surface rocks. A thick crust would also have been highly unstable, creating a massive overturn that would melt the entire crust. Since there is no trace of the first seven hundred million years of Earth's history, this might provide evidence to such an occurrence.

A thick, buoyant crust could not be easily broken up and subducted into the mantle, which is important for the operation of global tectonics. The lithospheric plates would float on the surface like pack ice in the arctic. This would make Earth an uninteresting place; there would be no mountains, no oceans, no volcanoes or earthquakes and NO LIFE! The continental crust averages twenty-five to thirty miles thick-and is up to forty-five miles thick in the mountainous regions. The oceanic crust is considerably thinner and in most places is only three to five miles thick. Like an iceberg, only the tip of the crust shows, while the rest is out of sight deep below the surface. The continental crust is twenty times older than the oceanic crust, which is no older than one hundred seventy million years. This is because the older oceanic crust has been consumed by the mantle and subduction zones spread around the world. Because of plate tectonics, as many as twenty oceans may have come and gone during the last two billion years.

Eight major and about half a dozen minor lithospheric plates act as rafts that carry the crust around on the sea of molten rock. The plates diverge at midocean ridges and converge at subduction zones where the plates are subducted into the mantle and remelted. The subduction zones that are located on the ocean floor are called deep-sea trenches. The plates and oceanic crust are continuously recycled through the mantle but the continental crust, because of its greater buoyancy, remains for the most part on the surface.

It is due to plate tectonics that life flourishes on Earth. It is even possible that there would not be active plate tectonics if Earth did not possess life. Lime-secreting organisms in the ocean remove carbon dioxide, an important greenhouse gas, from the atmosphere and store it in the bottom sediments. This keeps the Earth's surface temperature within the range needed for plate tectonics to operate, which in turn maintains living conditions on Earth.

The lithosphere, which includes the crust and its underlying plate, is generally between fifty and one hundred miles thick under the continents. Beneath the ocean floor it ranges from about five miles thick near spreading centers to about sixty miles thick at plate margins. Cracking open the continental lithosphere would be a major project, considering its great thickness. The best evidence for rifting of continents can be found in the African rift system, which has not yet fully ruptured. When it does, the present continental rift will be replaced by an oceanic rift as the area floods with seawater. This type of rifting is presently taking place in the Red Sea where Africa and Saudi Arabia are slowly drifting away from each other. The Salton Sea in southern California exists due to plates being pulled apart there, also.

The Rift Valley, sometimes known as the Great Rift Valley, is a gigantic trenchlike fracture in the Earth's crust. It extends from northern Syria across East Africa to southern Mozambique - a total distance of about 4,800 miles (7,700 kilometers) averaging twenty-five to thirty-five miles (forty to fifty-five kilometers) in width. It is typically bounded by high cliffs or tiers of cliffs that in places rise thousands of feet above the valley floor. The floor itself varies in elevation from 1,296 feet (395 meters) below sea level in the Dead Sea to 6,000 feet (1,800 meters) above the sea in Kenya. The overall contours of the Rift Valley can be traced on any map by noting the many bodies of water cupped in by the Jordan River Valley, the Dead Sea and the Gulf of Aqaba. Beyond that the great cleft is occupied by the Red Sea, with a branch extending east through the Gulf of Aden.

The Rift Valley enters Africa in the broad fan-shaped Danakil Depression, an area of intense heat and barren salt flats reaching some four hundred feet (one hundred twenty meters) below sea level. The valley floor rises to much higher elevations as it bisects Ethiopia and, in northern Kenya, contains the waters of Lake Turkana. Continuing across Tanzania and Malawi into Mozambique, the southern part of the Rift Valley is occupied by long, narrow Lake Malawi. To the north of Lake Malawi, the valley divides in two, with the Western Rift Valley branching away from the main Eastern Rift Valley. Curving northward along the border of Zaire, the Western Rift Valley contains a conspicuous chain of large lakes, including Lake Tanganyika and Lake Kiuu. Lake Victoria is located between the Western and Eastern Rift Valleys. It is the world's second largest freshwater lake. The lake was created by massive movements of the Earth's crust that resulted in the formation of the eastern and western branches of the Rift Valley. As the trenchlike rifts opened up, the plateau between them sagged in the center, forming the broad, shallow basin occupied by the lake. Kilimanjaro rises majestically above the plains of Tanzania near the Kenyan border. It is a towering, snow-covered summit with a maximum elevation of 19,340 feet (5,895 meters) above sea level located on the eastern side of the Eastern Rift Valley. At the highest elevations, there is the zone of permanent ice and snow that is responsible for the name Kilimanjaro, which in Swahili means "the mountain that glitters."

HOT SPOTS

More than half of the hot spots exist on the continents, with the greatest concentration, about twenty-five in all, in Africa, which has remained essentially stationary for millions of years. Hot spots are also numerous in Antarctica and Eurasia, indicating that these plates have very little movement. In contrast, rapidly moving plates such as the North and South American plates, hot spot volcanism is rare. Yellowstone National Park now contains the active hot spot in North America. Its track can be traced through volcanic rocks on the Snake River Plain for four hundred miles in southern Idaho. Over the past fifteen million years, the North American Plate slid southwestward across the hot spot, placing it under its temporary home at Yellowstone.

Several island chains in the Pacific Ocean are the result of hot spots, with the Hawaiian Island chain being the most visible. As the Pacific plate moves over a hot spot, each island is formed. A new island is presently being formed under the surface of the ocean and a sequence is seen when viewing a map of these islands. An abrupt change in the direction of plate movement is indicated by the change in direction of the chain of islands in the Pacific. The Emperor Seamount Chain began to form more than forty million years ago when the plate was moving northward. About twenty-five million years ago, the plate moved northwestward and started to form the Midway-Hawaiian Chain.

PLATE TECTONICS

Earth's outer shell is comprised of a dozen or so rigid mobile *tectonic* plates (Greek term TEKTON - to "build") composed of the upper mantle that are always in motion. This movement accounts for all geologic activity taking place on the Earth's surface. The plates ride on semimolten rocks of the Earth's mantle and carry the continents along with them like ships frozen in arctic pack ice.

When two plates collide, they create mountain ranges on the continents or volcanic islands on the ocean floor. The breakup of a plate creates new continents and oceans. The process of rifting (splitting) and patching of the continents has been going on for at least 2.8 billion years and FOUR supercontinents have formed and fragmented during this period. The first was Kenora, formed 2.3 billion years ago; the second was Amazonia, formed 1.5 billion years ago; the third was Baikalia formed 800 million years ago; and Pangaea which formed some 220 million years ago. A global mid-ocean ridge system that stretches forty thousand miles along the ocean floor like a seam on a baseball defines the outlines of the plates that cover the surface of the Earth. The mid-Atlantic Ridge magma which is pushed up from the interior of the Earth forces apart North America and Europe at the rate of one inch each year. In the Pacific Ocean, the same action is creating up to six inches each year of new Pacific Ocean plate. The pressure of magma being pushed up at mid-ocean ridges forces the ocean ridges further apart, forcing the ocean floor and the lithosphere upon which it rides away from the mid-ocean ridge. This generates about three cubic miles of oceanic crust every year. The Mid-Atlantic Ridge growth is now monitored regularly. The ocean floor closest to the underwater ridge is relatively young, while material farther away, on either side, is much older. This is what would be expected of hot molten matter being pushed up and solidifying as the Eurasian and North American plates are forced apart.

Paleomagnetism, the study of ancient, or fossilized, magnetism reveals that magnetic fields are generally in the north-south direction. As the material that is pushed up in the ridge solidifies, the material becomes magnetized. This is a continuous event and on average, the Earth's magnetic field reverses itself roughly every half-million years.

Matched fossils and plants on various continents verify an original single landmass. The Cape Mountains of South Africa are similar to the Sierra Mountains south of Buenos Aires in Argentina. Canada, Scotland and Norway have rock strata of the same type that was laid down in exactly the same order. The continents of Africa, South America, Australia, Asia and Antarctica showed evidence of contemporaneous glaciation in the late Paleozoic Era (570-245 MYA), around 270 million years ago. Deposits of glacial till (unsorted and unstratified glacial deposits which consists of rock, worn-down rock, etc.) indicate this along with the grooves in the ancient rocks excavated by boulders embedded in slowly moving masses of ice. The lines of ice flow were away from the equator and toward the pole, which would not be the case if the contents were situated as they are today since glacial centers do not exist on the equator. Coral reefs and coal deposits found in the North Polar regions indicate that a tropical climate once existed on them before their being relocated to the north.

India was originally connected to Australia and Antarctica and drifted for over 140 million years to slam into Asia creating the Himalaya Mountains from seabed being elevated. The Alps of northern Italy formed in much the same manner as the Himalayas when the Italian prong of the African plate was forced under the European plate. Since the European plate is only half as thick as the Indian plate, the Alps are only about half as tall as the Himalayas.

Much of the Himalayas that rise so majestically from the Indo-Gangetic Plain were formed beneath the sea millions of years ago. An ancient ocean named the Tethys Sea, after a figure in Greek mythology, once lay roughly where the Himalayas are now located. Rivers entering the sea carried in debris from the surrounding land, and marine animals died and contributed their skeletons to the sediments accumulating on the ocean floor. Eventually the sediments piled up in beds as much as six miles (ten kilometers) thick and were compacted into shale, limestone and several other kinds of sedimentary rock.

The plate forming the Indian subcontinent moved slowly toward the Asian mainland, and finally, about sixty-five million years ago, the two started to collide. The force of the collision wrinkled and folded the sedimentary rock layers of the former Tethys Sea, but in this case the wrinkles were of enormous size. The process was somewhat similar to two pieces of ice colliding on a frozen river and pushing up a rim of shattered ice at the point of the collision. In many places the heat and pressure of the collision melted the sedimentary rocks and changed them into gneisses, schists and other metamorphic rocks. This collision process continues today. The mountains are rising at a rate of three to four inches each year with natural erosion at the top almost equaling this amount. The net growth of the Himalayas amounts to only one to two inches over the course of a century.

The oldest evidence of life on Earth is micro fossils, the remains of ancient microorganisms, and stromatalites, the layered structures formed by the accumulation of fine sediment grains by colonies or primitive blue-green algae. Stromatalites were found in Western Australia 3.5 billion years old. Micro fossils found in South Africa are 3.3 billion years old. Another evident factor is that the fossils of a reptile extinct for nearly 200 million years were uncovered at only two locations on Earth, one on the Brazilian coast and the other on the west coast of Africa. These two plates are precisely where the continents apparently meshed as part of the ancestral supercontinent of Pangaea.

When the Paleozoic Era came to a close, *Gondawana* and *Laurasia* converged into the crescent-shaped supercontinent *Pangaea*, the major land feature on the Earth approximately 220 million years ago. The continent extended practically from pole to pole. The single ocean Panthalasia covered the rest of the planet. The sediments in the Tethys Sea, upon collision, were uplifted into mountain belts including the Ouachita and Appalachians of North America and the Hercynian Mountains of southern Europe. Dinosaurs, which were then the dominant form of life, could have traveled from Russia to Texas through Boston without getting their feet wet.

At the end of the Paleozoic Era the breakup of the continents began. During the Jurassic period that followed, South America began to separate from Africa like a zipper opening up from south to north. India was set fully adrift and headed toward southern Asia. Antarctica, still attached to Australia, swung away from Africa toward the southeast, forming the proto-Indian Ocean. North America drifted westward.

During the Cretaceous Period, the oceans rose several times, flooding the continents. One part of that inundation was the Western Interior Seaway, which stretched from the Gulf of Mexico well into Canada and divided North America. For tens of millions of years, the seaway retreated and advanced, its western edge once advancing within fifty miles of the Rocky Mountains. About 68 million years ago, the seaway withdrew for good, leaving behind the configuration of continent and surrounding oceans that exist today. The accumulations of marine sediments, eroded from the chain of mountainous highlands to the west, were deposited on the red beds of the Colorado plateau. The areas covered were filled with thick deposits of sediments that are presently exposed as impressive cliffs in the Western United States. During the two to three million years that followed both the environment and the animal and plant populations of the region remained stable. Then a giant asteroid slammed into the planet north of the Yucatan Peninsula 64.5 million years ago.

Much of western North America was assembled from oceanic island arcs and other crustal debris that were skimmed off the Pacific plate as the North American plate continued heading westward. Northern California is a jumble of crust assembled only a few hundred million years ago. A nearly complete slice of ocean crust, the type that is shoved up on the continents by drifting plates, sits in the middle of Wyoming. The Appalachians, which are 225 million years old, having been upraised by a continental collision between North America and northwest Africa near the beginning of the Triassic Period, were eroded to stumps by the Cretaceous Period. When a mountain ten thousand feet high vanishes over a period of forty million years, what has happened? Each million years it loses two hundred fifty feet, which means, each thousand years it loses three inches. The loss per year would be minimal - only three thousandths of an inch per year. There was a high degree of geologic activity around the rim of the Pacific Basin, and practically all the mountain ranges facing the Pacific and the island arcs along its perimeter developed during this time. When the Cretaceous Period came to a close, North America and Europe were no longer in contact except for a land bridge that spanned Greenland, the world's largest island. Greenland separated from North America during the early part of the Cenozoic Era. The Bering Strait between Alaska and Asia narrowed, creating a nearly landlocked Arctic Ocean.

Africa moved northward, leaving Antarctica as South America was leaving Africa moving westward. Antarctica at this time was still joined to Australia and India was continuing to narrow the gap between it and southern Asia. The Rocky Mountains, now 65 million years old, extend from Mexico to Canada and were pushed up during the Laramide orogeny (mountain building episode), which occurred from about 80 million to 40 million years ago. During the Miocene, a large part of western North America was raised, and the entire Rocky Mountain region was raised about a mile above sea level. Great blocks of granite rose high above the surrounding land, while to the west in the Great Basin and Range area, the crust was pulled apart and dropped in some places below sea level, such as Death Valley. The Great Basin is bounded by the Cascade Range and Sierra Nevada mountain ranges on the west, and the Rocky Mountains on the east, extending from Mexico to Canada. The Great Plains are east of the Rocky Mountains, also ranging from Mexico to Canada.

The meeting of the North American continent with the East Pacific Rise spreading center caused the sediments in the trench to be squeezed and forced up to form the coastal ranges of California. At about the same time, Baja, California was separated from the mainland to form the Gulf of California. Arabia had split off from Africa under similar circumstances to form the Red Sea. In South America, the mountainous spine of the Andes running along the western edge of the continent continued to rise throughout much of the Cenozoic Era due to the subduction of the Nazca plate underneath the South American plate. The Andes are rising about four inches (ten centimeters) per century. South America was temporarily connected to Antarctica by a narrow, curved land bridge. When Antarctica and Australia broke away from South America and moved eastward, they themselves separated. Antarctica moved toward the South Pole, while Australia continued in a northeasterly direction. Antarctica drifted over the South Pole and acquired a thick blanket of ice. Glaciers also grew for the first time in the highest elevations of the Rocky Mountains. At times, Alaska was connected with East Siberia at the Bering Strait closing off the Arctic Basin from the warm ocean currents, resulting in the accumulation of pack ice. A permanent ice cap did not develop over the North Pole until about four million years ago when Greenland also acquired its first major ice sheet. The polar seas have been considered the lungs of the oceans because they are the primary sites of oxygen intake. Bottom circulation of the Weddell Sea in the Antarctic and the Norwegian Sea in the Atlantic can be compared to blood circulation; within these currents oxygen is circulated to all parts of the ocean bottom to be consumed by the metabolism of organisms while they live and by their decay when they are dead.

The 300 million year cycle of convection currents in Earth's mantle causes periods of rapid mantle convections in which the supercontinents tend to break up raising the global temperature. During the times of low mantle convection, all the continents are assembled into a supercontinent. This causes a reduction of carbon dioxide resulting in lower global temperatures. Supercontinents usually have a more moderate temperature where plant and animal life can prosper.

According to the theory of plate tectonics, the destruction of old oceanic crust at subduction zones matches the creation of new oceanic crust at midocean ridges where lithospheric plates, along with their overlying sediments, are forced down into the mantle. The rate of seafloor spreading is not always the same as the rate of subduction. This results in a lateral motion of the associated midocean ridges. Most of the subduction zones are in the western Pacific, which explains why nowhere is the oceanic crust older than about 170 million years.

The convection motions in the mantle that drive continents around the face of Earth are powered by heat generated by the decay of radioactive elements. After the continents have been separated from each other, heat is more easily conducted through the newly formed ocean basins. When a certain amount of heat has escaped, the continents cease their separation and start their journey toward each other again. When the present continents have reached their maximum range, the crust of the Atlantic seafloor will age and become dense enough to sink under the landmasses of the Americas and Europe/Africa, beginning the closing of the Atlantic basin. As the convection patterns shift, all continents will rejoin into a super continent and a new cycle will begin.

The term "Continental Drift" is sometimes misleading. The continents are drifting but they are only the visible evidence of the plates because they appear above the ocean as dry land. The seafloor itself is the slowly drifting plate and the oceanic water merely fills in the depressions between continents. The southeastern portion of the Pacific Ocean, called the Nazca plate, contains no landmass at all. The Indian plate includes all of India, much of the Indian Ocean and all of Australia and its surrounding south seas. The American plate on the east extends from Scandinavia down the middle of the Atlantic Ocean to the latitude of Cape Horn at the southern tip of South America. On the west, the American plate extends from the South Pacific up the coast of Mexico, America and Alaska.

The various plates collide producing either mountain ranges or subduction zones, one plate sliding under another, ultimately to be destroyed as it sinks into the mantle. Subduction zones are responsible for most of the deep trenches in the world's oceans. Other plates shear or slide past one another, such as the most active region in North America - the San Adreas Fault in California. The sudden jerks as these two plates move against each other are the cause of many major earthquakes.

The oceanic crust is remarkable for its consistent thickness and temperature, averaging about four miles thick and not varying more than twenty degrees Celsius over most of the globe. As the rigid lithospheric plate carrying the oceanic crust descends into Earth's interior through the subduction process, it slowly breaks up and melts. Over millions of years, it is absorbed into the general circulation of the mantle. It is involved in the recycling process until it once again appears as magma pushing up either in a ridge or through volcanic action to begin its long trip over once again.

The Cenozoic Era, from 65 million years ago to the present, when volcanic activity was extensive, witnessed great outpourings of basalt on the Pacific Northwest area of the United States. These outpourings covered the states of Idaho, Washington and Oregon creating the Columbia River Plateau. Somewhat concurrently (80-40 MYA) the Rocky Mountains were pushed up. The Apennine Mountain Range that runs the length of central Italy was uplifted aproximately 20 million years ago. Three million years ago in this same Pacific Northwest area the northern part of the East Pacific Rise was consumed in a subduction zone located beneath the continent. As the fifty mile thick crustal plate was forced down into the mantle, the heat melted parts of the descending plate and the adjacent lithosphere plate, forming pockets of magma. The magma melted its way to the surface and formed the volcanoes of the Cascade Range - Mounts Baker, Rainier, St. Helens, Adams and Hood.

Between three and four million years ago, the Panama Isthmus, which separated North America from South America, was uplifted due to colliding oceanic plates. This halted the flow of cold water currents from the Atlantic into the Pacific, which together with the closing off of the Arctic Ocean from the warm Pacific currents, might have initiated the Pleistocene glacial epoch. This caused the formation of two polar ice caps. At two million years ago, an ice age overwhelmed the Northern Hemisphere.

During the last ice age ten thousand years ago, the northern landmasses were covered with ice sheets up to two miles thick. Under the weight of the ice, North America and Scandanavia began to sink. When the ice melted, the crust became lighter and began to rise. In Scandanavia, marine fossil beds have risen more than one thousand feet above the sea level since this last ice age.

The entire state of Alaska is a jumbled cluster of terranes (steeply tilted Paleozoic rocks). A large portion of the Alaskan Panhandle was once part of eastern Australia some five hundred million years ago. It broke off from Australia, traveled across the Pacific Ocean, stopped briefly at the coast of Peru, sliced past California, swiping some of the Mother Lode, and bumped into North America around one hundred million years ago.

The ice age caused the level of the world's oceans to be more than three hundred feet lower that they are today. The ice was created by snow that originated from the moisture taken from the oceans. Now that the water from the oceans was now residing in the glaciers and ice pack, the topography of the surface of the Earth has a much different appearance. The land areas that had previously been separated were now joined by necks of land. Australia was attached to Antarctica, Sri Lanka to India, Cyprus to western Asia and England to Europe. The most spectacular land bridge was that of Alaska to Siberia, uniting two continents, having its own name - Beringa, the lost land of the Bering Sea.

Mount Everest, in the Himalayan Mountain range in Nepal, is located at 27^0 59' N 86^0 56' E and has an elevation of 29,035 feet (8,850 meters) above sea level. Mount Chimborazo, located in Ecuador 1^0 35' S 78^0 45' W, has an elevation of 20,702 feet (6,310 meters) above sea level. Due to Earth's rotation causing an equatorial bulge, Mount Chimborazo's top peak is located 3,967.11 miles (6,384 km) from Earth's center. Mount Everest's top peak is only 3,965.75 miles (6,382.25 km) from Earth's center. Therefore, Mount Chimborazo is higher from the center of Earth than Mount Everest.

The present drift of the continents projects that the California coast will break off at the San Andreas Fault and proceed northward, where the plate on which it is riding will plunge down the Aleutian Trench. North America will reverse direction and head back toward Eurasia. South America will separate from North America and drift into the South Pacific. The African and Eurasian plates will continue to press against each other and the Mediterranean Sea, caught in the middle, will be squeezed dry. A new subcontinent will tear off Eastern Africa and probably drift into India. Australia will drift northward possibly colliding with Southeast Asia. The movements will eventually form the new single supercontinent which is named NEOPANGAEA. The process of continental breakup will begin again, forming new continents and new oceans that will have no resemblance to our present-day world. This continental movement causes the plates to thicken which will cause all plate movement to cease some two billion years from now. Volcanoes will no longer erupt, earthquakes will cease, and Earth will become a dead planet.

THE DRIFTING OF THE CONTINENTS

AGE MILLION YEARS		GONDWANA	LAURASIA
Pliocene	3		
			Opening of Gulf of California.
Miocene	11	Began spreading near Galapagos Islands. Opening of the Gulf of Aden.	Change spreading directions in Eastern Pacific.
			Birth of Iceland.
Miocene	22		
		Opening of Red Sea.	
Oligocene	35		
		Collision of India with Eurasia. Separation of Australia from Antarctica.	Spreading of Arctic Basin. Separation of Greenland from Norway.
Eocene	55		
Paleocene	65		
		Separation of New Zealand from Antarctica. Separation of Africa from Madagascar and South America.	Opening of the Labrador Sea. Opening of the Bay of Biscay. Major rifting of North America from Eurasia.
Cretaceous	135		
		Separation of Africa From India, Australia, New Zealand, Antarctica	
Jurassic	180		
			Begin separation of South America from Africa.
Triassic	230		
Permian	280		

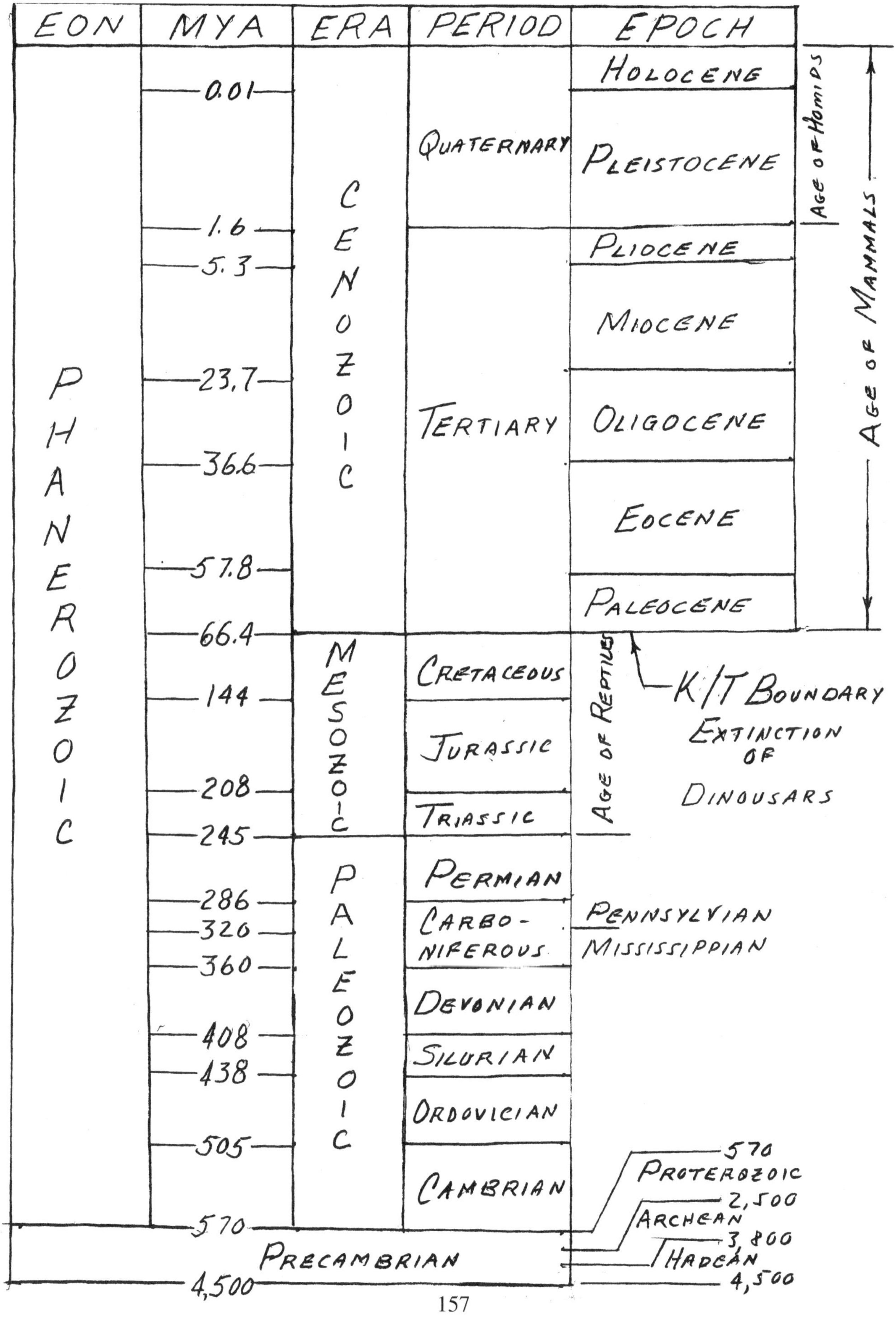

EON	MYA	ERA	PERIOD	EPOCH
PHANEROZOIC		CENOZOIC	QUATERNARY	HOLOCENE
	−0.01−			PLEISTOCENE
	−1.6−			PLIOCENE
	−5.3−		TERTIARY	MIOCENE
	−23.7−			OLIGOCENE
	−36.6−			EOCENE
	−57.8−			PALEOCENE
	−66.4−	MESOZOIC	CRETACEOUS	
	−144−		JURASSIC	
	−208−		TRIASSIC	
	−245−	PALEOZOIC	PERMIAN	
	−286−		CARBO- NIFEROUS	PENNSYLVIAN MISSISSIPPIAN
	−326−			
	−360−		DEVONIAN	
	−408−		SILURIAN	
	−438−		ORDOVICIAN	
	−505−		CAMBRIAN	
	−570−			
	−4,500−	PRECAMBRIAN		

AGE OF HOMIDS

AGE OF MAMMALS

AGE OF REPTILES

K/T BOUNDARY
EXTINCTION
OF
DINOUSARS

570
PROTEROZOIC
2,500
ARCHEAN
3,800
HADEAN
4,500

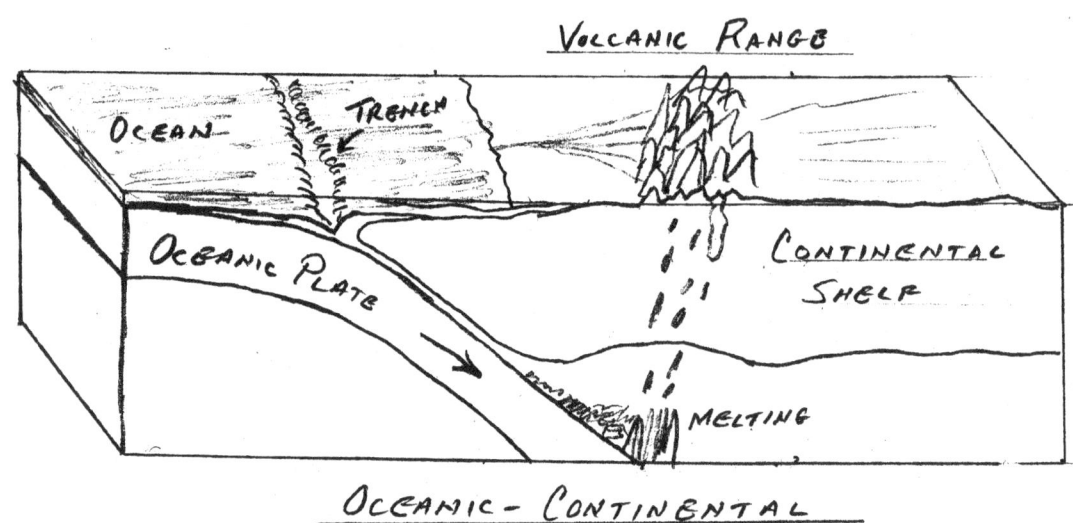

VOLCANIC RANGE

OCEAN

TRENCH

OCEANIC PLATE

CONTINENTAL SHELF

MELTING

OCEANIC - CONTINENTAL

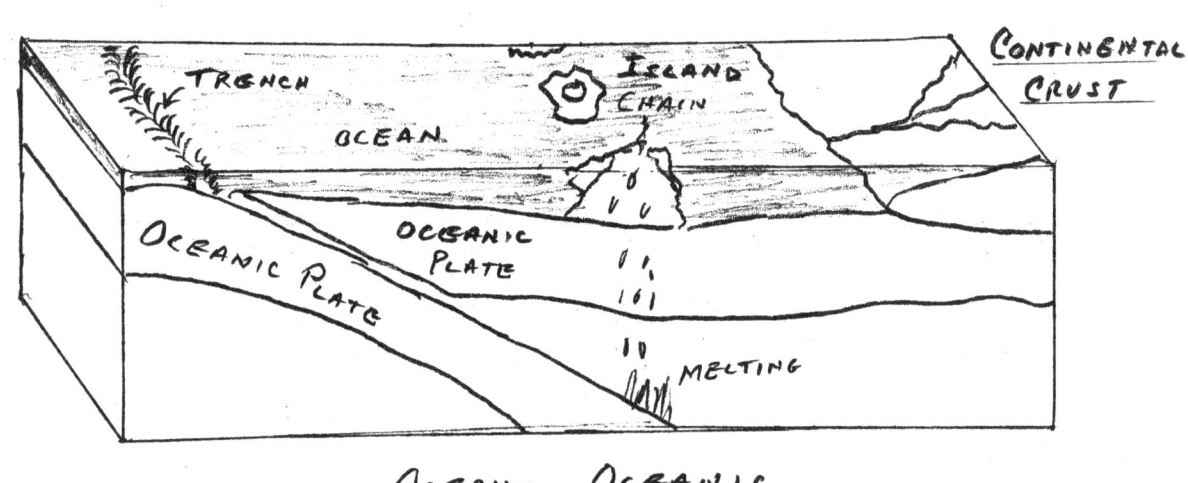

CONTINENTAL CRUST

TRENCH

OCEAN

ISLAND CHAIN

OCEANIC PLATE

OCEANIC PLATE

MELTING

OCEANIC - OCEANIC

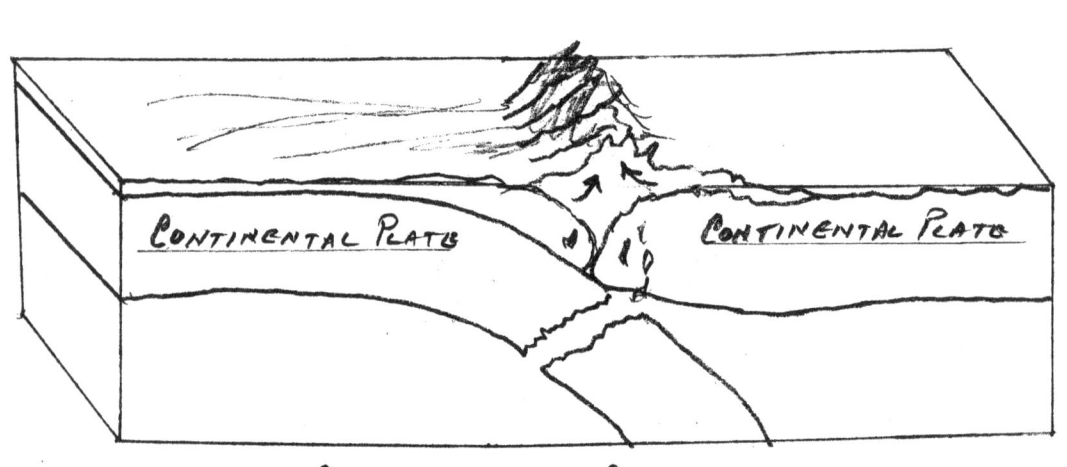

CONTINENTAL PLATE

CONTINENTAL PLATE

CONTINENTAL - CONTINENTAL

158

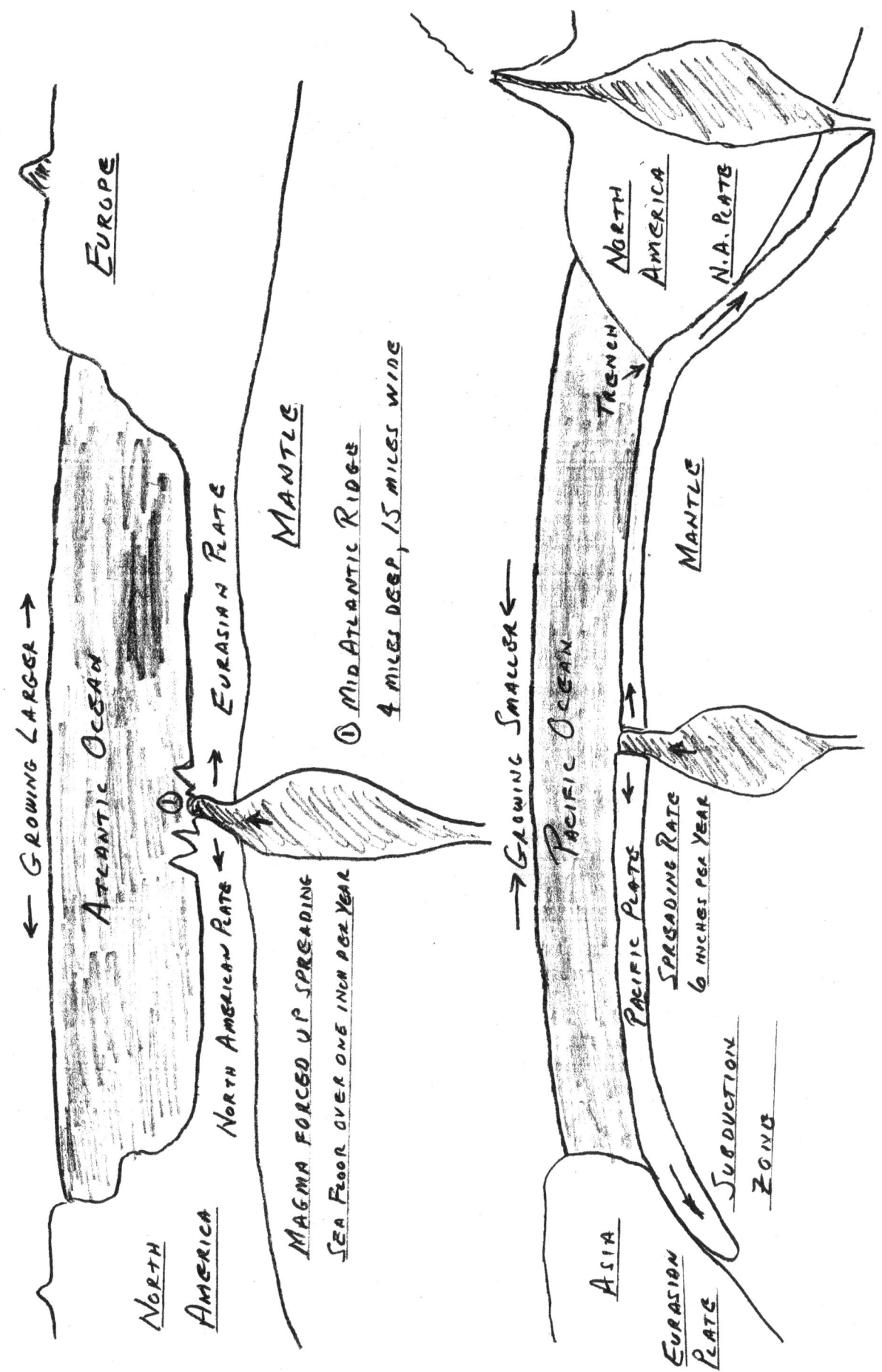

EUROPE

GROWING LARGER →

ATLANTIC OCEAN

NORTH
AMERICA

NORTH AMERICAN PLATE

EURASIAN PLATE

MANTLE

① MID ATLANTIC RIDGE
4 MILES DEEP, 15 MILES WIDE

MAGMA FORCED UP SPREADING

SEA FLOOR OVER ONE INCH PER YEAR

NORTH
AMERICA

N.A. PLATE

← GROWING SMALLER ←

PACIFIC OCEAN

TRENCH

MANTLE

PACIFIC PLATE

SPREADING RATE
6 INCHES PER YEAR

SUBDUCTION
ZONE

ASIA

EURASIAN
PLATE

159

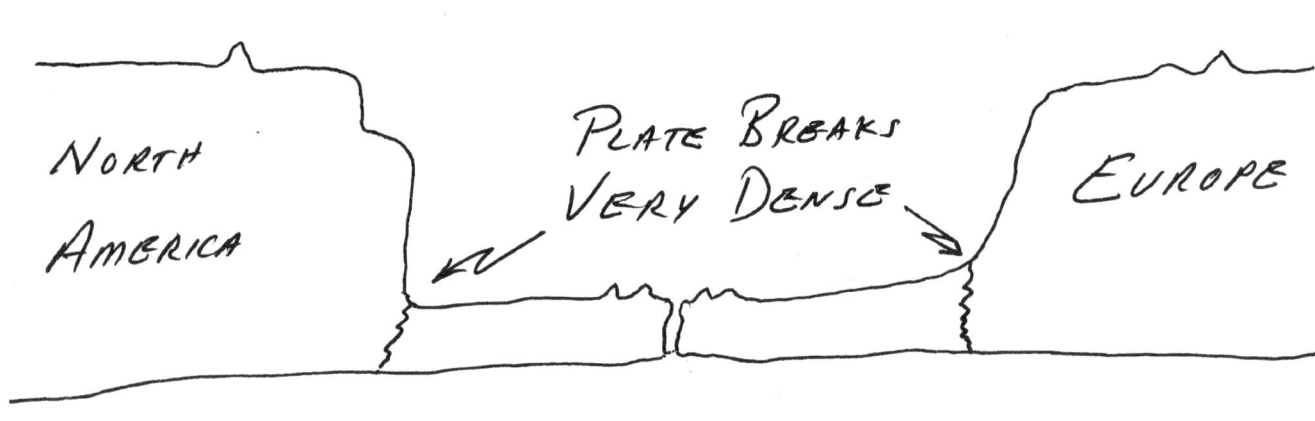

NORTH AMERICA

PLATE BREAKS VERY DENSE

EUROPE

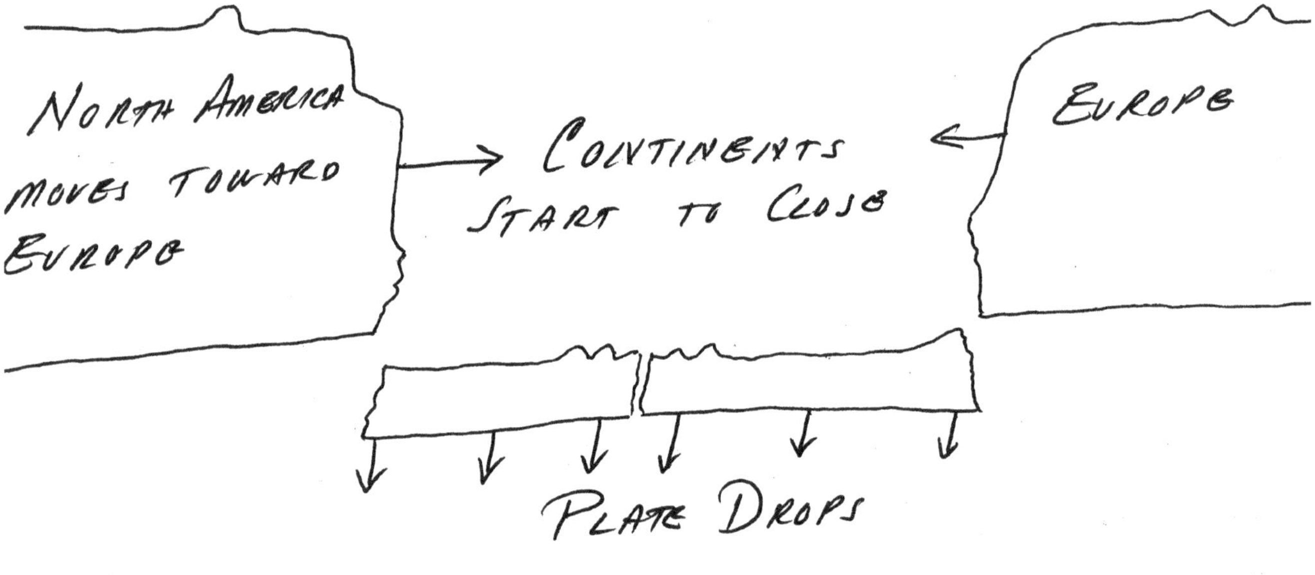

NORTH AMERICA MOVES TOWARD EUROPE

CONTINENTS START TO CLOSE

EUROPE

PLATE DROPS

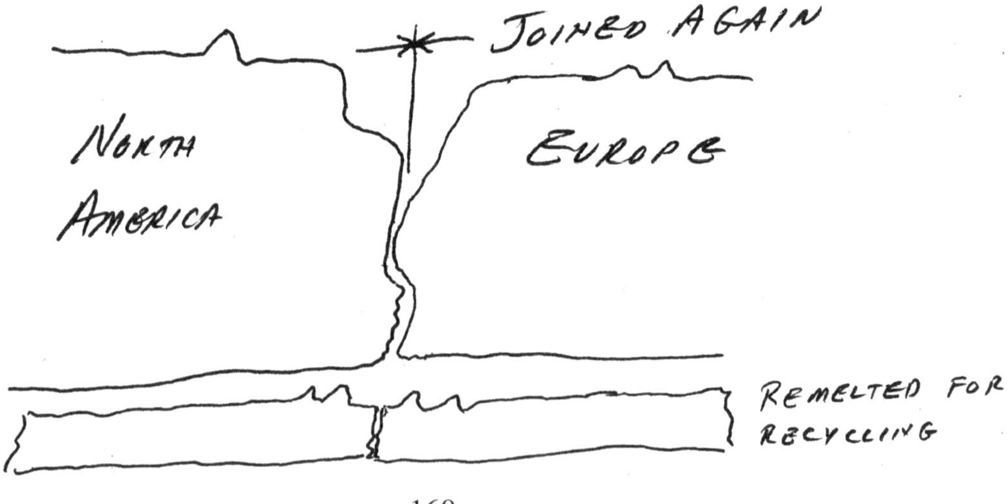

JOINED AGAIN

NORTH AMERICA

EUROPE

REMELTED FOR RECYCLING

PANGAEA

North America

Asia

Europe

LAURASIA

South America

Africa

GONDWANA

India

Antarctica

Australia

PANGAEA

MATCHING GEOLOGIC PROVINCES

GONDWANA

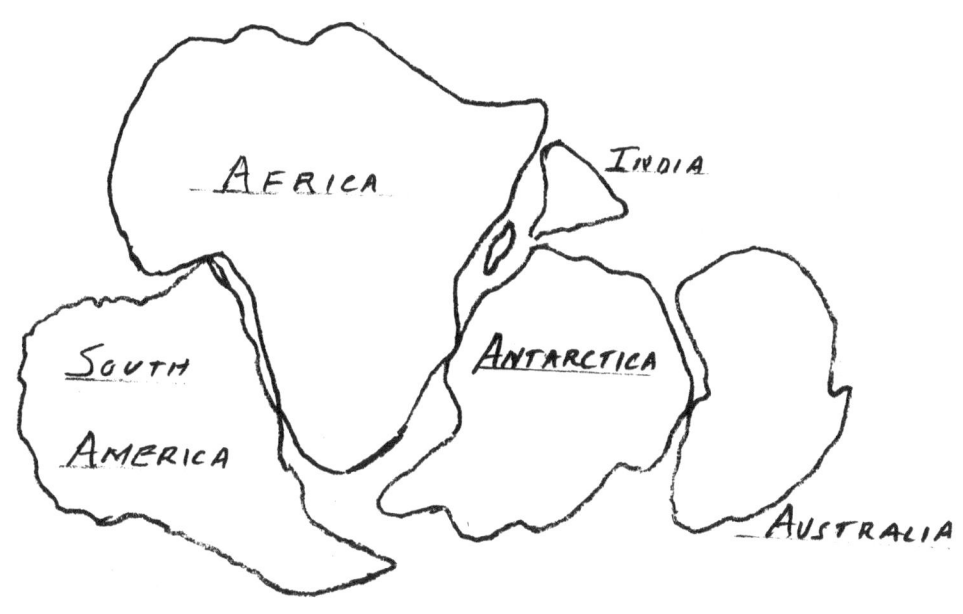

COMBINATION DURING THE PALEOZOIC 430 MYA

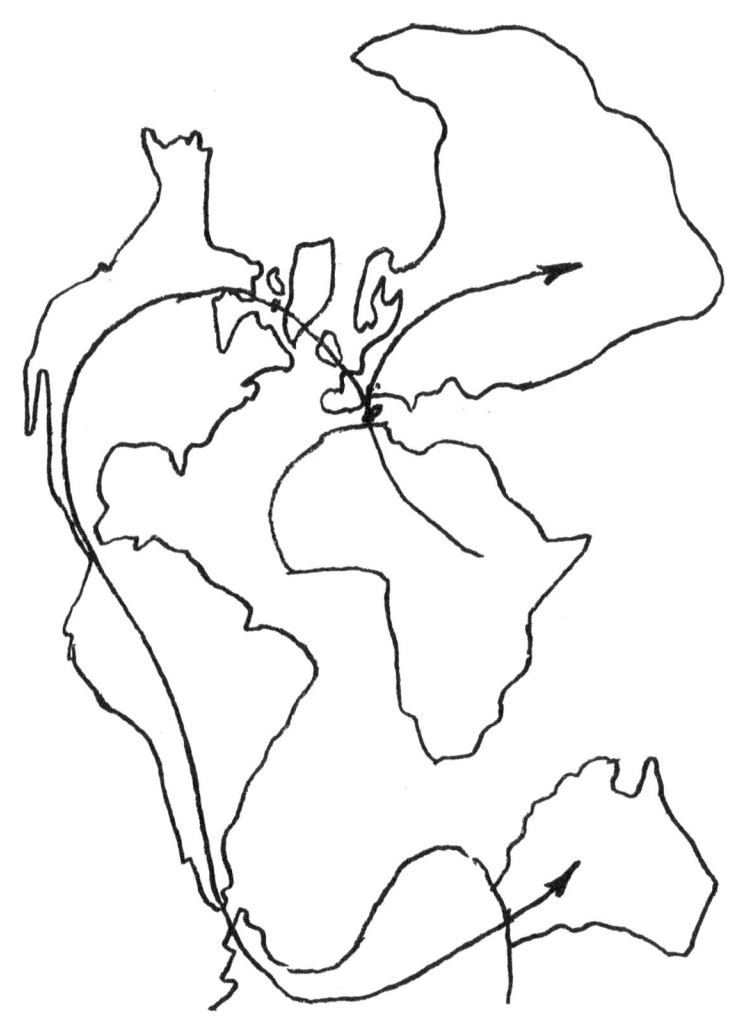

POSSIBLE ROUTES FOR MARISUPALS (DINOSAURS)
FROM AFRICA TO EUROPE/ASIA AND NORTH AMERICA
ALL THIS WAY TO AUSTRALIA.

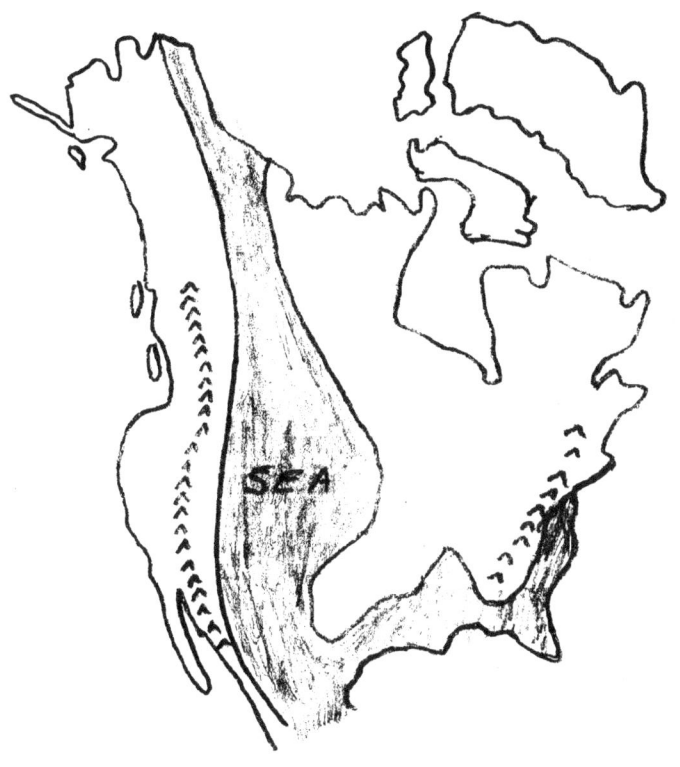

WESTERN CRETACEOUS INTERIOR SEA IN NORTH AMERICA
FILLED WITH THICK DEPOSITS OF SEDIMENT

THE MESOZOIC ERA IS KNOWN FOR ITS EROSION OF PREVIOUSLY
FORMED MOUNTAIN RANGES SUCH AS THE APPALACIANS, WHICH
WERE ERODED TO STUMPS. LARGE INLAND LAKES FORMED
AND THE SEAS INVADED THE INTERIORS OF MOST CONTINENTS.
MUCH OF RUSSIA, WESTERN SIBERIA AND ASIA WERE UNDER
WATER, ALONG WITH PORTIONS OF WESTERN NORTH AMERICA
AND PARTS OF SOUTH AMERICA.

165

THE DRIFT OF INDIA   140 MYA - 40 MYA

COLLIDES WITH ASIA - FORMS HIMALAYA MOUNTAINS

Plate Boundaries

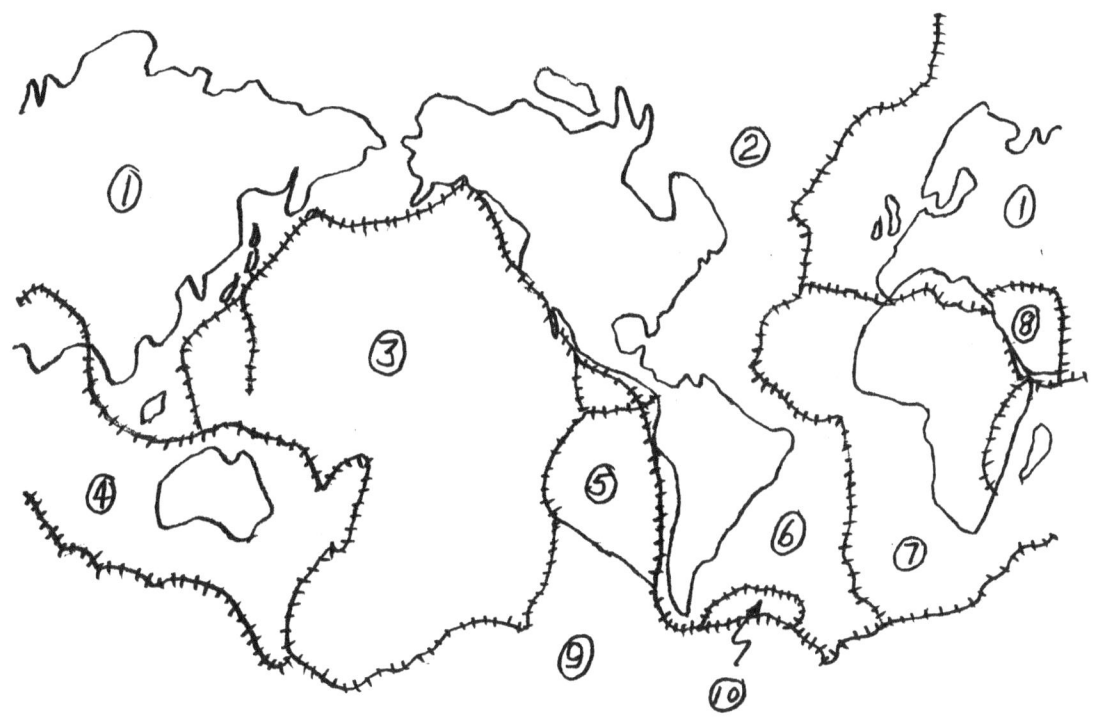

1. Eurasian
2. North American
3. Pacific
4. Indian Australian
5. Nazca
6. South American
7. African
8. Arabian
9. Antarctic
10. Scotia

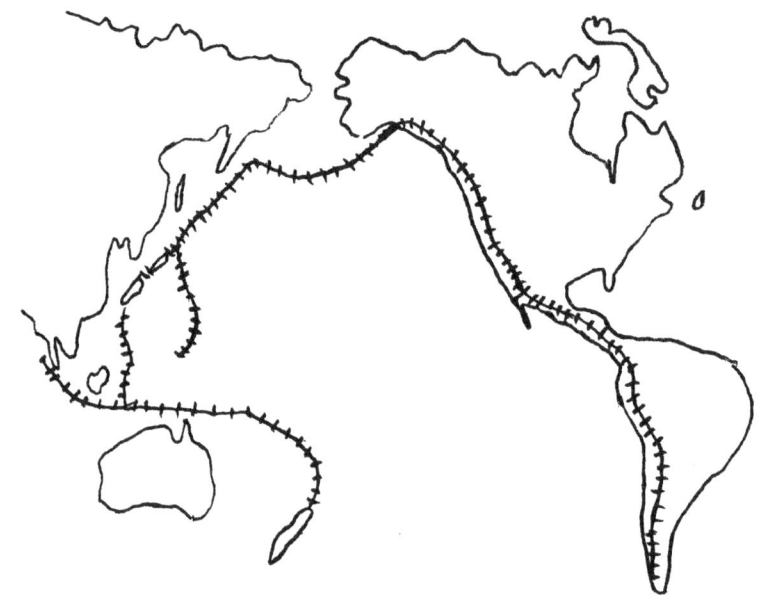

RING OF FIRE ++++++++++

RING OF SUBDUCTION ZONES
SURROUNDING THE PACIFIC OCEAN

East African Rift System

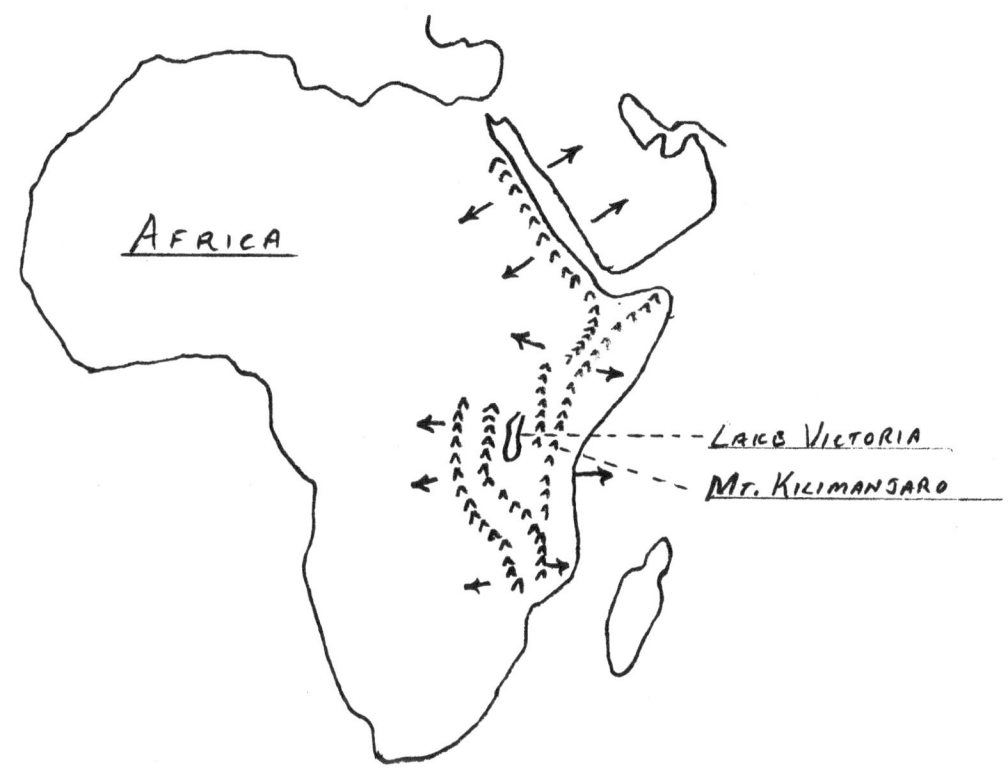

Africa

Lake Victoria

Mt. Kilimanjaro

The continent is being pulled apart by plate tectonics. The Red Sea is expanding separating Africa from Arabia.

A Rift is the center of an extensional spreading center where continental or oceanic plate spreading occurs.

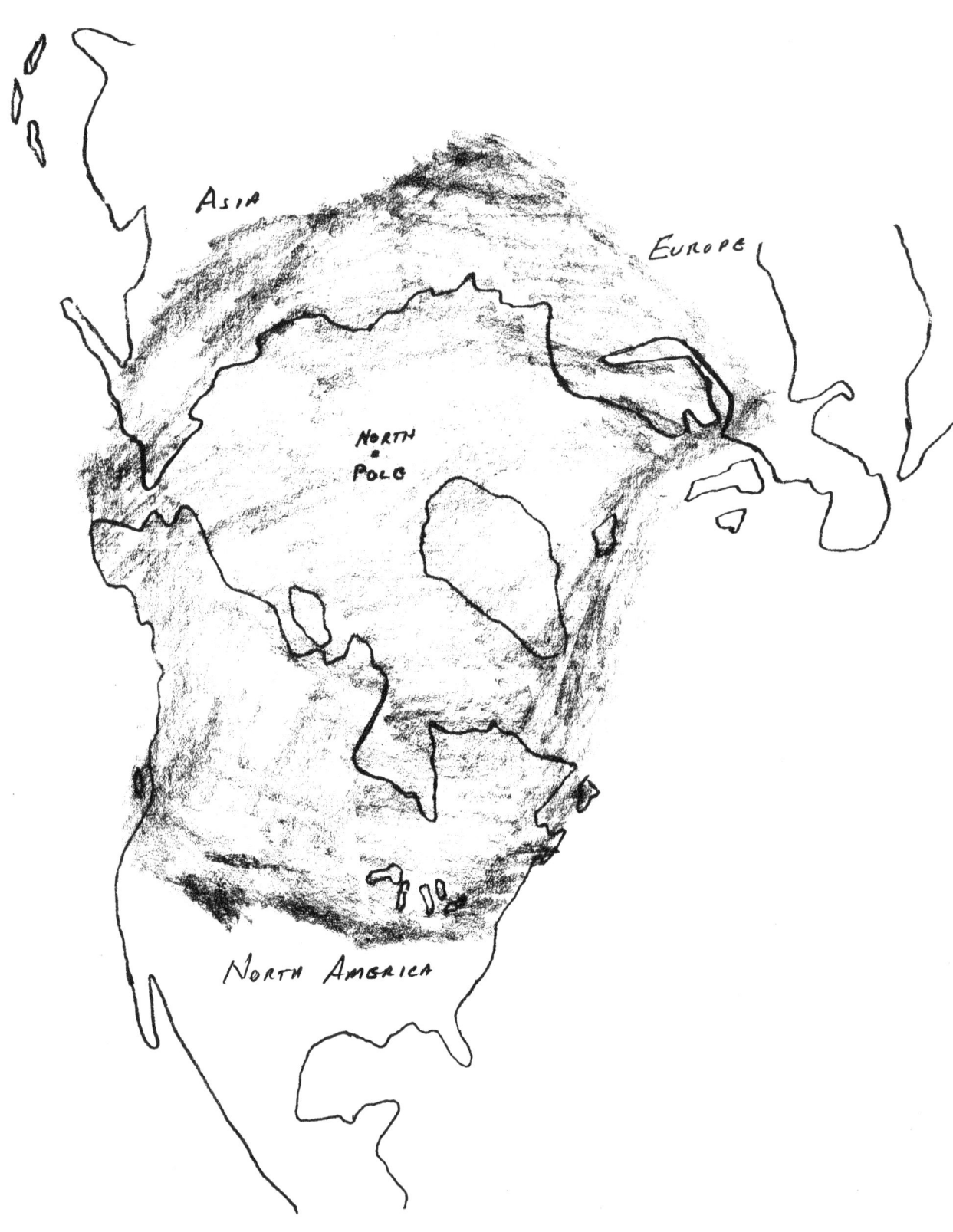

ASIA

EUROPE

NORTH POLE

NORTH AMERICA

THE EXTENT OF GLACIATION DURING THE LAST ICE AGE

THE MEDITERRANEAN SEA

Around 6.3 million years ago the whole blue Mediterranean Sea was a hot, dry, deep salty desert basin, the deepest and widest desert basin the world ever knew.   The separation of Europe, Asia and Africa created this valley of over a mile deep covering 969,100 square miles. The Caribbean Sea covers 971,400 square miles for comparison. The narrow passage of the Straits of Gibralter, the *Pillars of Hercules*, was sealed by a natural dam one mile high and almost twenty miles long. The Atlantic Ocean spilled over this dam as Niagra Falls does today. The evaporation in this basin more than made up for the inflow of these falls plus the rivers Nile, Rhone and others, causing it to dry out every few thousand years, adding salt to the floor. The weather would be extreme - hot desert winds blew dust and salt across the enormous desert basin. This was no single monotonous landscape, but a whole varied region, crossed and marked by volcanic mountain ranges, their low foothills sloping to flat valleys and deeper basins. Salty marshes and lakes would form and dry again as rainfall and rivers from the continents flowed and changed. Seawater from the dam in the west also continued to enter and evaporate. The residue eventually spread into an uniform layer of sea salt about eight hundred feet thick covering the floor of the entire basin.

The original granite floor of the Nile, located eight hundred miles south from the sea, was six hundred feet below the present level of the Mediterranean Sea. The Nile, Rhone and other rivers were merely waterfalls into this giant basin. This area went through many wet and dry cycles before the dam at Gibralter broke. Due to the constant action of plate tectonics, the material of the dam continued to spread and weaken. It could have also been at the end of an ice age when the water used for the ice was returned to the oceans increasing their depths by hundreds of feet. This combination could have caused a break in the dam at Gibralter. This flood would be the greatest Earth would ever witness with unbelievable forces. It would take over seventy years for the filling of the Mediterranean Sea from the Atlantic to complete and become stable some 5.3 million years ago. Plate tectonics will close the Straits of Gibralter again in the future and the sea will again dry up after several thousand years or so because the inflow of rivers cannot keep up with the rate of evaporation. Many more feet of salt will be added to the basin floor as it again becomes a very inhospitable basin.

THE WATER CYCLE

The oceans cover about 70 percent of Earth's surface with an average depth of about two and one-half miles, amounting to nearly two hundred fifty million cubic miles of water. Each day one trillion tons of water rain down on the planet, most of which falls directly into the oceans. The movement of water on the planet is one of nature's most important cycles; for without the movement of water over the land and back to the oceans, there would be no life as we know it.

The average journey water takes from the oceans to the atmosphere, across the land, and back to the oceans again requires about ten days. The journey is only a few hours in the tropical coastal regions, but as long as ten thousand years in the polar regions. This is what is known as the hydrologic cycle - the water cycle.

During a glacial period, water becomes locked upon the land as ice, which lowers the sea levels (the water for the ice has to come from somewhere). As the glaciers melt, the sea levels rise and the amount of change in sea level can be calculated because the area of maximum ice coverage is known - the area and the thickness. The Antarctic ice sheet alone contains enough water to raise sea levels throughout the world by about seventy meters (two hundred thirty feet). If this were to happen, the North Pole ice cap, Greenland and all the glaciers of Earth would also melt causing the sea levels to rise another thirty meters (almost one hundred feet) to a total of over one hundred meters (three hundred thirty feet) above present levels.

The dates of sea level changes are well documented by radiocarbon samples of terrestrial organic matter and near-shore marine organisms obtained from drilling and dredging off the continental shelf. These data show that about thirty-five thousand years ago the sea level was near its present position. Gradually it receded and by eighteen thousand years ago it had dropped nearly one hundred thirty-seven meters (four hundred fifty feet). It then rose rather rapidly to within six meters (twenty feet) of its present level. These fluctuations caused the Atlantic shoreline to recede one hundred to two hundred kilometers (sixty to one hundred twenty-five miles), exposing vast areas of the shelf. Early man probably inhabited large parts of the continental shelf now more than one hundred meters (three hundred thirty feet) below sea level.

If continental and ocean basins did not exist and the Earth's surface was smooth, water would completely surround the planet as a uniform layer measuring approximately twenty-four hundred meters (seven thousand nine hundred feet) deep.

HISTORY OF THE DEEP CIRCULATION IN THE OCEAN

50 Million Years Ago: The oceans could flow freely around the world at the equator. A rather uniform climate with warm oceans existed even near the poles. Deep water in the oceans was much warmer than it is today and there were only alpine glaciers on Antarctica.

35 - 40 Million Years Ago: The equatorial seaway begins to close. There is a sharp cooling of the surface and of the deep water in the south. The Antarctic glaciers reach to sea with glacial debris. The seaway between Australia and Antarctica opens. Cooler bottom water flows north and flushes the ocean. The snow limit drops sharply.

25 - 35 Million Years Ago: A stable situation exists with possible partial circulation around Antarctica. The equatorial circulation is interrupted between the Mediterranean Sea and the Far East.

25 Million Years Ago: The Drake Passage between South America and Antarctica begins to open.

15 Million Years Ago: The Drake Passage is open; the circum-Antarctic current is formed. Major sea ice forms around Antarctica, which is glaciated, making it the first major glaciating of the Modern Ice Age. The Antarctic bottom water forms. The snow limit rises.

3 - 5 Million Years Ago: Arctic glaciation begins.

2 Million Years Ago: An ice age overwhelms the Northern Hemisphere.

DIMENSIONS OF DEEP OCEAN TRENCHES

TRENCH	DEPTH (MILES)	WIDTH (MILES)	LENGTH (MILES)
Aleutian	4.8	31	2,300
Japan	5.2	62	500
Java	4.7	50	2,800
Kuril-Damehata	6.5	74	1,400
Marianas	6.8	43	1,600
Middle America	4.2	25	1,700
Peru - Chili	5.0	62	3,700
Philippine	6.5	37	870
Puerto Rico	5.2	74	960
South Sandwich	5.2	56	900
Tonga	6.7	34	870

MAGNETISM

MAGNETOSPHERE

The magnetosphere is a region around Earth defined by the Earth's magnetic field, filled with plasma and radiation, and shaped like a comet with a long tail pointing away from the Sun. It is about one hundred twenty-five thousand miles (two hundred thousand kilometers) across, and the tail is millions of miles long.

The magnetosphere contains distinct regions coupled with one another through complicated mechanisms of energy, momentum, mass and wave transfer. Electric currents millions of amperes strong flow in these regions and strongly distort the Earth's magnetic field. The hot plasma in the boundry layers and the plasma sheet is mainly of solar-wind origin and is composed of hydrogen ions, some helium and heavier ions, and electrons. The less hot plasma-sphere, an extension into space of the Earth's ionosphere, is composed mainly of hydrogen ions, oxygen ions and electrons.

VAN ALLEN BELTS

There is a doughnut-shaped zone of radiation surrounding the Earth at the equator that consists of an inner belt and an outer belt. This radiation, consisting of high-energy protons and electrons, was discovered in 1958 by the American physicist James A. Van Allen. The radiation zone extends from about two hundred forty kilometers (about one hundred fifty miles) in altitude to over fifty thousand kilometers (about thirty-one thousand miles). The lower limit of the zone is determined by the absorbing effect of the Earth's atmosphere. The outer limit is highly variable, depending on the trapping capability of the geomagnetic field - sometimes out to eighty thousand kilometers (fifty thousand miles).

The maximum intensity of energetic protons occurs at an altitude of about thirteen thousand kilometers (eight thousand miles). The electron concentration has two maximums: in an inner belt at an altitude of about twenty-five hundred kilometers (fifteen hundred fifty miles); and an outer belt at about twenty-two thousand kilometers (thirteen thousand six hundred seventy miles). The maximum electron concentrations are high enough to cause severe radiation damage to unprotected electronic instruments in satellites and are of great concern with regard to the safety of manned spacecraft.

EARTH'S MAGNETIC FIELD

Geomagnetism is the natural magnetism of the planet Earth, which acts as a giant magnet with two opposing poles. This magnetic field also serves as a barrier to some solar and cosmic radiation that would otherwise be harmful to life on Earth. The magnetic field may be described as that of a short bar magnet, or *dipole*, at the center of Earth. The axis of the dipole that best approximates the Earth's field is inclined at about eleven degrees to Earth's axis of rotation. The *geomagnetic poles* - the points where the axis of the imaginary dipole meets the Earth's surface - are located at approximately 74.9^0 N latitude, 101.8^0 W longitude in the Northern Hemisphere and 68.7^0 S latitude, 143.0^0 E longitude in the Southern Hemisphere. The geomagnetic pole in the Northern Hemisphere has a south polarity at the present time while north polarity exists in the Southern Hemisphere location. The magnetic field of Earth is not stable - it varies both in intensity and location. The source of the magnetism, the Earth's core, is not solid itself and is subject to variations. From this the magnetic poles tend to wander and even reverse themselves periodically.

EARTH'S POLE REVERSALS

The latest pole reversal period occurred between 2.4 million years ago and 200 thousand years ago that included at least five reversals. The polarity reversal of Earth's magnetic field is a physical reality. Over the last 170 million years, the Earth's magnetic field has reversed at least three hundred times. Our magnetic field is electromagnetic; the field is induced by the flow of charged particles with Earth's molten iron core. Random motions with the fluid can result in a reversal of the Earth's magnetic field whenever a certain critical configuration is reached. Other reversal periods were 0.7, 1.7, 0.9 and 1.7 million years long. This was initially found alongside the spreading of the Mid-Atlantic Trench and is now monitored by the seafloor spreading of this trench.

ELECTROMAGNETISM

An electric current produces a magnetic field. A moving magnetic field creates an electric field that accelerates charge within a wire to cause a current to flow by induction. Another method is to move a current-carrying wire near a wire that has no current causing a current to be started in this neighboring wire without moving either. All modern generators are based on induction by this changing magnetic field creating an electric field - one cannot exist without the other. You cannot have an electric current without creating magnetic field and likewise, a magnetic field will induce a current in any neighboring wire. Charge is carried by electrons, protons and other elementary particles. Charge alone is sufficient to produce both electric and magnetic phenomena. Only charge has been found. Magnetic poles do not exist independent of each other. Magnetic fields do exist.

AURORA

Charged particles headed toward Earth, protons and electrons from the solar wind of the Sun, can be trapped by Earth's magnetism. In this way the Earth's magnetic field exerts electromagnetic control over the particles, collecting them into the shape of the Van Allen Belts. The charged particles often escape from the magnetosphere near the North and South magnetic poles, where the field lines intersect the atmosphere. Their collisions with the air rip apart some atmospheric molecules, creating a spectacular light show. Called an *aurora*, this colorful display results when atmospheric molecules, excited by collisions with the charged particles, fall back to their ground state and emit visible light (photons). Many different colors are produced because each type of atom or molecule can take one of several possible paths as it returns to its ground state, causing the emission of the photons. Aurora are the most brilliant in the Arctic and Antarctic circles. In the Arctic, the spectacle is called the Aurora Borealis, or northern lights. In the Antarctic, it is called the Aurora Australis, or southern lights.

MAGNETOSPHERE
OF EARTH

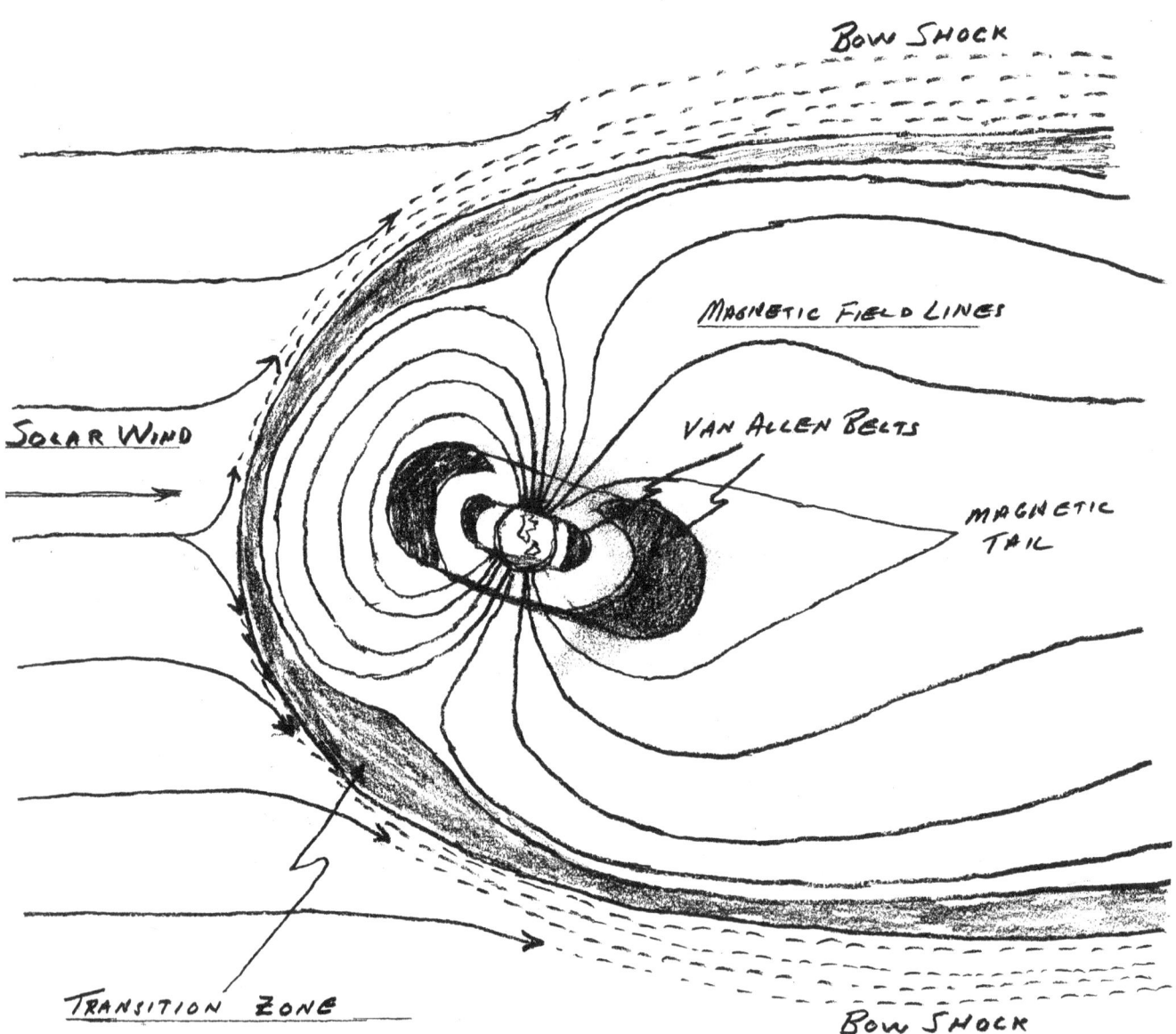

BOW SHOCK

MAGNETIC FIELD LINES

VAN ALLEN BELTS

MAGNETIC TAIL

SOLAR WIND

TRANSITION ZONE

BOW SHOCK

JUPITER AND SATURN

ALSO HAVE MAGNETOSPHERES

EXAMPLES OF MAGNETIC FIELD DISTRIBUTIONS

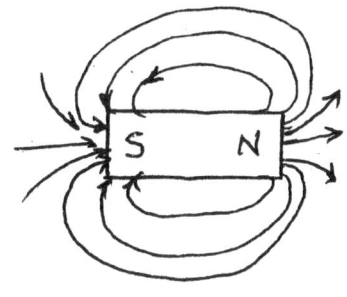

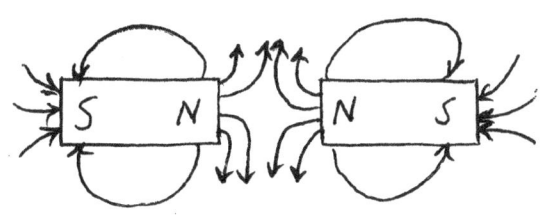

LIKE POLES REPEL

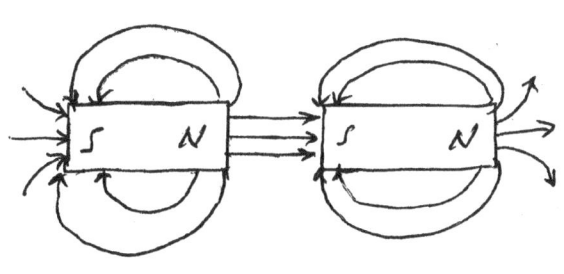

UNLIKE POLES ATTRACT

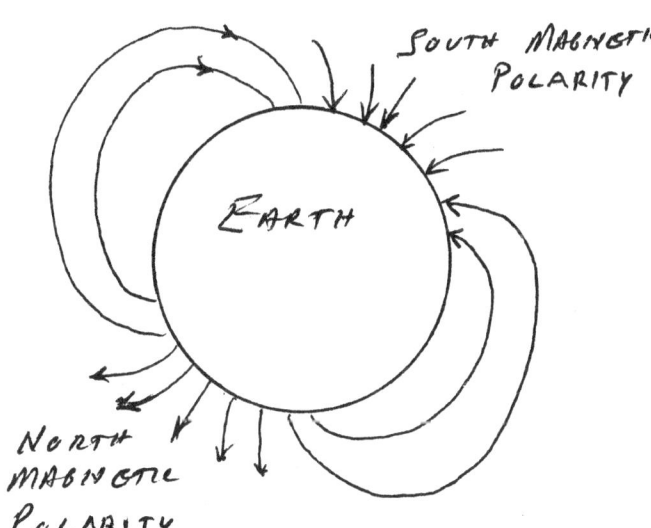

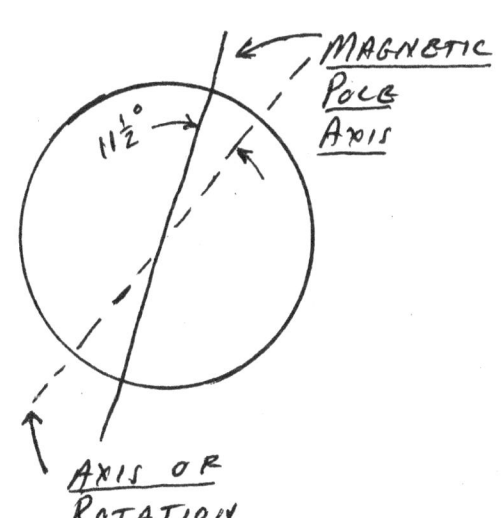

ATMOSPHERIC PRESSURE

ALTITUDE		PRESSURE
FEET	METERS	POUNDS/IN^2
Sea level	Sea level	14.696
10,000	3,000	10.2
20,000	6,000	6.8
30,000	9,000	4.5
40,000	12,000	2.8
50,000	15,000	1.8

HYDROSTATIC PRESSURE

DYNE = produces an acceleration of 1 cm/sec^2 on a mass of 1 gram.
One million DYNE/cm^2 = 1 BAR
One-tenth of a BAR = 1 DECIBAR

DECIBAR: Hydrostatic pressure exerted on one centimeter of surface by one meter of ocean water depth increases by one decibar per meter of depth. This is equal to having one pound per square inch added for each 2.25 feet of depth.

COMPRESSIBILITY: Forces the molecules of ocean water closer together which increases the density. If compressibility of seawater was zero, the real ocean level would be ninety feet higher than the present level.

OCEAN DEPTH		HYDROSTATIC PRESSURE		
2.25	feet	1 PSI		
22.5	feet	10 PSI		
225	feet	100 PSI		
2,250	feet	1,000 PSI	½	TON/IN^2
4,500	feet	2,000 PSI	1	TON/IN^2
6,750	feet	3,000 PSI	1½	TON/IN^2
9,000	feet	4,000 PSI	2	TON/IN^2
18,000	feet	8,000 PSI	4	TON/IN^2
36,000	feet	16,000 PSI	8	TON/IN^2

HERE WE COME

Don't ask me about the meaning of life! I don't even know how to fix the toaster.

Perhaps millions of ages before the commencement of the history of mankind, would it be too bold to imagine, that all warm-blooded animals have arisen from one living filament, which the great first cause endured with animality? What a magnificent idea of the infinite power of THE GREAT ARCHITECT! THE CAUSE OF CAUSES! PARENT OF PARENTS!

> Erasmus Darwin (1731-1802)
> Author *Zoonomia* (1794-1796)
> (Grandfather of Charles Darwin)

Charles Darwin (1809-1882) received his degree as a clergyman from Christ's College, Cambridge in 1831. He published *The Origin of Species* in 1859, *The Descent of Man* in 1871 and is buried in Westminster Abbey.

We are made of the ash of stars. Stars burn hydrogen, the most common element in the Universe. The ash is carbon, oxygen, nitrogen and other elements of life.

> Chet Raymo, Author
> *Biography of a Planet*

Rome - History	753 BC - 476
Republic	509 BC - 476

Early Middle Ages	500 - 1000*
High Middle Ages	1000 - 1300*
Renaissance	1300 - 1600

*Dark Ages	400 - 1300 Feudalism, Roman Catholic Domination

BASIC REQUIREMENTS

An examination of life on Earth reveals just how delicately our existence is balanced on the scales of chance. There is a long list of indispensable prerequisites for the survival of our species. First, there must be an abundant supply of the elements which make up the raw materials of our bodies: hydrogen, carbon, nitrogen, oxygen, phosphorous, sulfur and calcium. Second, there must be no risk of contamination by other chemicals which are poisonous; we would not want an atmosphere of methane or ammonia as found on many other planets. Third, we require a rather narrow range of temperatures so that our body chemistry can proceed at a correct pace. Without special clothing it is doubtful if humans could survive for long outside the temperature range of 5^0 - 40^0 C (41^0 - 104^0 F). This temperature is attainable in orbits that range between eighty-eight to ninety-five million miles from the Sun because glaciation begins to dominate the higher elevations at ninety-four million miles. Fourth, a supply of free energy is required, which in our case is supplied by the Sun. It is very important that this energy supply remains stable and is not subject to large fluctuations, which not only require the Sun continue to burn with uniformity, but that the Earth's orbit be nearly circular to avoid drifting toward or away from the solar surface. A fifth requirement is that the Earth's gravity be strong enough to prevent the atmosphere from evaporating into space, but weak enough so that we may move about easily and fall over every now and then without hurting ourselves.

A closer inspection shows that the Earth is endowed with still more amazing conveniences. Without the layer of ozone above the atmosphere, deadly ultraviolet radiation from the Sun would destroy all living creatures on the surface - the only safe place to live would be in the oceans. In the absence of a magnetic field, cosmic subatomic particles would inundate the Earth's surface. Considering that the Universe is full of violence and cataclysms, our little corner of the cosmos enjoys a peaceful existence.

The status of these "coincidences" changed dramatically when it was discovered that life on Earth is not static, but continually evolving. It then became possible, on the basis of Darwin's theory of evolution, to turn the problem upside down and ask, not why the Earth is so well suited to life, but why life is so well adapted to the Earth. Mutation and natural selection supplied the answer. Organisms that by random change find themselves slightly more attuned to the prevailing conditions have a selective advantage in the survival stakes, and will tend to proliferate at the expense of their less well adapted neighbors. Had gravity, for example, been stronger, it would have favored the development of smaller, squatter creatures with stronger bones. A higher ambient temperature would encourage the development of cooling fins and other means of heat control. In many ways there is nothing very special about the Earth after all, as far as life is concerned. Had the conditions been different, we would have been different.

All forms of chemical life are essentially electromagnetic in nature; that is, the forces which control the chemical processes in our bodies are the electric and magnetic forces which act between atoms. In addition there is gravity which anchors us to our system. This does not mean that we would have evolved to fit any conditions whatever, because there are certain absolute limits and requirements without which no life at all is possible. It is doubtful if life can exist on a planet with no atmosphere, such as the Moon, or with a temperature well above that of boiling water. It is also hard to imagine life around a Sun with erratic habits; many stars flare up unpredictably and others explode. By appreciating that the Sun is just a typical star we can view life on Earth in a more cosmic perspective. Stars come in all varieties of size, mass and temperature and our Sun is merely a small to medium sized object compared to the range of sizes that exist.

As you journey through this book, you will observe that all objects, animate and inanimate, are composed of compounds. These compounds consist of molecules and elements. Elements are atoms; atoms are protons, neutrons and electrons; protons and neutrons consist of quarks; electrons are electromagnetic forces; quarks and electrons are electromagnetic - *waves*. All matter is basically high-energy waves (fields) accumulated into larger and larger objects or substances.

Is it possible for life to exist anywhere in the Universe? Of course - all that is needed is a situation similar to that on Earth. We realize that liquid water, moisture in some form, is one demanding ingredient because boiling water would evaporate and frozen water would not allow the process necessary for life to begin. To many, fiction is easily accepted and facts are not because there is more information required when dealing with facts. We will assume life evolved on the Moon. Since there is no atmosphere and gravity is about one-sixth that of Earth, the Moon-man may have developed in this manner:

1. Assume a quadruped in the general appearance of man that walks, can run and has arms.
2. Eyes would be required - quantity unknown.
3. The mouth, nose and ears are not required due to no atmosphere. Atmosphere is required for both breathing and transmitting sound waves; talking and hearing.
4. A valve in the hand may consume nourishment since nothing grows on the surface - possibly a mineral intake from the ice caps.

5. Size and weight could be up to six times as that on Earth to duplicate movements and weights as Earthlings experience. People could be thirty-six feet tall and weigh between one thousand to twelve hundred Earth pounds and be as agile on the Moon as we are on Earth.
6. The remainder of the civilization and its development is given to your imagination to complete.

SEQUENCE OF EVENTS

Photosynthesis is the fixing of carbon dioxide and water into starch by chlorophyll-containing organisms using sunlight. The by-product of this micro-organic process is oxygen. The organisms responsible on Earth are blue-green algae - cyanophytes. They were the dominant form of life, along with bacteria, for two billion years on this planet. They still live here. The very earliest forms did not even have an oxygen component atom. Evidence has been found of these single-cell creatures in the oldest rocks and it has been concluded they existed when Earth's atmosphere had no oxygen. They did not contain DNA molecules in their nuclei to control chemistry and reproduction. They were truly simple life forms, not far removed from amino acids and chains of molecules wandering in the direction of life. They were enough to do the job.

The growth and multiplication of the blue-green algae was incredibly slow. The ultraviolet bath on the primordial Earth was fierce, there was no oxygen available and the life form was inefficient. It caught on fast by geologic time intervals and evolved into a more complex form. About one and a half billion years ago, nuclei appeared in the blue-green algae, and they assumed more or less the form in which they are found today. The new and improved cyanophytes did their job with spectacular efficiency, if not great speed. They were far more efficient at converting carbon dioxide to oxygen through photosynthesis than their ancestors. In a billion generations, a billion billion generations of these unthinking organic field hands, Earth's atmosphere wound up seven hundred million years ago with five percent oxygen and much less carbon dioxide than was originally present.

There was enough oxygen at least for other forms to survive and multiply. In one hundred million years, life on Earth had changed from single cells to complex forms - muilticellular organisms. Hard tissue developed an anchor for muscles in the more evolved forms and the hard tissue left a fossil record that dates back six hundred million years. Life on Earth was off and running.

The surface of Venus has ninety times the air pressure than that of Earth, ninety atmospheres of carbon dioxide. The carbon dioxide absorbs the infrared, trapping it like a heat lamp making it extremely hot. Since we are known as the sister planet of Venus, where is Earth's carbon dioxide? It is in the rocks. Shellfish and plankton long ago converted carbon dioxide into shells. Shells from the dead creatures spread through the ocean, forming sediments which eventually became rocks, such as limestone. If you took the carbon dioxide out of the rocks, Earth would have very nearly the same atmospheric pressure as Venus. Under these conditions, most of the atmosphere would be carbon dioxide, Earth would be extremely hot and would be unable to support life. We would just be another Venus.

Life was established by Earth's forces, ignited by lightning. Not from a single branch of lightning, but a planet-wide lightning storm burning the oceans for millions of years. After the birth of the first cell, the development of the cell advanced during the next several million years. These later cells invented sex and the Earth's diversity expanded a hundredfold as now two genetically different cells could unite and fashion from their genetic abilities a radically new cell. These new cells also invented the habit of eating living cells, and thus deepened the community of Earth not only with the intimacy of sexual bonding, but also with the intimacy associated with environmental predator-prey relationships. These steps led to the later development of the first multicellular animal.

Four million years ago humans stood up on two limbs and by two million years ago they began using their free hands to shape Earth's materials into tools. One and a half million years ago these hands were controlling fire, shaping the Sun's energy that had been stored in sticks, to advance their own projects. Around thirty-five thousand years ago cave paintings began to appear. Twenty thousand years ago, these humans became aware of seeds and seasons and twelve thousand years ago they began consciously shaping these patterns by domesticating plants and animals.

A secure supply of food enabled populations to surge. The first Neolithic villages to sustain human groupings of more than a thousand people were Jerico, Catal Hiiyuk and Hassuna ten thousand years ago. Jerico is located 820 feet below sea level and its walls are nine thousand years old. Catal Hiiyuk (Huyuk) is located in Anatolia - modern Turkey. Hassuna in Mesopotamia was located in the middle Tigris region - modern Iraq. As time passed other villages arose as craftwork evolved and language expanded. Harappa and Mehenjo-Daro of northwest India were the earliest Neolithic developments in this area of the world. Religions appeared and evolved to try to explain the reasons for what had happened - each area had its own story of the cause of their creation. Five thousand years ago, the human venture mutated into a new way of life, the urban civilization as existed in Sumer.

As industrial humans multiplied into the billions to become the most numerous of all of Earth's complex organisms, as they decisively inserted themselves into the environmental communities throughout the planet, drastically reducing Earth's diversity and channeling the majority of the Gross Earth Product into human social systems, a momentous change in human consciousness was in process. Humans were involved in a process that began over thirteen and a half billion years ago. When the ancestors of all the Native Americans came to the Americas it was a world that had never heard language, known tools or seen any version of human society. It was their New World, the prehistoric New World. Australia has almost the same history. The oldest dates for human occupation of Australia are now somewhat earlier than the oldest credible dates for America, about forty or fifty thousand years. The first Australians were also homo sapiens, as all humans are today in all the continents. The reason is pretty clear; the falling sealevels of glacial times came in the time of our earliest ancestors. They could manage to travel by canoe or alongshore. That was enough to get us to where we are today, on all continents, where the earlier forms of humanoids could and did know only the wider area of the land-rich hemisphere.

During the last ice age, which ended ten thousand years ago, the oceans of the world were over two hundred and fifty feet shallower. The water was required to form the ice that covered most of the upper half of the Northern Hemisphere that had depths of two miles or more. This created many land bridges throughout the world in which humans and animals could migrate to new or better areas for survival. Archaeologists believe that Civilization began at the end of the last Ice Age. Nearly ten to twenty billion different species have existed on Earth since life first arose. More than 99.9 percent of these species are extinct. Man is a species and will also become extinct at some time in the future.

SPECIES	DATE APPEARED	LOCATION
Homo Habilis	1.9 - 1.6 Million Years Ago	Tanzania
Homo Erectus	1.7 MYA - 250,000 years ago	Java - Indonesia
Homo Sapiens - Neanderthal	200,000 - 30,000 years ago	Europe, Asia, Africa
Homo Sapiens - Sapiens	100,000 years ago	Europe

The last Ice Age ended ten thousand years ago. The Mesolithic peoples began food gathering some seven thousand years ago. Neolithic people began food processing in the Fertile Crescent in the Middle East around the Sahara Desert from Ur (inland) to Palestine (coastal). This spread to the Balkan Peninsula by 5000 BC, to Egypt and Central Europe by 4000 BC and then to Britain and India by 3000 BC. The Neolithic cultures of Middle America and the Andes are found as independent developments. The Neolithic Age in China was 6500-1700 BC and was developed independently from that in the Middle East.

The four earliest civilizations (Mesopotamian, Egyptian, Indian and Chinese) arose between 3100-1500 BC, each in the valley of a great river system. Located immediately north of Sumer lay the narrow region of Akkad which was inhabited by Semites who had absorbed Sumerian culture. Appearing late in the fourth millennium BC, the Akkadians were the earliest of the Semitic peoples who filtered into Mesopotamia and Arabia. The oldest sailing boat known is represented by a model found in a Sumerian grave of about 3500 BC. Following this date wheeled vehicles appear in the form of animal drawn war chariots. Pack animals were relied on for the transport of goods overland. The potter's wheel was also first used in Sumer after 3500 BC. About 3100 BC metal workers discovered that copper was improved by the addition of tin. The resulting alloy, bronze, was harder than copper and provided a sharper cutting edge. The advent of civilization in Sumer is associated with the beginning of the Bronze Age in the West, which in time spread to Europe, Egypt and Asia. The Bronze Age lasted until about 1200 BC, when iron weapons and tools began to replace those made of bronze. The Hittites are credited with being the first people to work iron from local deposits. Not until after 1200 BC did iron metallurgy become widespread. Why civilizations arose in these locations and not elsewhere is a question whose answer is not known.

Archaeology may discover, in the following centuries, that the major races of humans evolved independently. What could come to be discovered is that the three distinct races of today - Caucasian, Asian, African - are the results of at least three different evolutions. Only time will tell. We do not have enough information to be exact in any hypotheses at this time although present findings of a new human species leads to this possibility. Stay tuned.

The geological record shows that very few species have existed longer than twenty million years, the average being about five million years. The composition of humans is:

Oxygen	65 percent	$_8O^{16}$
Carbon	18 percent	$_6C^{12}$
Hydrogen	10 percent	$_1H^1$
Nitrogen	3 percent	$_7N^{14}$
Calcium	1.5 percent	$_{20}Ca^{40}$
Phosphorus	1 percent	$_{15}P^{30}$
Sulfur	0.25 percent	$_{16}S^{32}$

"I have in the past compared the two elements, carbon and silicon. They share certain important properties, yet differ in a property that is crucial for life. Thus carbon dioxide is a gas in the atmosphere and dissolves readily in all the waters of the Earth - the places from which life on Earth derives its carbon. Silicon dioxide, however, is the stuff of sand, most rocks and quartz. So I end by saying ... and that's why silicon is good for making rocks, but to make life requires carbon. That sounds like disparaging silicon. But now let me say something more, that I think begins to approach what one might mean by an aspect of mind in a stone. If silicon were not good for making rocks, there would be no place in the Universe in which carbon could make life. For silicon makes the surface layers of plants, the universal abode of life. What I am getting at is the marvelous fitting together of all aspects of this life-breeding Universe.

"Let me give another example: Of the 92 natural elements, 99 percent of all living matter is made of just four: hydrogen, oxygen, nitrogen and carbon. I think that must be whenever life arises in the Universe, because only these four elements possess the unique properties upon which life depends. Now something wonderful: Life to persist indefinitely on any planet in the Universe must come to depend upon the light from its star. So it is that all life on Earth runs on the light of our star, on sunlight. And what makes sunlight? Those same four elements that constitute life on Earth also makes the sunlight on which that life runs.

"Humankind then takes a great place in cosmic evolution, one of transcendent worth and dignity, in which our purpose is to know and create and try to understand. We are an intrinsic part of the Universe. Much of the history of our galaxy is bound up in us. The carbon, nitrogen and oxygen that form the bulk of our bodies was made in the deep interiors of former generations of stars that have died. The salts of the ancient seas circulate in our blood. We see this Universe not from the outside, but from the inside: Its suff is our stuff. I once wrote, 'A physicist is the atoms way of knowing about atoms.' In our knowing, the Universe comes to know itself."

George Wald

13.5 BYA	Big bang.
12.5	Galaxies forming.
12.2	Oldest quasar - designation PK52000+330.
11	Element making boom; many quasars.
10	Average age of solar elements.
9	Average age of spiral arm elements of the Milky Way.
6.7	Universe contained one-quarter of present volume.
6	Sterilization of many galaxies when the gases were lost and they were no longer able to breed new stars.
4.65	Early presolar supernova created by the explosion of a large short-lived star.
4.45	Earth complete as a result of storm of cosmic impacts.
4	Universe has one-half of present volume. Chemical evolution of single cell life on Earth begins.
3.9	Cosmic impacts on Earth abating. Photosynthesis.
3.5	Living reefs begin as ocean basins fill.
2.9	Gold boom; gold belts formed in old continental rocks.
2.8	Continent making as lighter density matter is pushed to top as plates are formed. Plate tectonics evolved and volcanoes are formed as plates subducted.
2.5	Iron born. Oxygen precipitated iron from seawater laying down more than 90 percent of world's mineable iron by 1.8 BYA.
2.3	Supercontinent of *Kenora* formed and first of many ice ages occurred.
2.2	Mountain building begins as *Kenora* crushes together.
2.1	Host cells; bacteria gathering.
2	Uranium reactions in what was later to be Western Africa.
1.9	Additional asteroid and comet impacts bring metals from outer space to Earth. Chromium and platinum were deposited in areas that later became South Africa. Iron and nickel were deposited in what is now Ontario, Canada.
1.8	Oxygen revolution. Purple bacteria appeared as first cells which lived in deep water.
1.7	Proto-animals; single-celled evolved.
1.6	Blue-green algae present.
1.5	Supercontinent of *Amazonia* formed. First multicellular proto plants, fungi and proto animals evolve. Mountain building of period completed as continents drifted - North America, Africa and Australia.

1.4 BYA	Mountain building pulse - microcontinent attached to eastern edge of North America, the collision creating the Appalachian Mountains.
1.3	Plants - microseaweed and multicelled organisms.
1	Seaweed on Earth as life invented sex as a more efficient means of reproduction. Mars becomes defunct as geological action ceases causing its frozen crust to thicken.
950 MYA	Ice Ages - onset of glaciation. Proto animals adapt to sex to reproduce more freely.
850	Copper boom - mountain building formed deposits in Africa and Asia.
800	Supercontinent of *Baikalia* formed and promptly fell apart.
770	Ice Age returns.
670	Coldest of all Ice Ages returns. Animals with multi-cells evolved - jellyfish. Made hollow bodies with internal organs enclosed in shells.
650	*Gondwandaland* assembling. Mountain building.
600	Early hard animals, similar to earthworms and insects.
570	Many hard animals such as shelled animals and sea plants.
510	Vertebrate animals and primitive fish.
460	Europe hits North America. An ocean callled *Iapetus* was converted into a chain of mountains - Greenland, Norway and Scotland.
440	Ordovician catastrophe; Ice Ages - Africa glaciated.
435	Land plants.
425	Jawed fish and life ashore began. Predatory fish and ancestors of the bony land animals also emerge.
400	Sharks appear.
395	Insects appeared on ground.
370	Fransnian catastrophe - tidal waves wiped out limestone reefs of the world. Amphibians appear.
350	Euramerica hits *Gondwandaland*.
330	Winged insects.
313	Reptiles evolved from amphibian ancestors.
310	Polycosaurus became the first large animal to walk on land. First trees appear.
300	Siberia hits Europe from the east.
290	Ice Ages. *Gondwandaland* extended across South Pole lowering the sea level due to glaciation.
260	Coal making reached an all time peak.
256	Therapsids replaced Polycosaurus.

245 MYA	Permian terminal catastrophe. Ninety-six percent of all species of marine animals were annihilated and the reefs and seabeds were sterilized. On land, all of the large mammal-like reptiles perished.
235	Dinosaurs and flowers are prevalent.
225	Pterosaurs (flying reptiles) and giant dinosaurs present.
220	*Pangaea* assembled. Africa is east of North America. China formed.
175	Brachiosaurs - long necked plant eaters.
170	Petroleum boom. Earth's climate entered a new "greenhouse" phase that was better for creating oil than coal.
150	Birds with feathers for warmth appear as descendants of dinosaurs.
145	Kimmeridgian turnover - drainage of the continental margins.
125	Marsupial mammals and appearance of flowers, etc. Modern mammals produced by egg laying.
117	Brachiosaurs extinct.
114	Floral revolution and first placental mammal.
100	India separates from Antarctica.
95	Cenomanian turnover; pre-primate.
93	Continents flooded; petroleum peak; Gulf of Mexico, Venezuela and North Africa lay under tropical seas. Arabia and Iran accumulated richest oil fields.
85	South America separated from Africa.
70	Coal boom. Earth's climate switched from "greenhouse" to "ice house" conditions which was better for creating coal than oil.
69	Fossil primates.
65	Cretaceous terminal catastrophe and the dinosaurs became extinct. The oceans were dead for ten thousand years. Crocodiles, small reptiles and some small furry animals survive.
60	North America separated from Europe and early Rocky Mountain Ranges formed.
55	New mammals - early horses, elephants and whales.
50	Australia separates from Antarctica; kangaroos and koalas appear.
45	India hits Eurasia at one hundred kilometers per million years.
40	More new mammals - list nearly complete.
37	Eocene terminal turnover - cosmic impact; seasons begin; penguins in Antarctica.
35	Early cats and dogs. New World monkeys.
30	Circumantarctic current; Japan separates from Eurasia; Arabia breaks away from Africa. Pigs and bears appear.

29 MYA	Sea level falls as Antarctica accumulates ice. Hang-nose primates.
25	Whales growing. Saber-tooth tigers.
24	Grass evolved from sea plants.
21	Apes and monkeys split.
20	Old World interchange; deer appear. Reunion of Africa and Eurasia.
19	Early antelopes.
16	Oregon eruptions build Columbia River Plateau.
15	Miocene disruption - cosmic object falls on Europe. Antarctica goes into permanent deep freeze.
13	Eastern Antarctic deep freeze; volcanic activity as ice sheet grows until 11 MYA.
9	Northern glaciers; mountain building.
6.6	West Antarctic deep freeze. Mastadons.
6.3	Mediterranean dryout - glacier forming drops sea level. The French Rhone and Egyptian Nile become waterfalls into the Meditarranean.
5.3	Mediterranean normalized. Atlantic waterfall at Gibralter excavated a channel deep enough to refill the Mediterranean.
5	Himalayan uplift - India pushes into Asia.
4.5	Andean uplift - western subduction of South America.
3.7	Modern horses in North America.
3.5	Red Sea ocean-like; early cattle.
3.25	Current ice ages begin. Amount of ice first exceeded present stock of the Arctic and Greenland ice sheets.
3	American interchange; arc of volcanic islands join North America and South America creating Panamanian land bridge, Central America. This opened the route between the two Americas for animal immigration each way. Trees and forests grew taking land from the large animals. This caused the demise of the large animals because the grassy areas were less and the formation of forests of trees caused limitation to their movements.
2	Man.

Hadean Era	4.5 - 3.8 BYA (Billion Years Ago)
Archean Era	3.8 - 2.5 BYA
Proterozoic Era	2.5 BYA - 570 MYA (Million Years Ago)
Paleozoic Era	570 - 245 MYA
Mesozoic Era	245 - 65 MYA
Cenozoic Era	65 MYA - Present

Ice Ages - Ancient	2.3 BYA, 950 MYA, 770 MYA, 670 MYA, 440 MYA, 290 MYA
Current	3.25 MYA, 2.4 MYA, 800,000 Years Ago, 550,000 YA, 430,000 YA
Most recent	72,000 YA, 62,000 YA, 58,000 YA, 28,000 YA, 18,000 YA, 14,000 YA, 10,900-10,300 YA
False	115,000 YA, 95,000 YA
Little	1530-1850 - Glaciers advanced in North America and Europe.

7000 BC	Neolithic revolution - Egypt and Near East.
3500	Emergence of civilization in Sumer - city-states appear.
3100	Cuneiform writing on clay by Sumerians. Menes unites Egypt.
2800-2370	Old Sumerian Period.
2613-2181	Old Kingdom of Egypt - Pyramid Age.
2500-1500	Indus Valley civilization.
2370-2315	Sargon.
2370-2150	Akkadian Empire.
2113-2006	Neo-Sumerain Period.
2050-1800	Middle Kingdom of Egypt.
2000-1200	Aegean civilization of Greece.
2000-1000	Indo-Europeans invade Italian peninsula. Latins settle in lower Tiber Valley.
2000	Achaean Greeks invade Peloponnesus. Hittites enter Asia Minor.
1792-1750	Hammurabi.
1760	Hammurabi rules lower Mesopotamia. Babylonian Empire.
1700-1450	Zenith of Minoan culture.
1700-1122	Shang Dynasty - China's first civilization.
1595	Hittites sack Babylon.
1570-1090	New Kingdom (Empire) in Egypt.
1500-1000	Early Vedic Age and caste system in India.
1500	Emergence of Hinduism and Judaism.
1450-1200	Mycenaen Age; Hittite Empire.
1402-1363	Akhenaton (Amenhotep IV)
1250	Destruction of Troy.
1150-750	Greek Dark Ages, Homeric Age.
1122-271	Chou Dynasty - China's "Classical Age."
1020-922	United Hebrew Kingdom

1000-500	Later Vedic Age, caste system expanded.
922-721	Hebrew Kingdom divided into Israel and Judah.
800-600	Foundation of Hinduism.
800	Carthage founded in North Africa by Phoenicians. Independent kingdom of Kush in Africa.
753	Founding of Rome.
750-338	Hellenic Age in Greece.
745-612	Assyrian Empire.
700	Kush conquers and rules Egypt.
660	Legendary beginning of the line of Japanese Emperors.
628	Zoroaster.
604-539	Chaldean Empire - Nebuchadnezzar.
582-497	Pythagoras.
563-483	Buddha.
551-479	Confucius.
550-330	Persian Empire.
539-469	Parmenides.
509	Beginning of Roman Republic.
500	Buddhism and Confucianism; Nok culture in Africa.
470-399	Socrates.
461	Golden Age of Greece.
451	Rome arranges calendar to present twelve months.
450	Laws of Twelve Tables of Rome.
427-347	Plato.
400-347	Eudoxus.
384-322	Aristotle.
356-323	Alexander III (the Great).
342-270	Epicuras.
336	Hellenistic Age of Greece.
331	Empire of Alexander the Great.
325-265	Euclid.
322	Mauryan Empire of India.
310-230	Aristarchus of Samos.
300	Romans build Appian Way.
295	Library founded at Alexandria.
287-212	Archimedes.
276-194	Eratosthenes.
221	End of Chou Dynasty in China. Ch'in Dynasty begins.

202	Han Dynasty in China.
150	Classical period of Mesoamerican culture begins with Olmecs.
100	Taoism. Rule by Yamato clan in Japan.
69-30	Cleopatra.
47	Alexandrian Library burned in battle between Caesar and Ptolemy.
46	Julian calendar introduced.
27	Augustus Emperor of Rome.

End of BC

23	Later Han Dynasty in China.
43	Beginning of Roman occupation of England.
70	Rome destroys Jerusalem ending Hebrew state.
96	Beginning of reigns of Five Good Emperors of Rome.
100-178c.	Claudius Ptolemy
200	Funan dominates border states of Southeast Asia.
220	Fall of later Han Dynasty in China. Collapse of Kushan State in India.
272	Alexandrian Library burned by Roman Emperor Lucius Domitus Aeroeliones.
300	Rise of Ghana in Africa.
320	Chandra Gupta I establishes Gupta Empire in India.
330	Founding of Constantinople.
378	Visgoths defeat Romans at Battle of Adrianople.
391	Alexandrian Library burned by Christian Emperor Theodosius I.
395	Theodosius I divided Roman Empire between his two sons.
407	Rome withdraws troops from England.
476	Traditional date of fall of Roman Empire.
481	Clovis I becomes ruler of small Frankish Kingdom.
532	BC and AD notations added to Catholic calendar; Dionysius Exiguus.
500	Shintoism.
570-632	Muhammad (Mohammed).
589	Sui Dynasty in China begins.
600	Cities begin to arise in interior highlands of Peru.
618	T'ang Dynasty in China begins.
622	Hijra; Muhammad and followers flee from Mecca to Medina.
640	Alexandrian Library burned by Muslims commanded by Caliph Dinar.
661	Founding of Umayyad Dynasty.
689-741	Charles Martel, ruler of United Frankish Kingdom.
711	Muslim forces from North Africa invade Spain.

714-768	Pepin, son of Charles Martel, rules Frankish Kingdom.
732	Battle of Tours as Muslims defeated in France.
750	Beginning of Abbasio Dynasty; high tide of Islamic power and civilization.
794	Heian Period in Japan begins.
800	Charlemagne, Pepin's son, crowned Emper of France. Peak of power of Frankish State and Carolingian Empire.
900	Toltecs create new power in Valley of Mexico.
907	Last T'ang Emperor deposed.
910	Reconquista; Christians begin reclaiming Spain from the Muslims.
960	Sung Dynasty in China begins.
988	Vladimir accepts Orthodox Christianity for Russia.
1000	Orthodox splits from Roman Catholic Church. Peak of development in Ghana. Incas settle the Cuzzo Valley in the Andes. Vietnam achieves independence from China. Kingdom of Kanem in Africa.
1019	Peak of Kievian Russia.
1050-1123	Omar Khayyam.
1054	Crab Nebula outburst. Greek and Latin churches mutually excommunicate each other.
1066	Norman conquest of England.
1095-1291	Period of nine Christian Crusades, militarily a failure.
1127	China divided between southern Sung Dynasty and Northern Ch'in.
1162-1227	Genghis Khan (Temujin)
1185	Kamakura Shogunate in Japan.
1200	Peak of Khmer Empire in Southeast Asia. Kingdom of Benin in Africa.
1206	Delhi Sultanate established. Mongol leader Temujim recognized as Genghis Khan.
1215	Magna Carter.
1216-1294	Kublai Khan.
1225-1274	Thomas Aquinas.
1275	Marco Polo at court of Kublai Khan.
1281	Mongol invasion of Japan repelled by *Kamikaze* ("The Divine Wind").
1300	Aztec confederacy in Mexico.
1304-1374	Petrarch
1325	Tenochtitlan founded.
1336	Ashikaga Shogunate in Japan.
1348	Black Death sweeps Europe.

1368	Ming Dynasty in China is established.
1392	Yi Song-gye founds Yi Dynasty in Korea.
1400	Quattrocento of Italian Renissance.
1400-1468	Johannes Guttenbeerg.
1430	Portuguese arrive in African kingdom of Kongo.
1450	Commercial revolution begins.
1452-1519	Leonrdo da Vinci.
1453	Constantinople falls to Ottoman Turks.
1464	Peak of power in African Kingdom of Songhai.
1473-1543	Nicholas Copernicus.
1475-1564	Michelangelo.
1483-1546	Martin Luther.
1492	Columbus reaches the New World.
1500	Protestant Reformation. High Renaissance in Italy.
1509-1564	John Calvin.
1517	Luthur issues ninety-five theses.
1519	Cortez arrives in Mexico.
1520	Ottoman power peaks under Suleiman the Magnificent.
1527	Beginning of Mughul Empire in India.
1531	Pizzaro conquers Peru.
1546-1601	Tycho Brahe.
1548-1600	Giordano Bruno.
1564-1642	Galileo Galilei.
1571-1630	Johannes Kepler.
1582	Gregorian Calendar.
1588	Spanish Armada defeated.
1603	Tokugawa Shogunate in Japan.
1613	Michael Romanov established Romanov Dynasty in Russia.
1628	Mughul Empire reaches height under Shah Jahan.
1637	Japan expels all Europeans.
1642-1727	Issac Newton.
1644	Manchus invade Ming China and establish Ch'ing Dynasty.
1656-1742	Edmund Halley.
1688	Glorious Revolution in England.
1700	Age of Reason; Industrial Revolution begins in England.
1731-1802	Erasmus Darwin.
1757	Battle of Plassey begins domination of India by British East India Co.
1775	American Revolution begins.

1781	William Herschel discovers Uranus.
1788	First English colony in Australia.
1789	French Revolution begins. U. S. Constitution adopted.
1790	Metric system established in France.
1804	Napoleon declares himself Emperor.
1809-1882	Charles Darwin.
1814-1874	Anders Jonas Ångström.
1815	Napoleon defeated and exiled to St. Helena.
1822	Brazil achieves independence.
1822-1895	Louis Pasteur.
1824-1907	William Thompson (Lord Kelvin).
1829	Greece achieves independence from Turkey.
1830	July Revolution in Paris enthrones Louis Philippe.
1831-1879	James C. Maxwell.
1840	Johann Galle found Neptune.
1846-1914	George Westinghouse.
1847-1931	Thomas Edison.
1848	Communist Manifesto published.
1852-1931	Albert Michaelson.
1854	Japan opens ports to the West.
1856-1943	Nikola Tesla.
1857-1894	Heinrich Hertz.
1858-1947	Max Planck.
1859-1906	Pierre Curie.
1859	Darwin publishes *On the Origin of the Species*.
1865-1923	Charles Steinmetz.
1867-1934	Marie Curie.
1879-1955	Albert Einstein.
1882-1945	Robert Goddard.
1885-1962	Niels Bohr.
1885	French gain control over Indonesia.
1889-1953	Edwin Hubble.
1894	Sino-Japanese War begins.
1898	Spanish-American War.
1899	Boer War begins.
1900	Boxer rebellion in China.
1900-1958	Wolfgang Pauli.
1901	Commonwealth of Australia formed.

1914-1918   World War I.
1930   Clyde Tombaugh found Pluto.
1939-1945   World War II.
1991   Dismemberment of Soviet Union into independent states.

WORLD POPULATION GROWTH

YEAR	POPULATION	INCREASE	TIME	PER YEAR
0	200 Million	----------------	--------------	
1000	300 Million	100 Million	1000 years	100,000
1300	396 Million	96 Million	300 years	320,000
1500	480 Million	84 Million	200 years	420,000
1600	500 Million	20 Million	100 years	200,000*
1700	640 Million	140 Million	100 years	1,400,000
1804	1 Billion	360 Million	104 years	3,460,000
1900	1.65 Billion	650 Million	96 years	6,771,000
1927	2 Billion	350 Million	27 years	13,000,000
1960	3 Billion	1 Billion	33 years	33,000,000
1974	4 Billion	1 Billion	14 years	71,400,000
1987	5 Billion	1 Billion	13 years	77,000,000
2000	6 Billion	1 Billion	13 years	77,000,000

* Result of fourteenth century plagues.

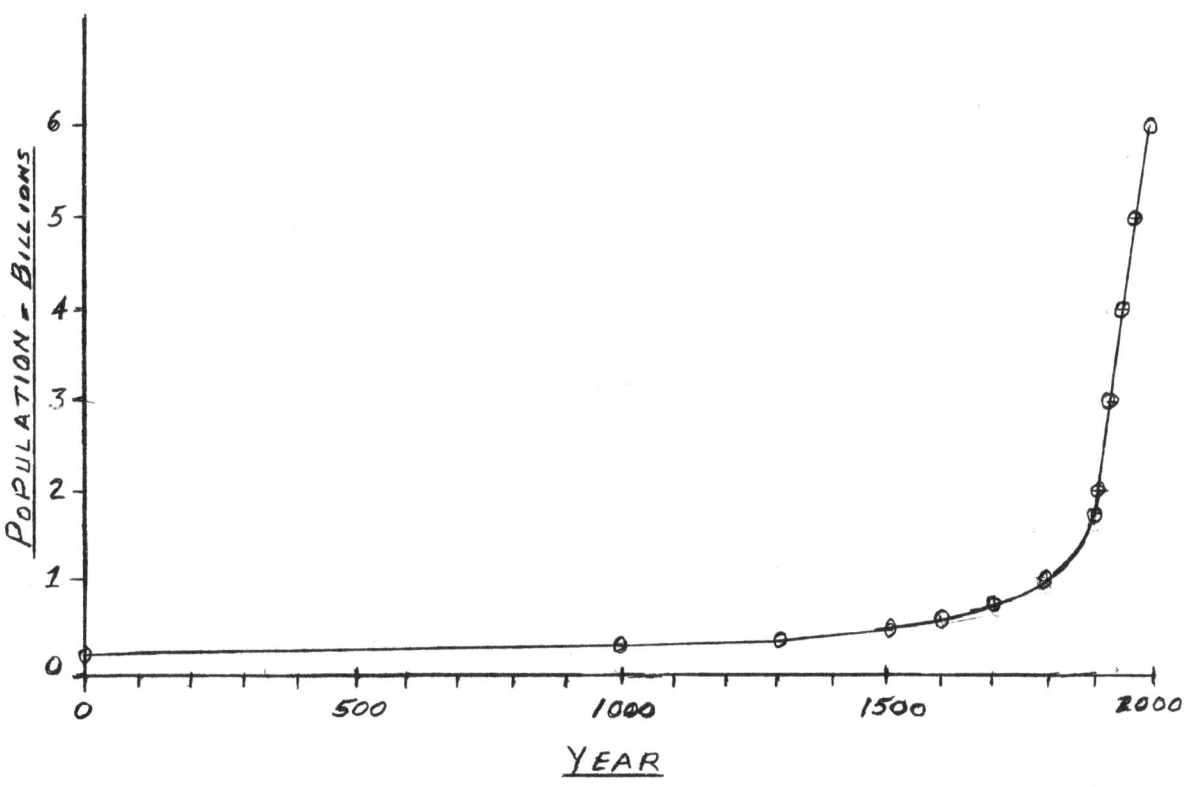

201

ARCHAEOLOGY

If Columbus' objective had been China in 1492 instead of India, what would he have named the natives he encountered in the Caribbean? CHINANS?

Archaeology is that science or art which is concerned with the material remains of man's past. There are two aspects to the archaeologist's concern. The first of these is the discovery and reclamation of the ancient remains; this usually involves field excavation or at least surface collecting. The second concern is the analysis, interpretation and publication of the findings. The general term used to describe any one of the things made by man is "artifact."

The Stone Age is identified through the discovery of pebble tools, core tools, flake tools and blade tools. The late Stone Age also consisted of paintings and drawings on the walls of caves. This was also the beginning of the creation of art products - figurines of animals, humans and other objects. The new Stone Age was characterized by implements whose edges were sharpened by grinding instead of flaking. All other discoveries are usually later than this period. Due to the accuracy of radiocarbon dating, artifacts are easily placed in their proper time in history. Science has advanced to the state that identification of diseases is possible and causes of death and dates of plagues are known. In the many cultures that occurred without the knowledge of the written language, archaeology is the only method to factually reconstruct their history.

Let us study these hypotheses: Corn could have been domesticated simultaneously all over Mesoamerica in 3000 BC; corn could have been domesticated independently in both the north and the south at the same time (3000 BC); corn was not domesticated in Mesoamerica at all but, rather, traded from somewhere else, such as South America or even Mesopotamia. These are just a few of the infinite number of possible hypotheses that could be cited to explain the archaeological facts.

Richard "Scotty" MacNeish's (1919-2001) experience told him that unburned corncobs could be preserved for millennia in an arid cave environment but would decompose rapidly when exposed to moisture. In 1962 he concentrated his search to dry caves and skipped the wet riverbeds. The geneticists surmised that maize had been domesticated as highland grass; accordingly, MacNeish decided to confine his search to the highlands. He searched the Tehuacan Valley of Pueblo State, Mexico, looking inside dry caves and overhangs. After personally searching thirty-eight caves, he tested Coxcatlan Cave, which had preserved six petite corncobs more primitive than any previously discovered. Subsequent radiocarbon assay placed their antiquity at about fifty-six hundred years. These findings bolstered MacNeish's theory about the origins of corn in southern Mexico. From these origins, corn could have been traded throughout North and South America as a common domesticated plant for all the natives.

CHINESE EXPLORATIONS

Unknown to the European (Portuguese) exploration of the western coast of Africa during the fifteenth century, the Chinese were trading along Africa's east coast. Theirs was a more respectful and less violent campaign. They mounted expeditions of thousands of men in fleets of junks each five times or more the size of the Portuguese caravels. They conducted peaceful trade backed by this show of force and are recorded to have resorted to violence only on three occasions in a century of explorations. But the Chinese folded their sails following the death of the adventurous Emperor Yung Lo. China abandoned its extensive naval expeditions into the Indian Ocean after 1440. By the time Da Gamma reached India the Chinese anti-exploration faction had made it a crime to build an oceangoing junk and had burned the ship's logbooks. Some of these logbooks were thought to have contained accounts of voyages extending across the Pacific as far as to the coast of the Americas. These burnings were ordered on the grounds that they contained "deceitful exaggerations of bizarre things." This, by the way, was just what Western critics said of Marco Polo's account of China.

EGYPTIAN EXPLORATIONS

In about 600 BC, to facilitate trade, the Egyptian Pharaoh Necho commissioned a Phoenician expedition which circumnavigated Africa in three years; twenty-one hundred years before Vasco da Gamma of Portugal.

EUROPEAN EXPLORATIONS

Spain - Columbus: 1492-1493; the Caribbean Islands of the New World.
Portugal - Vasco da Gamma: 1497-1499; around Cape of Good Hope to
            India.
Spain - Magellan: 1519-1522; circumnavigate the globe east to west.

CHINESE INVENTIONS

INVENTION	DATE INVENTED	WESTERN DEVELOPMENT
Astronomical observations*	2400 BC	4,000 years later
Acupuncture	1500 BC	-------
Plow	600 BC	2,200 years later
Circulation of blood	600 BC	1,800 years later
Cast Iron	400 BC	1,700 years later
Compass	400 BC	1,500 years later
Steel	200 BC	2,000 years later
Paper	200 BC	1,400 years later
Suspension bridge	100 BC	2,000 years later
Wheelbarrow	100 BC	1,300 years later
Seismograph	130	1,400 years later
Observed supernova	7-7-185	-------
Modern geology	200	1,500 years later
Porcelain	300	1,700 years later
Stirrups	300	300 years later
Steam engine	500	1,200 years later
Paddle wheel for boat	500	1,000 years later
Mirror	500	1,500 years later
Matches	577	1,000 years later
Chess	600	500 years later
Discover solar wind	600	1,400 years later
Brandy and whiskey	700	500 years later
Clock - mechanical	725	585 years later
Printing - block	800	700 years later
Printing - movable	1045	400 years later
Crab Nebula outburst	1054	-------
Metal type - Korea	1234	-------

*Based on the equator and the poles.

The Great Wall of China is the only man-made structure on Earth that can be seen from the Moon.

ROMAN ENGINEERING

The Empire's needs required a communication system of paved roads and bridges as well as huge public buildings and aqueducts. As road builders, the Romans surpassed all previous peoples. Constructed of layers of stone and gravel according to sound engineering principles, their roads were planned for the use of armies and messengers and were kept in constant repair. The earliest and best-known main Roman highway was the Appian Way. Running from Rome to the Bay of Naples, it was built about 300 BC to facilitate Rome's expansion southward. It has been said that the speed of travel possible on Roman highways was not surpassed until the early nineteenth century.

In designing their bridges and aqueducts, the Romans placed a series of stone arches next to one another to provide mutual support. At times several tiers of arches were used, one on top of the other. Fourteen aqueducts, stretching a total of two hundred sixty-five miles, supplied some fifty gallons of water daily for each inhabitant of Rome. These were proudly described by Rome's superintendent of aqueducts as "a signal testimony to the greatness of the Roman Empire," to be contrasted with "the idle pyramids or all the useless, though famous, works of the Greeks."

The largest Roman domed structure is the Pantheon, the oldest important roofed building in the world that is still intact. The massive dome rests on thick round walls of poured concrete that could not be weakened by window openings. The size of the dome remained unsurpassed until the twentieth century. The Romans had invented concrete and it became the primary ingredient of their construction. Roman buildings were built to last, and their size, grandeur and decorative richness symbolized the proud imperial spirit of Rome.

TROUBLES

Poverty in time of trouble is something riches cannot buy.

Chinese peasant

I asked a Burmese man why the women, after centuries of following their men, now walk ahead. He said there were many unexploded mines since the war.

The absence of evidence is no evidence of absence.

Martin J. Rees

Ideas to scientists are like mistresses to philanderers. If one has many, one does not mind losing a few.

I will debate a person on a subject in which equal knowledge exists. I will not debate a person on a singular subject that presents many avenues of approach when that person subscribes totally to a single avenue. I will not debate a person when the opposite is true - I know when I'm outclassed!

MASS EXTINCTIONS

Several mass extinctions have occurred on Earth since its formation over four and one-half billion years ago. The first life to evolve on this forming planet did so without our present atmosphere. The first cells were *bacteria* whose by-product of living and dying was *oxygen*. This process proceeded for the first two and one-half billion years after the formation of the first bacteria, producing enough for our atmosphere to eventually become saturated with oxygen. All other life forms that had evolved during this time with no oxygen died - gassed by oxygen. This was the first mass extinction.

The stage was now set for new forms of life using oxygen to evolve. During the Cambrian Period - 530 MYA - the first sign of a backboned fish appeared. This fish was the ancestor to all species with backbones. The evolution process was the development of legs from the fins allowing certain fish to "walk" on the surface of the oceans. The next step was for these fish to venture out of the oceans onto the land that now was protected by the ozone layer, allowing life to occur not only in the oceans, but on the open ground surface in the atmosphere. The evolution of land animals now began. Sharks were evolving with no bones, just cartilage for spines, during the Carboniferous Period.

Earth entered a long warm phase of so-called greenhouse conditions some four hundred thirty million years ago. A catastrophe struck the seabed three hundred seventy million years ago when giant tidal waves cleared the shores and wiped out the limestone reefs of the world. During the Permian Period -280 MYA- the evolution continued as fish, insects, reptiles and mammal-like reptiles (human ancestors) formed to populate the Earth. This was the time of the completion of the formation of Pangaea; all of the continents that had been separated five hundred seventy million years ago had now come together again resulting in one single landmass and one massive ocean. All the seas had disappeared and all climates had changed to that dictated by the location of this massive landmass. It took ten to twenty million years for the seas to be lifted and dried, eliminating all life in them. Volcanic activity increased tremendously due to the movement of plate tectonics as time passed. Sulfur dioxide was released in massive quantities into the atmosphere causing the extinction of most of the remaining land-borne life. Later, carbon dioxide dominated the atmosphere creating a greenhouse effect. The Permian Terminal Catastrophe of two hundred forty-five million years ago was the most deadly disaster on record. At the end of the Permian Geological Period, ninety-six percent of all species of marine animals were annihilated, and the reefs and seabeds were sterilized. The cause of this disaster is not known, but the results are known.

If the cause had been an asteroid or comet collision, the result would have been equivalent to a worldwide magnitude twelve earthquake.

During the Jurassic (180 MYA) and the Cretaceous (135 MYA) Periods the dinosaur era began. This was also the beginning of the breakup of Pangaea. The separation of South America from Africa began, followed by the separation of Africa from India, Australia, New Zealand and Antarctica. Deserts were appearing and mountains were being formed, creating new flowers. Turtles made their entrance along with forty foot reptilian sea monsters. All the animals roamed freely over this one gigantic land and were about to begin a trip of their own as the plate tectonic activity of the continental drift began to separate the landmasses one more time. The resulted effect was that the remains of species of this period are found in many continents today because they rode these separating continents without even knowing they were moving.

From this time until sixty-five million years ago the continents continued their journeys. North America was separating from Eurasia; Africa was separating from both Madagascar and South America; New Zealand began separating from Antarctica; the opening of the Bay of Biscay began; and the Labrador Sea opened. India was separating from Australia to begin its trip to join Asia. The Cretaceous Impact by an asteroid in the Yucatan Peninsula during this time caused the extinction of seventy-six percent of life on Earth and created the tremendous shock wave that traveled through and around the Earth causing the volcanic action resulting in the Decan Traps of India. The Decan Trap lava flow, when completed, had created an additional land depth of eight thousand feet. The gasses produced from these continual eruptions were added to the dense atmosphere created by the impact, leaving the Sun blocked out for a year or so with the resulting consequences. This was the last mass extinction to date. After the dust settled, animals started their evolution and humans could now evolve because the massive dinosaur control of Earth had ended.

A CAUSE OF MASS EXTINCTIONS

An asteroid or a large comet that collides with Earth will cause a mass extinction. Comets, Greek meaning "longhaired," consist of: a nucleus which consists of dust and rock particles; a coma, Greek meaning "hair," consists of volatiles which are frozen around the nucleus; a long tail which flows from the coma. The coma (volatiles - substances that evaporate readily) are mostly methane, ammonia, water and other compounds of hydrogen combined with carbon, nitrogen or oxygen atoms. The tail will always point away from the Sun due to the force of the solar wind. The composition of the coma and tail would be considered a vacuum on Earth. Some asteroids could be the remains of a volatile-free comet, a rock. An asteroid is defined as a subplanet - Greek, "like a star." Both asteroids and comets were active in the formation of Earth and each has an orbit different from the other.

There are three possible collision paths that an asteroid or comet would take: head on, overtaking or intermediate. Earth's speed in orbit is 66,588 miles per hour, almost 30 kilometers per second. Asteroid collisions would probably be in an overtaking situation with a relative velocity of 10 to 15 kilometers per second (22,370 to 33,555 miles per hour). Comets, traveling at 30 kilometers per second (76,110 miles per hour), usually impact head-on or from an intermediate path. The relative velocity of comet impacts could approach 130,000 miles per hour. A comet with the same mass as an asteroid would produce four to nine times as severe an impact as the asteroid due to the increased collision speed.

The comet or asteroid would part the atmosphere on Earth like a super projectile leaving a hole through the air as wide as itself. Even though air rushes back into the hole at the speed of sound, it would take tens of seconds for the hole to disappear. Upon striking Earth, the object would decelerate to zero velocity only after depositing most of its kinetic energy into the material it encounters - Earth - which contains a total mass of material many times the mass of the incoming object.

The oceans have an average depth of seven kilometers (4.33 miles), so the ultimate result would be much the same whether the object struck land or water. It would create a hole as it passed through the water and then excavate a crater several kilometers deep which would be from 50 to 100 kilometers (31 to 62 miles) wide in the Earth's crust. The water around the object would vaporize at once, doubling the water vapor content in the atmosphere. The earthquake from such an impact would release 100 billion times more energy than did the 1906 San Francisco earthquake which had a Richter reading of over eight. A tsunami one kilometer (two-thirds of a mile) high would roll across the ocean floor at 1,000 kilometers per hour (620 miles per hour) and destroy everything hundreds of kilometers inland from the coasts. These are the side effects.

A ten kilometer (6.2 mile) object would move about 200 cubic kilometers (50 cubic miles) of rock, more than a thousand times more than the amount of material excavated in over ten years to create the Panama Canal. Upon impact this material would be heated instantly, spraying it sideways and upwards from its center. Matter heading upward would encounter no resistance from either the ocean or the atmosphere; each would have been pushed aside enough for the material to rise unobstructed. The pulverized particles would each acquire a ballistic trajectory and orbit our planet. Some of the particles would fly off into interplanetary space while most, held by Earth's gravity, would fall back onto the top of the atmosphere at a point far from the hole from which they had emerged. The heavier ones would fall through the atmosphere at points all around the globe, but the lighter ones would float like oil on water, suspended in the stratosphere as fine-grained, low mass dust grains. The eventual result would be death for many species of life on Earth.

A long period of darkness would result from the dust lifted 20 to 40 kilometers (12.5 to 25 miles) into the atmosphere, above the tropopause, usually 10 kilometers (6.2 miles) high, that puts a lid on all weather patterns. Any dust that enters the stratosphere, over the tropopause, will take months to settle to the surface because it cannot be brought down by rain, since water vapor does not rise that high. This dust would arise from the vaporization of many cubic kilometers of matter at the site of the impact.

The effects would be significant surface darkening over many weeks, subfreezing land temperatures persisting for up to several months, large perturbations in global circulation patterns and dramatic changes in local weather, including the precipitation rates. As long as large numbers of dust particles remained in the upper atmosphere they would absorb all sunlight, heating the upper atmosphere layers but cooling layers near the surface, which would no longer receive the Sun's light and heat. The dust particles would eventually rain out of the upper atmosphere, colliding with each other and joining together to form larger particles, which cannot remain suspended in the rarefied air, and would sink toward the surface.

These months would be unpleasant for life as we know it. For a number of weeks the intensity of sunlight on the Earth's surface would fall to below one percent of its normal value, low enough to halt photosynthesis among the phytoplankton in the oceans. It has been calculated that somewhere between ten and one hundred days without sunlight would be enough to kill all the phytoplankton. Since these tiny plants form the bottom of the entire food chain in the seas, this loss would in turn cause the disappearance of smaller marine animals unable to survive for a few weeks without food.

The other significant effect this dust produces, the cooling of the Earth's entire surface, is the great damaging effects on life. The average surface temperature would decline in a few days after the dust entered the atmosphere by some forty degrees Celsius to -30^0 C (by seventy degrees Fahrenheit to -22^0 F). This would exist primarily over the landmass, but storms over the oceans would create adverse conditions. We would now enter three months of darkness.

Recovery to the normal surface temperature of the Earth would occur in six months or so. Within a decade or two, many plants would again grow with reduced diversity. Best off would be the benthic (bottom-dwelling) forms of sea life. From the ocean depths could arise the next forms of intelligent life. The sharks of today, which have already survived for two hundred million years with few changes, could evolve into the world leaders of tomorrow.

YUCATAN PENINSULA IMPACT

The Cretaceous impact of sixty-five million years ago disrupted the global ecosystem and led to a major mass extinction. This impact of a fifteen kilometer (nine and one-third mile) object released more than one hundred million megatons of energy and excavated a crater, Chicxulub in Mexico, approximately two hundred kilometers (one hundred twenty-four miles) in diameter. Among the environmental consequences were devastating wildfires and dramatic short-term disturbances in the climate which produced some ten trillion tons of fine dust which was injected into the stratosphere. Major mass extinction of the dinosaurs and almost all land based life occurred. The dinosaurs were involved in their own extinction process at this time due to the increasing evolution of forests and other natural changes. Major mass extinctions occur at intervals of many millions of years according to fossil records.

If the asteroid that created this impact had safely crossed the Earth's orbit an hour earlier, some of the dinosaurs would have survived a little longer. The evolution process would have been entirely different and human existence would have been unlikely. Perhaps the most terrible of the "terrible lizards" was named *Stenoychosaurus Inequalis*, whose large braincase, stereo vision and opposing thumbs hint at the potential for becoming a race of intelligent life. The story of life on Earth is not to be understood without reference to events of this kind, which repeatedly cleared the principal actors from the stage and gave their understudies their chance to advance. Major impacts probably occur every few hundred thousand years, on average, and stupendous ones every fifty million years or so.

TUNGUSKA

An explosion occurred in Central Siberia June 30, 1908 at 60^0 55' North and 101^0 57' East that was the equivalent of a ten megaton nuclear detonation. It was heard at a distance of one thousand kilometers (620 miles). It exploded eight and one-half kilometers (5.25 miles) above the ground. Trees were uprooted and blown radially outward thirty to forty kilometers (19 to 25 miles) from the blast. Two thousand square kilometers (772 square miles) was devastated by fire. It is not known but is assumed that a comet traveling the opposite direction of Earth's orbit collided with our atmosphere and caused the blast. The head-on collision must have occurred around sixty kilometers per second (over 130,000 miles per hour). The mass of the comet has been estimated to weigh thirty million kilograms (33,000 tons). Its nucleus would have measured forty meters (120 feet) in diameter. It was noted that Comet Encke changed its orbit in 1908 and the change could have been induced by the loss of a fragment of the size of the Tunguska object.

EARTH'S CLIMATE
OR
WHY I DON'T WANT TO BE A WEATHERMAN

The Earth's climate is changing. This same statement was true four and one-half billion years ago and has been true for each year since. Predictive climate based on cycles is impossible because too many factors about Earth are changing independent of each other. Each change by one may be increased, decreased or canceled by one or several other changes that may be happening at the same time.

To begin this list of variables we must start with the Sun. The first influence involved is Earth's orbit about the Sun. The orbit is not a circle, but an ellipse with the Sun closer to one end than the other. The ellipse changes shape over time resulting in different amounts of energy (sunlight) reaching different parts of Earth. This cycle is approximately 95,800 years. If this were the only variable, we could predict the weather for 95,800 years from the present because we would expect the conditions to be the same again. Sorry, we still have a few more effects of the Sun to list.

The second of the Sun's influence on Earth's climate and weather is the angle of our tilt as we orbit the Sun. As you look at a globe representing Earth, you notice that it is positioned at an angle. The globe is not constructed in this manner for you to view it easier, it is showing the actual inclination (tilt) of the axis between the North Geographic Pole and the South Geographic Pole, presently a 23.45 degree tilt from the vertical, which Earth spins as it orbits the Sun. The statement is *presently* because this angle of tilt steadily varies between 21.39 degrees and 24.36 degrees. The Sun is directly overhead at this latitude (23.45^0) North during the northern summer - the Tropic of Cancer - and overhead at this same latitude South during the southern summer - the Tropic of Capricorn. The total width of the tropics is two times 23.45 degrees (north and south of the equator) for a total of 46.90 degrees. At the lower range the tropics would total only 42.78 degrees but would be 48.72 degrees at the higher range of the tilt of our axis. This produces an area of 5.94 degree variation of the tropics during this cycle of 41,000 years. The northern and southern hemispheres vary between being cooler and warmer during this cycle which affects the climate and weather. This 41,000 year cycle is independent of the 95,800 year cycle.

Another of the Sun's influences on Earth's climate and weather is precession, the wobble of the Earth like a top. The precession period is 25,735 years and again, does not coincide with either the 95,800 or 41,000 year cycles of orbit and tilt. These three influences also help create the erratic cycles of the Ice Ages. Other Sun actions that affect Earth are the Sun spots that appear in approximated eleven year cycles. Another variable is an apparent pulsating Sun, growing larger and then smaller, on a seventy-six year cycle. These cycles may create or add to the severity of drought cycles. Solar flares also present many electrical problems on Earth. All of these influences of the Sun are in effect and ever changing.

Our solar system has a system of erratic asteroids whose orbits were converted into highly elongated ones due to Jupiter's gravity. These are the Apollo asteroids that are orbiting on the outer edge of the asteroid belt that may intersect Earth's orbit. The solar system's orbit within the Milky Way galaxy crosses a belt of asteroids within the galaxy on a thirty-three million year cycle. These asteroids present collision problems to Earth.

Another variable to add to the above is the continuous drifting of the continents, plate tectonics. This continuous movement of the landmasses adds another problem. The landscape is continually changing location - moving away, moving toward, moving north, moving south - always moving. This changes where and how much sunlight is reflected each day. These movements create another variable, ocean currents which create our winds and carry storms that are born in Africa across the Atlantic Ocean to be deposited on the East Coast of the United States as hurricanes. As currents change, directions change and this phenomenon persists in all our oceans and large bodies of water. A couple of these are known as El Nino and La Nina. Another possible variable of which there is no cycle in creating changes in Earth's climate and weather is MAN. The requirements to maintain life is a precarious balance of carbon dioxide and oxygen. Man may not be helping matters.

So, why would anyone want to be a weatherman? The changes are each happening on their own timetables with no consideration for the others. The variables in long range forecasting are many and there may be others we have not yet had to consider. Earth has never been the same in any two times in its existence and never will - there are enough variables to assure this.

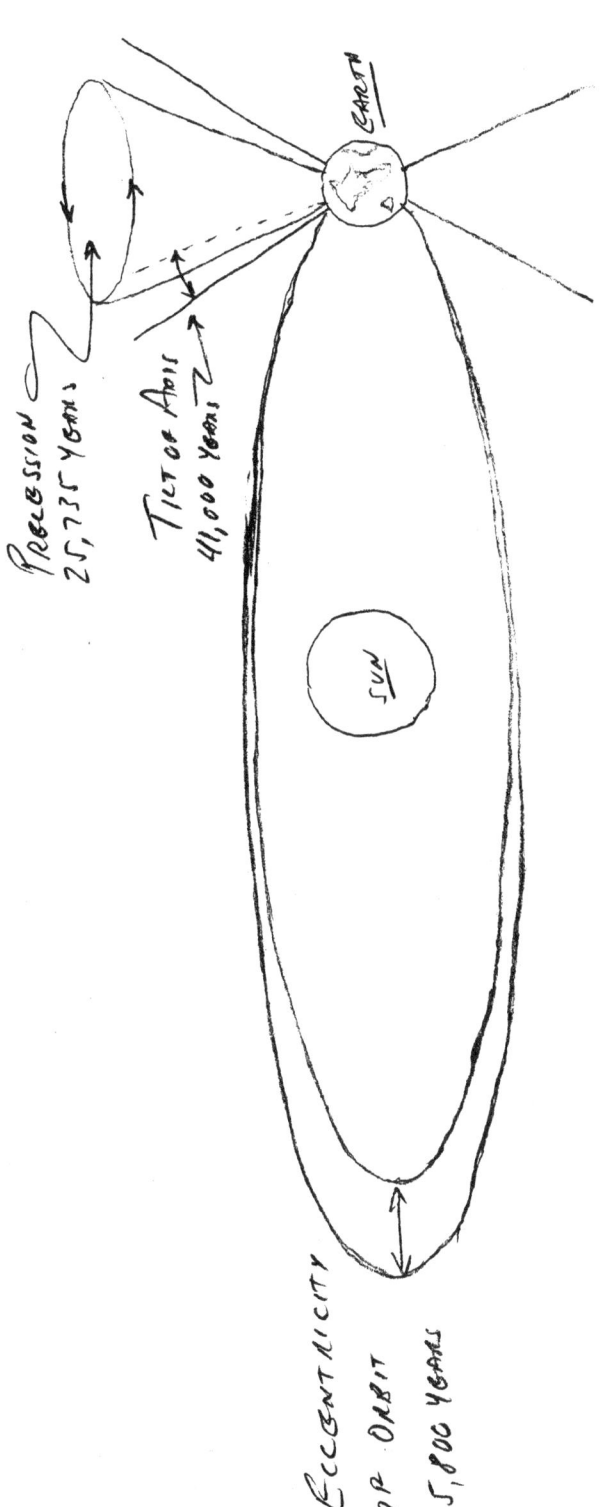

Precession
25,735 years

Tilt of Axis
41,000 years

Earth

Sun

Eccentricity
of Orbit
95,800 years

Variations in Earth's Orbit

217

OZONE

Ozone in the high atmosphere plays an important role with respect to life on Earth and the structure of the atmosphere. When the intense sunlight reaching the high atmosphere breaks oxygen molecules into two oxygen atoms, most of these atoms reassemble, not as common oxygen, but as molecule of *ozone* containing three oxygen atoms. Ozone is destroyed by being turned back into oxygen in a different set of reactions. There was no ozone layer during the early formation of Earth. Only following the development of an atmosphere was this formation possible. As the ultraviolet (UV) radiation split the water into its two components, two hydrogen atoms and one oxygen atom, it caused the formation of the three atom molecules of ozone.

Ozone's structure allows it to absorb a certain kind of ultraviolet sunlight that would otherwise reach the surface of the Earth and affect living material. If there were no ozone in our atmosphere, there would be no life on the surface - all life would be limited to the oceans. The radiation of most concern is called UV-B, radiation between the wavelengths of 280 and 320 nanometers ($280 - 320 \times 10^{-9}$ meters). Ozone also provides another important role in the high atmosphere. By absorbing ultraviolet radiation, ozone deposits the heat associated with these wavelengths into that level of the atmosphere, creating a layer much warmer than those immediately below. The stable region so created is the stratosphere and it is in this stable layer that disturbing changes are occurring. Chlorine, an effective chemical catalyst that can change ozone into normal oxygen, is appearing in rapidly increasing concentrations in the atmosphere.

Each day ozone is created during daylight hours by reactions driven by intense sunlight. Each day, a fraction of all the ozone in the stratosphere is destroyed by reactions with chemicals occurring naturally in the stratosphere. The amount created is more or less fixed, while the amount destroyed increases as the total amount of ozone increases. The amount of ozone builds up until the amount created equals the amount destroyed and an approximate equilibrium is reached. The introduction of chlorine into the atmosphere and its conversion of ozone into oxygen is disturbing this balance by causing the destruction of ozone to exceed he creation of ozone. If this were to continue, the ultraviolet wavelengths would begin to reach Earth unimpeded and cause life on the surface to become impossible to continue. The range for ozone is its location in the stratosphere fifteen to fifty kilometers (10 to 30 miles) above sealevel. If all the ozone were to be brought down to sealevel, the layer would only be three millimeters thick (one-eighth of an inch). This very small percentage of our atmosphere is one of our most important.

OXYGEN

Oxygen accounts for twenty-one percent of the volume of our atmosphere. Oxygen is produced by living organisms. If this volume were more than twenty-one percent, fires would burn continuously - you could not put them out. A volume of less than twenty-one percent would cause humans and other complicated forms of life to asphyxiate. Earth is a "living" entity and can be killed. Nitrogen accounts for almost seventy-eight percent of the remaining volume of our atmosphere.

AVERAGE TEMPERATURE OF THE EARTH

NO SUN

The internal core temperature of Earth would not be able to warm the surface. The atmosphere would freeze solid and the Earth would be covered with a layer of nitrogen and oxygen snow thirty feet thick at minus twenty degrees Celsius (minus five degrees Fahrenheit). The oceans would be ice.

WITH THE SUN BUT NO EARTH ATMOSPHERE

The effect would be the same as with NO SUN due to the fact that all energy that was received would be reflected, causing no absorption of heat.

GREENHOUSE EFFECT

PROCESS - INFRARED TRAPPING
EFFECT - CLIMATE HEATING

The increase of carbon in the atmosphere is due to the increase of carbon dioxide being released because of the burning of fossil fuels, the great increase in the global population (each of us exhale carbon dioxide after each breath is drawn in) and other natural releases. Studies of the concentration of carbon dioxide in the air, from air samples trapped at different depths in polar ice and taken in air samples from the Mauna Loa Observatory, show an increase from about 280 parts per million by volume (PPMV) around the year 1750 to 375 PPMV by the year 2000 - an increase by one-third. Carbon dioxide traps infrared radiation and this increase tends to increase climate heating.

The gases in the atmosphere absorb much of the infrared frequencies. When sunlight hits Earth and is reflected (radiated back to space), this blanket of infrared absorbing gases obstructs the way. As a result the Earth has to warm up some to achieve the equilibrium between the sunlight coming in and the infrared radiation emitted out to space. This causes the temperature of Earth's surface to be thirteen degrees Celsius (55.4 degrees Fahrenheit). The adding of the greenhouse gases would cause more infrared radiation to be absorbed and add to global warming.

The burning of fossil fuels and the destruction of the forests places seven million tons of carbon dioxide into the atmosphere each year. Trees remove carbon dioxide from the atmosphere and convert it to wood. The global temperature change between the Ice Age and an interglacial interval is only three to six degrees Celsius (five to eleven degrees Fahrenheit). Anything higher would be new. As the Earth warms, sea levels rise. Seawater expands as it warms causing the glaciers and polar ice to melt. The sea level could rise one meter by the year 2100. Polynesia, Melanesia and the Indian Ocean will be submerged. Coastal cities and major river cities will have problems.

Sea levels the last two hundred million years have varied from above two hundred meters of present sea level to as low as two hundred meters below. High levels cause flooding of the continents and droughts occur during the Ice Ages because most of the water is converted to ice leaving smaller amounts for rain.

Another infrared trapping substance is methane. Modern measurements show that the atmospheric concentration of methane is increasing about twice as fast as that of carbon dioxide. This concentration increased from about 650 parts per billion by volume (PPBV) around the year 1400 to almost 1800 PPBV by 2000.

EL NINO

El Nino, or "the Christ Child," was named because of its appearance during the end of the year in a cyclic pattern. When a great barometric low is located over Indonesia and Australia, the Western Pacific is warm and that warm water maintains the powerful center of rainfall in the west that India experiences as the annual monsoon. By contrast, the sea surface off the coast of Peru during this period is relatively cold; it is made up of water from the deep ocean that wells up to the surface near South America. The temperature difference between east and west affects the atmosphere as well: The sea surface temperature gradient drives air along the ocean surface from the cooler east out to the warmer west - the trade winds. The winds in their turn set up a feedback loop, pushing warm surface water along in front of them, piling up warm water in the west, moving it out of the way of the upwelling cold water in the east. In this manner the winds maintain the conditions that set them blowing in the first place.

If the warm water extended eastward, the rain would also move east; the temperature difference between the west and the east would drop in turn. The Pacific low-pressure zone would now be found to the east of its ordinary position over Indonesia and Australia; the winds would weaken, allowing even more warm water to flow east and slowing the rise of cold water from the deep. Feedback, again, working in the opposite direction. As long as warm water continued to spread over the eastern Pacific Ocean, the El Nino pattern would maintain itself; it would be dry in the west and wet in the Central Pacific and the Americas. As long as these conditions continued, the Pacific low-pressure zone would be displaced to the east.

The problem starts when the temperature of the water off the coast of Peru starts to rise. Eventually it causes droughts in the western Pacific and excessive rainfall in the eastern Pacific.

LA NINA

La Nina, or "little sister," follows El Nino as the jet stream moves accordingly. The eastern Pacific becomes cooler as rains enter the Pacific Northwest.

ACID RAIN

In the eastern half of the United States, somewhere between ten to fifteen million tons of sulfur get poured into the atmosphere each year. For oxides of nitrogen the possible range is wider, with emissions ranging from ten to twenty million tons annually. The burning of fossil fuels produces over ninety percent of the sulfur and nitrogen that returns to Earth as sulfuric and nitric acids. About two-thirds of the acidity in acid rain in the East derives from sulfur, the rest from nitrogen. Seventy percent of the sulfur and thirty percent of the nitrogen comes from electric power plants and smelters. Transportation, essentially car exhaust, accounts for another forty percent of the nitrogen. Effects include lake pollution, the killing of forests and the erosion of marble statues and buildings.

What is happening is that the by-products of our economic life are altering the chemical composition of the atmosphere by increasing the amount of carbon dioxide and a number of other rarer compounds. These compounds together alter the radiative properties of the atmosphere. The net result is that the lower atmosphere and the surface of the Earth are going to retain more heat. Within a relatively short time of a few generations or so, the entire planet could be a lot warmer.

One of the easiest predictions to make of the greenhouse effect is that it will produce a rapid rise in sea level. The rise in temperature will warm both the ocean and the land. The oceans will expand as they warm. One model calculates that the ocean's height would rise by eighty centimeters (thirty-two inches) as they reached a new thermodynamic equilibrium. Add the melting water from the great continental glaciers that remain and the upper limit of the rise of the sea level could exceed three meters (ten feet).

The ozone is peculiarly vulnerable to a set of reactions involving the element chlorine. Chlorine can attack a molecule of ozone, capturing one of its three oxygen atoms to form chlorine monoxide. When that molecule encounters another single atom of oxygen, another reaction occurs, producing one molecule of ordinary molecular oxygen and one atom of chlorine, which is then free to break up another ozone molecule. Chlorofluorocarbons could destroy between seven and thirteen percent of the ozone layer within a century.

Recent studies have also revealed that water droplets in fog or clouds can be much more acidic than the water in rain and that these tiny drops, some as powerful as battery acid, can cause damage to leaves and needles beyond that created by exposure to pollutant gases. This discovery helped explain why trees at higher altitudes, which are frequently covered in clouds, appeared particularly susceptible to damage.

HEAT TRANSFER

Heat always passes from warmer to cooler bodies. If several bodies of different temperatures are in thermal contact, those that are warm will become cooler and those that are cool become warmer - they tend to reach a common temperature.

CONDUCTION: The molecules at the heat source are caused to move more rapidly. These molecules and free electrons collide with their neighbors and cause them to move faster. This process continues until the increased motion has become transmitted to all of the molecules and has caused the entire body to become hot. The conduction of heat is accomplished by electron and molecular collision.

Solids whose molecules have a "loose" outer electron conduct heat and electricity well. Metals have the "loosest" outer electrons and are the best conductors of heat and electricity for this reason. Silver is the best conductor, followed by copper, aluminum and iron.

INSULATOR: These are poor conductors. Examples are wool, wood, straw, paper, cork and styrofoam.

CONVECTION: Heat is transmitted by currents in liquids and gases. When air comes in contact with a hot surface, it is heated and then rises, being replaced by cooler air. After the warm column of air rises, it cools and then descends repeating the process.

Warm air rises because it expands and becomes less dense than the surrounding air and is buoyed upward like a balloon. The buoyancy is upward because the air pressure below a region of warmed air is greater than the air pressure above. The warm air rises because the buoyant force (upward) is greater than its weight (gravity). This is the reason helium is located in the higher elevations of our atmosphere - gravity. Its atomic mass is the least and therefore affected less by gravity than neon, argon, krypton, xenon and radon. This is the reason *radon* is so dangerous when it is concentrated. It is heavy and accumulates at ground level. GASES HAVE WEIGHT!

Because warm air rises, we know the air in a room will be warmer near the ceiling than at the floor. We would also think the atmosphere would be warmer with increasing altitude. Why aren't the mountains warm and green and the villages below cold and snow covered? The reason this will not work is because as warm air rises in the atmosphere it is rising in lesser and lesser atmospheric pressure. This causes the air to expand. The rapid expansion of air is a cooling process. Compression heats air - expansion cools it.

Convection is a means of heat transmission in all fluids, whether liquids or gases. If a fluid is heated from below, its molecules increase in speed and rise, permitting cooler fluid to come to the bottom. In this way, convection currents keep the fluid stirred up as it heats.

RADIATION: The transfer of energy by the rapid oscillations of electro-magnetic fields in space. These are radio waves, television waves, infrared waves, light waves, ultraviolet waves and x-rays. These are basically the same phenomenon - electromagnetic waves that transmit energy at different levels.

Strictly speaking, heat is not transmitted in the radiation process. The thermal energy of a radiating body is transformed, at the instant of radiation, into radiant energy. Thermal energy from the Sun is *transformed* into radiant energy through space and is *transformed* into thermal energy when it strikes the surface of an object.

A very good absorber of radiation reflects very little light and appears dark. A perfect absorber reflects no radiation and appears perfectly black, and since it is absorbing all the radiation frequencies, it will become hot. A very poor absorber of radiation reflects most of its incident radiation and appears to mirror the incoming light - reflecting the frequencies instead of absorbing them. It will remain cooler because it is not absorbing the frequencies, it is reflecting them.

The ability of an object to be an absorber or a reflector is due to its molecular and atomic contents. The molecules of a dark object are chemically different from the molecules of a reflective object. These objects are composed of different atoms. Atoms reradiate photons from their electron shells in accordance to the atom's energy levels. Incoming photons may have more energy than the atom can reradiate causing the atom to absorb the energy, which transfers itself to heat. This would tell us that these molecules have color - they are darker than the reflective surface.

THE CARBON CYCLE

Plants replenish oxygen by utilizing carbon dioxide that is the primary source of carbon for photosynthesis, the basis for all life. Carbon dioxide acts as a thermostat that regulates the temperature of Earth. Too little, and Earth cools; too much and Earth would heat up. The oxygen in the upper atmosphere promoted the formation of the ozone layer, blocking out the ultraviolet rays which allowed life to exist on land. Without the ozone layer life could only exist in the oceans.

Carbon 14 is a radioactive isotope of carbon that has a half-life of about 5,730 years. The Earth's atmosphere contains much carbon, mostly in the form of carbon dioxide gas, with small traces of carbon 14. Most atmospheric carbon is the nonradioactive isotope, carbon 12. The ratio of carbon 14 to carbon 12 is virtually constant in the atmosphere. As a plant absorbs carbon dioxide from the air in the process of photosynthesis, the carbon 12 stays in the plant while the carbon 14 is converted into nitrogen. (*Cosmic rays knock neutrons out of gases in the atmosphere. The free neutron enters a nitrogen atom changing it to a carbon 14 atom. Beta decay of a neutron releases an electron and the neutron becomes a proton producing a nitrogen 14 atom.*) Thus, in a plant, the ratio of carbon 14 to carbon 12 is smaller than the ratio in the atmosphere. Even when the plant dies or is consumed by an animal or person, this ratio will continue to decrease. From this a method of dating objects called *carbon 14 dating* has been developed.

Suppose an object has been discovered in which the ratio of carbon 14 to carbon 12 is only about half the ratio found in the atmosphere. This means the object ceased living about 5,700 years ago.

RADIOISOTOPES
FOR
GEOLOGIC DATING

RADIOACTIVE PARENT	HALF-LIFE (YEARS)	DAUGHTER PRODUCT	ROCKS, MINERALS COMMONLY DATED
Uranium 238	4.5 Billion	lead 206	zircon, uraninite, pitchblend
Uranium 235	713 Million	lead 207	zircon, uraninite, pitchblend
Potassium 40	1.3 Billion	argon 40	muscovite, biotite, hornblend, glauconite, sanidine, volcanic rock
Rubidium 87	47 Billion	strontium 87	muscovite, biotite, lepidolite, microline, glauconite, metamorphic rock
Carbon 14	5,730	nitrogen 14	all plant and animal materials

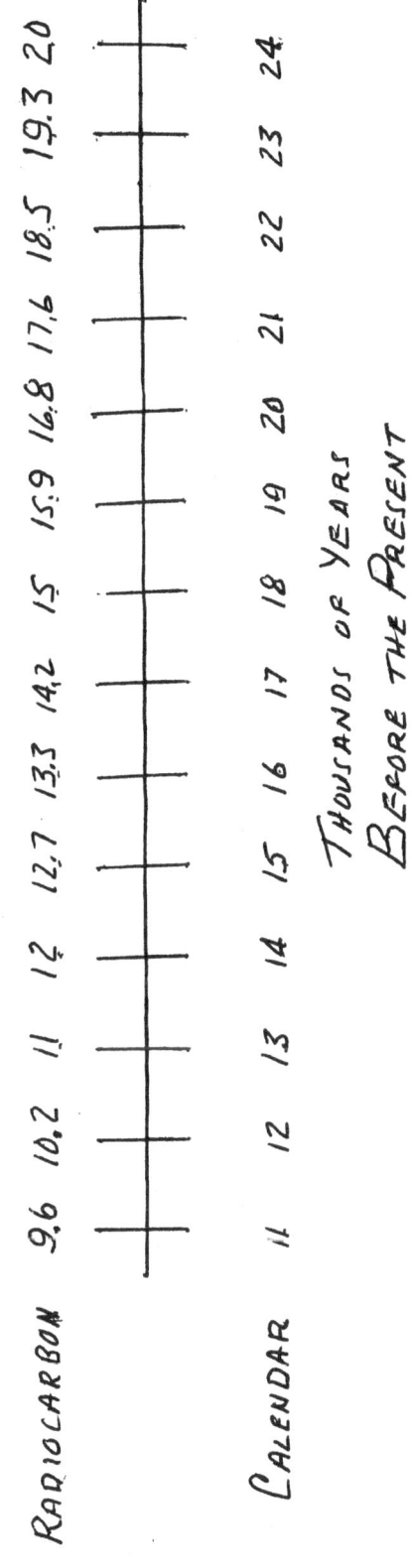

Archaeological Dating

Matching Radiocarbon Dates to the Calendar

RADIOCARBON 9.6 10.2 11 12 12.7 13.3 14.2 15 15.9 16.8 17.6 18.5 19.3 20

CALENDAR 11 12 13 14 15 16 17 18 19 20 21 22 23 24

Thousands of Years
Before the Present

LITTLE THINGS

The world is composed of four elements - earth (fall), water (fall), fire (rise) and air (rise).

Aristotle

I assert that the elements are not four as Aristotle says, but that they are innumerable and indefinitely divisible.

Anaxagoras of Clazomenae in Iona

I agree with Aristotle that there are only four elements - earth, water, fire and air.

Empedocles of Acragas in Sicily

During the fifth century BC, Leucippus of Milteus and his student Democritus of Abdera in Thrace saw innumerable individual pieces of matter. They were the first, explicit, thoroughgoing materialists. Democritus named these individual pieces of matter "atoms," which meant "indivisible."

Principle of uniformitarianism - the laws of nature do not change with time.

Thales of Milteus was exploring electrostatic charges in 600 BC. Walk across a new carpet and then touch a piece of metal - zap! Electrostatic *DIS*charge.

Only one millionth of a billionth of an atom is solid nucleus. This applies to everything.

As early as 1898, the American geologist Thomas Crowder Chamberlain was speculating that atoms were "complex organizations and seats of enormous energies" and that "the extraordinary conditions that reside in the center of the Sun may ... set free a portion of this energy."

In the span of forty years, by 1938, humankind had progressed from ignorance of the very existence of atoms to an understanding of the primary thermonuclear fusion process that powers the Sun.

The elementary particles apparently provide the key to some of the fundamental mysteries in early Cosmology ... And Cosmology, it turns out, provides a sort of testing ground for some of the ideas of elementary particle physics.

> Murray Gell-Mann, Nobel Prize in 1969
> Discovered the quark.

A bar of gold, though it looks solid, is composed almost entirely of empty space: The nucleus of each of its atoms is so small that if one atom were enlarged a million billion times, until its outer electron shell was as big as greater Los Angeles, its nucleus would still be only about the size of a compact car parked downtown. The electron shells would be zones of insubstantial lightning, each a mile or so thick, separated by many miles of space. Nor, to return to the old classical metaphor, does a que ball strike a billiard ball. Rather, the negatively charged fields of the two balls repel each other; on the subatomic scale, the billiard balls are as spacious as galaxies, and were it not for their like electrical charges they could, like galaxies, pass right through each other unscathed.

> Timothy Ferris
> *Coming of Age in the Milky Way*

The electron sphere is ten thousand times larger than the sphere of the nucleus. If the nucleus sphere were one inch in diameter, the electron sphere would be:

10,000 inches in diameter
833.33 feet in diameter
277.8 yards in diameter (138.9 yards radius)

If the Sun were to be considered a nucleus, Pluto would be orbiting in the electron sphere.

ATOMIC STRUCTURE

The fact that the mass of an atom is concentrated in the nucleus makes the nucleus extremely dense. The principle building block of the nucleus is the nucleon, which in turn is composed of fundamental particles called quarks. When the nucleon is in an electrically neutral state, it is a neutron. When it is in an electrically positively charged state, it is a proton. All protons are identical; they are copies of one another. Likewise with neutrons; each neutron is identical to every other neutron. The known atomic nuclei are composed from one to about two hundred seventy nucleons.

The nucleus is surrounded by clouds of negatively charged particles called electrons and this combination results in an *atom* - the nucleus plus the electrons. Electrons make up the flow of electricity in electrical circuits and bind to the outer electron shells of other atoms to make molecules and compounds. An electron is about two thousand times lighter than a nucleon (proton or neutron) and contribute very little mass to the total mass of the atom. Electrons are negatively charged and repel other electrons, but are attracted to the positively charged protons in the nucleus. An electron in one atom is identical to any electron in or out of any other atom. The number of protons in the nucleus is electrically balanced by an equal number of electrons whirling in the surrounding clouds. The basic atom is electrically neutral in cases where the proton and electron counts are equal.

As with the solar system, the atom consists of empty space. The main characteristic that distinguishes atoms from one another is the number of electrons in orbit about the nucleus - generally the same as the number of protons in the nucleus. Helium has two electrons orbiting two protons and two neutrons; uranium, the heaviest natural element, has 92 protons and 146 neutrons packed into its central nucleus with 92 electrons in orbit. The uranium atom occupies little more space than the single proton and electron of the hydrogen atom. The average diameter of an atom is 10^{-8} centimeter (0.00000001 centimeter), which is also called one Ångstrom - Å.

Matter exists in four states: solid, liquid, gaseous and plasma. In the solid state, the atoms and molecules vibrate about *fixed* positions - ICE. The liquid state occurs when the molecular vibration is increased enough to shake the molecules apart and cause them to wander throughout the material in *nonfixed* positions - WATER. The gaseous state occurs when more energy is put into a material causing the molecules to vibrate at even greater rates, the result being that they *break away* from each other - STEAM. Heating steam to over two thousand degrees Celsius causes the atoms themselves to be *shaken apart*, making a gas of free electrons and bare nuclei - PLASMA. All substances can be transformed from any state to another in going from one state to the other, whether up or down the ladder.

PARTICLE	SPIN	MASS	
Photon*	1	0	
Graviton*	2	0	
Electron's neutrino*	½	0	
Electron*	½	1	
Muon's neutrino	½	0	
Muon	½	206.77	
Meson - Pion	0	273.1	
Kaon	0	966.4	
Eta	0	1074	
Baryons - nucleons			
proton*	½	1836.10	Rotation 10^{22}/sec
neutron[#]	½	1838.63	Beta decay when alone

Note: *Only particles with infinite lifetimes.
[#]Stabilized when with proton.

CLASSICAL HELIUM ATOM
+ {Proton}   O {Neutron}   - {Electron}

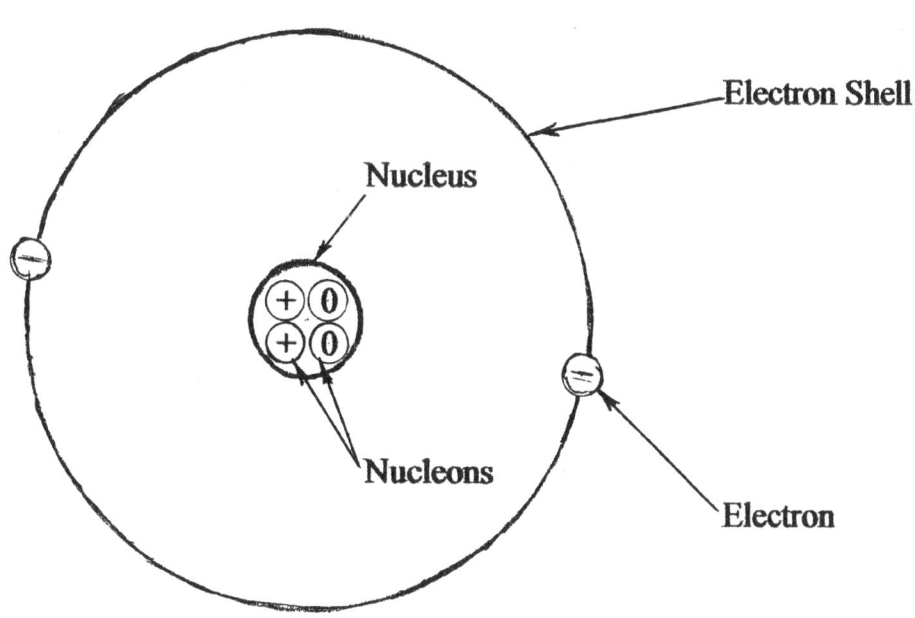

NUCLEUS CONFIGURATION

A convenient way to visualize the parts that are assembled to create the protons and neutrons of the nucleus of an atom is to think of a group of tennis balls (spheres) with split seams, each able to fit one over the other. These tennis balls are each solidly colored - red, blue or green. Each is marked either "up" or "down." These tennis balls represent *quarks* - the smallest of the building blocks of the nucleon. The nucleus of an atom consists of a specific number of protons (positive charge) and neutrons (neutral charge). The neutron is neutral because internally it consists of a proton (positive charge), an electron (negative charge) and a neutrino. The number of protons in the nucleus determines what the element will be - such as carbon, oxygen, potassium or uranium.

Since protons have a positive charge, we cannot just fill up a bucket with them - "like" charges *repel*, which causes each proton to want to push away from any other nearby proton. Since they are required to be close together in the nucleus, I have developed a model that takes care of this problem by using the neutrons to insulate protons from each other. How are the nucleons assembled to solve this problem? We will build this assembly. These tennis balls, quarks, are each representing a quantum of energy - spheres that are in motion - spinning at a frequency in the gamma ray range, a million billion billion times a second. Since these quarks would be hard to handle, we will use our tennis balls.

To build a proton, we will select three tennis balls (quarks) - one red, one blue and one green. The artistic will notice that these are considered our primary colors and when placed together form the visual white! This sounds so scientific and probably had deep thinkers involved. I'm sorry to say that this is not the case - we have used all the Greek alphabet and names for scientific identifications and have conveniently chosen simple English terms to name the newly discovered particles of the Universe. Now that we have our feet back on the ground, we will assemble these balls one over the other to form a single ball three layers thick.

You will notice that the balls are also marked randomly "up" or "down" (another use of our newly discovered English notation system). Other directions that are available are: strange, charmed, bottom and top - we won't use these but I knew you had to know that others are available for special purposes. Now back to business: Select two balls that are marked "up" and a third one marked "down." It doesn't seem to matter which color has "up" or which has "down" as long as the three different colors are marked with two colors being marked "up" with the third color being marked "down." The proton is constructed by holding a ball marked "up," placing a ball marked "down" over it and then placing the third ball marked "up" over the other two. You now have a proton. This nucleon is the nucleus of the hydrogen atom. Take a break - you have done a lot!

Now that you are back from your break we will immediately build a neutron. We don't waste time around here. We will generally proceed as we did with the proton except that we notice that each tennis ball is such that it can also fit over any other collection of tennis balls (our proton). This time we choose three colors again, but with a difference. We now want two "down" tennis balls and only one "up" tennis ball. We are going to make a neutron and neutrons are different from protons. We will construct this neutron over the proton we assembled. Place one of your "down" tennis balls over the proton we just made before you took your break. The astute will notice that we are now in a process of alternating "up" and "down" tennis balls (quarks) in a sequence. Continue to assemble the neutron by placing an "up" tennis ball over the "down" tennis ball and then finishing the operation by placing a "down" tennis ball over the last "up" tennis ball. We now have a neutron (two "downs" and one "up") over a proton (two "ups" and one "down") which is a nucleus consisting of one proton and one neutron - deuterium (heavy hydrogen). This is a product of the fusion process - we got there without so much noise as a hydrogen bomb explosion or having to go to the Sun to get one - it is too hot there. We want to build something closer to home - a helium nucleus.

A point of interest: The "down" quarks (tennis balls) weigh a little more than the "up" quarks (tennis balls) causing the neutron to weigh more than the proton. One "up" quark has a mass number of 611.19 and weighs 0.556×10^{-24} grams while the "down" quark has a mass number of 613.72 and weighs 0.558×10^{-24} grams. (You had already figured this out because you remembered that a neutron consists of a proton, electron and a neutrino - it had to weigh more than a single proton.)

To make a helium atom, we set aside the proton and neutron assembly we completed and produce another proton in the same manner in which we assembled the first proton. Now construct another neutron over the second proton and - VOILA! - we have made a helium nucleus when we combine our two assemblies. This helium nucleus also runs around the Universe under another name at times - an alpha particle. We will explain that later in this book. To complete our helium atom, we must have an equal number of electrons to match the number of protons (two) to produce a neutral electrical charge for the atom. All we have to do is to find a couple of electrons floating around and attach them outside our nucleus and we have built a helium atom. To place our atom in proper perspective, the sphere of the electron field is ten thousand times the diameter of our nucleus. Since our nucleus measures about six inches in diameter, the electron sphere (distance from the nucleus) would be 833 yards (2,500 feet). As we look at our assembly, we notice two things: the neutrons act as insulators between the protons; there is a lot of distance (empty space) between our six inch nucleus and the spinning sphere of electrons 833 yards away, or a diameter of 1,666 yards (5,000 feet - almost one mile). We notice that the space between the nucleus and the electron field of the atom is made of NOTHING!

This NOTHING is showing us two of the four fundamental forces of nature; the Strong Nuclear Force and the Electromagnetic Force. When we constructed our nucleus, we used glue (gluons) to hold our tennis balls together. This quanta of gluons is the strong nuclear force that holds all these parts of the nucleus together, called *Quantum Chromodynamics*. That is an expression that you can use on your friends - you can really forget this big word because "gluons" works nicely. A gluon is a quantum.

Now let us put two of these helium atoms we have assembled together. You notice they can only join with their outer electron spheres meeting each other. This joining together of atoms is due to the electromagnetic force. The electrons are the messengers of the electromagnetic force. This means that the closest one nucleus can be to the other is almost *one mile* and the limits of these two electron fields would be almost *two miles* wide. We have just built more of NOTHING! From this we deduce that the world is made of a lot of nothing, but a lot of force (energy), $E = MC^2$. This force (energy) does have mass (weight to us on Earth).

Since all matter is made of "nothing," the joining together of trillions of these atoms just to make a speck of dust, how is it that certain things weigh so much! We used tennis balls to construct our helium atom consisting of a nucleus six inches in diameter with an electron sphere with a diameter of one mile to illustrate a true large scale example that we can comprehend and realize the tremendous volume of this "nothingness." Another ratio that is accurate is with the Sun's diameter acting as a nucleus, the diameter of the orbit of Pluto, eight trillion miles, would be the electron shell - now *there* is a lot of "nothing." Let's quit playing around and get to the reality of these measurements. The diameter of a hydrogen nucleus is 2.406×10^{-13} centimeters (or you may write 0.0000000000002406 cm). To place the electron field in its proper relationship of being ten thousand times larger - 10,000 or 10^4 - we immediately know this to be 2.406×10^{-9} centimeters (0.000000002406 cm). See how much larger the electron shell is than the nucleus, or was it easier to visualize this using tennis balls? These are the dimensions of the lightest atom - hydrogen. The hydrogen atom has an atomic weight of 1.00797. This number tells us two things: It consists of only one proton and has one electron spinning in orbit about the nucleus; and the weight, the number used by scientists in determining many things, density being one. Examine the periodic table in the Appendix to observe the flow in the growth of the atomic weights of the elements (atoms). The heaviest natural atom, uranium, has an average diameter of 14.88×10^{-13} centimeters for its nucleus and a diameter of 14.88×10^{-9} centimeters for its outer electron field (shell). It has an atomic weight of 238.03 with a nucleus consisting of 92 protons and 146 neutrons. There are 92 electrons in several shells orbiting this nucleus. There is quite a difference between these two atoms - hydrogen and uranium in about everything except in *size* - the uranium atom is not really that much larger than the hydrogen atom.

Now back to what we were supposed to be discussing - where does this weight come from? A proton weighs 1836 times more than an electron which has been assigned the value of one (1). A neutron weighs slightly more at 1838. The density of nucleons is 2.3×10^{14} grams per cubic centimeter, which to you and me is the same as saying that a cubic inch of nucleons (this is nucleus material only - no electrons or the "nothingness") weighs 4.15 *billion tons per cubic inch*. Now we know where the weight is centered. This is no different than our solar system in which the Sun has 99.865 percent of the mass of the entire solar system. Just for fun let's see how many nuclei - that is what we call this nucleus filled with protons and neutrons - would fit into a ball (sphere) one centimeter in diameter. This ball would hold 71.3×10^{39} hydrogen nuclei but only 301×10^{36} uranium nuclei because the uranium nucleus is just a tiny bit larger than the hydrogen nucleus. Yep, that little 39 above the ten (10) means that 39 zeroes have to be added after the 71 if you want to see how many this number represents. In a one inch diameter sphere there would be 1.17×10^{42} hydrogen nuclei but only 4.94×10^{39} uranium nuclei. See what happens when we remove the electron spheres from our atom leaving only the nuclei. Oh yes, the one centimeter diameter sphere would weigh 132 million tons and the one inch diameter sphere would weigh 2.156 billion tons. Now we know where the mass (weight on Earth) originates. Don't you think it is time for another break.

The classical diagram of an atom has a lot of circles - a large circle for the electron field, a smaller circle for the nucleus, and inside this nucleus circle are more circles. The small circles inside the nucleus are marked (+) for protons and (neut or 0) for neutrons. Little circles are spotted on the large electron field circle with a (-) to signify that these are electrons. Easy to draw but hard to explain where everything really is. I looked at the bundle of neutrons sitting there, lazily collected together with no problems with each other while the protons were doing all they could to get away from each other. My model of the nucleus was going to be a little different and maybe present a more feasible and elementary assembly. Since all atoms have a minimum number of neutrons equaling the number of protons (except for the hydrogen nucleus of only one proton), the assembly of our nucleus can be made and dropped into a bucket with no problem. Any extra neutrons may be thrown in the bucket on top of our assemblies for good measure - this way only friendly neutrons are next to each other, the troublesome protons are encased in the neutral neutrons.

Think you are smart now, don't you? We are just getting started, so hold on. The proton is positive - how do we know this to be true other than by experiment which always proves this to be true. Why is the neutron always neutral? We physicists know because of a lot of mathematics and fancy equations, but here it is in a simplified explanation. Remember our "up" and "down" quarks. Our proton consists of two "up" quarks and one "down" quark. The neutron was assembled using two "down" quarks and one "up" quark. Now here comes the fancy footwork simplified: Assign the "up" quark a charge of +2/3 and the "down" quark a charge of -1/3. For you that have to know, the other quarks charges are: charmed (+2/3); strange (-1/3); top (+2/3) and bottom (-1/3). Now back to business. Let us add the values together and see what we come up with:

proton = [2 (+2/3) + 1(-1/3)] = [(4/3) - (1/3)] = (3/3) = 1 (positive);
neutron = [2 (-1/3) + 1 (+2/3)] = [- (2/3) + (2/3)] = 0 (neutral).

See how simple that was when all the mathematics and fancy equations are left out!

This raises a question on Beta Decay - a free neutron decaying into a proton, electron and a neutrino - what about all these quarks, can it happen? You betycha! The mass of the proton is 1836 and the neutron's mass is 1838 - what about that difference of two (2)? That is one good thing about having a problem - you now have something you have to work yourself out of. Let's give it a go. We now know that the "down" quarks weigh more than the "up" quarks. We covered this near the beginning of this subject and all the mathematics are shown in the Appendix. However, we will summarize the findings here and now. We know the mass of the neutron is two (2) greater than the proton. We will now call on the Weak Force - radioactivity. A free neutron (one that is not in a nucleus or closely associated with a proton) has a life of ten to fifteen minutes (twelve minutes is always a safe number to use) before the transformation (radioactivity) begins. One of the "down" quarks of the neutron (-1/3 charge) is converted to an "up" quark with its (+2/3) charge, making the ex-neutron of two "down" and one "up" now a proton of two "up" and one "down" quark system. "This is fine for you to try to tell me," you think, "but what happened to that mass of (2) that the neutron HAD." Simple, beta decay causes a release of an electron (mass 1) as the neutron is converted to the proton and the other mass of one (1) was used in the energy expended by the radioactivity process. What's your next question? The charge of the neutron, neutral, has not changed because the neutral neutron is now a proton (positive charge), an electron (negative charge) and a neutrino (no charge) which has been set free to roam the Universe. There has been a change (conversion) but without a change in charge because the proton (+1) and the electron (-1) cancel each other out for a neutral hydrogen atom. All that occurred to the atom that released the neutron was that it became an isotope of the original atom - no protons or electrons of this atom were involved. This is a continuous process in uranium and some other heavy atoms that have an abundance of extra, active neutrons.

ELECTRONS

Now that we are so proficient in the assembly of a nucleus, we will move a long way out from the nucleus to a place where it is just as busy - the electron field. The classical picture of the electron is one of circles drawn around a solid ball, the nucleus. Each of these circles has one or more little circles drawn on itself with little "minuses" (-), indicating electrons, drawn inside. It appears that a lot of little "minus" (denoting negative charge) balls are going around the big ball (nucleus) with (+)'s and (0)'s inside it. This description will suffice for the novice, but we know better.

Each electron, if we were able to weigh it, would almost be considered massless - the proton is 1836 times as massive. Now where is this electron located? Let's go back to our classical drawing of the small circle of the nucleus and the larger circle of the electron's path around it. These circles on the paper are really hollow spheres - balls. The sphere of the electron field is ten thousand times larger than the nucleus' sphere. We know from our discussion of the nucleus how far away this electron field, or shell, is located. The classical picture was derived from the model of our solar system, but Pluto is solid and the electron is not.

Our electron is an energy field shaped like a sphere containing this energy. The electron is in motion, and I mean fast. This energy is spinning in this field at a rotation speed of 10^{15} times a second. Yes, that is 1,000,000,000,000,000 times each second - this is in the visible light range. The closer the electron is to the nucleus, the faster it spins and the more energy it contains. This will be explained later when we get into quantum mechanics. This energy creates the appearance of a solid ball called a "shell." Electrons normally appear in pairs in atoms and are very close together - they might have a wreck some time if they ran into each other at these speeds. Science has given us an out by Wolfgang Pauli's Exclusion Principle which states that these two cannot occupy the same shell, each has to have its own orbit. It is resolved by the fact that the two electrons in the orbitals have opposite spins - one has spin "up" and the other has spin "down." This satisfies the requirements of the Pauli Exclusion Principle.

Electrons, in addition to their mass and electric charge, have this intrinsic quantity of angular momentum called "spin," almost as if they were tiny spinning balls. Associated with the spin is a magnetic field like that of a tiny bar magnet lined up with the spin axis. Scientists represent the spin with a vector. For a sphere spinning "west to east" the vector points *north*, or "up." The vector points *south*, or "down," for the companion which must have this opposite spin. In a magnetic field, electrons with spin "up" and spin "down" have different energies. In an ordinary electric circuit the spins are oriented at random and have no effect on the current flow. Utilizing the polarization of spin (magnetism) is leading to great advances in memory storage in the computer industry. Electron spins restricted to spin "up" and spin "down" will be used as bits in memory storage [bits are either a one (1) or a zero (0)]. The electrons provide their own energy, not requiring an outside source such as a battery or other forms of electricity to maintain their momentum. They will always retain their *spin* which is either a (1) or a (0).

These orbitals these electrons create are called SHELLS. Elements, each atom, contain from one to seven shells. These shells are numbered and lettered outward from the nucleus and divided into SUBSHELLS. The outermost shell of all atoms bears the special name VALENCE, and the electrons in that shell are called VALENCE ELECTRONS. These electron shells would appear as onion skins do on an onion. All electrical activity occurs in the valence shell. The geometry of its electron orbitals determines how a material behaves chemically, or structurally, and indicates whether or not it is an electrical conductor. Electrons account for the physical and chemical properties of the elements and their characteristic spectra. The electrons actually have almost no influence on the nucleus.

What do these electrons really look like? These days physicists are pretty sure it is not a particle at all. It is not exactly a cloudlike field of energy, either. Nor is it a wave, but it might be all three at once somehow. It appears to be what is required for it to perform its job at the time of need - a particle, a wave or a shell.

ALPHA PARTICLE - α

Alpha particles consist of two protons and two neutrons that have been ejected from a heavy unstable nucleus. They are equal to the nucleus of a helium atom and, upon ejection, easily capture two stray electrons and become helium atoms. Alpha particles have a penetration depth of 0.006 inch - a sheet of paper.

BETA PARTICLE - β

A fast moving electron is ejected when a neutron is converted into a proton in a nucleus due to beta decay. Carbon 14 - $_6C^{14}$ - six protons and eight neutrons becomes nitrogen - $_7N^{14}$ - seven protons and seven neutrons through this beta decay. The ejected electron enters the second shell neutralizing the atom resulting in a neutrally charged atom. Where we had six protons and six electrons in Carbon 14, we now have seven protons and seven electrons in Nitrogen 14. Beta particles have a penetration depth of 0.06 inch (1.5 millimeters) of aluminum.

BETA DECAY

An isolated neutron, one that has no proton companion, has a life of about fifteen minutes before decaying into a proton, electron and neutrino. The proton and electron stay joined as a new hydrogen atom.

GAMMA RAY - γ

High energy electromagnetic radiation emitted during nuclear reactions or decay. These frequencies originate in the nucleus and can penetrate two inches (five centimeters) of aluminum. Gamma rays are found in the frequency range of 3×10^{18} to 3×10^{24} Hz.

CHANGES:
Alpha particle: -4 mass; -2 atomic number; *consist of 2 protons and 2 neutrons.*
Beta decay: +1 atomic number; *neutron conversion to proton, electron, neutrino.*
Electron capture: -1 atomic number; *a proton and an electron combined to form a neutron* due to compression by gravity forming a neutron star - reverse beta decay.

FISSION

Nuclear fission involves the delicate balance within the nucleus between the attraction of nuclear forces and the repulsion of electrical forces. In nearly all nuclei the nuclear forces dominate. This is rarely the case in uranium. If the uranium nucleus is stretched into an elongated shape, the electrical force may push it into an even more elongated shape. If the elongation passes a critical point, nuclear forces give way to electrical ones, and the nucleus separates. This is "fission."

The absorption of a neutron by a uranium atom is apparently enough to cause such an elongation. The resultant fission process may produce any of several combinations of smaller nuclei.

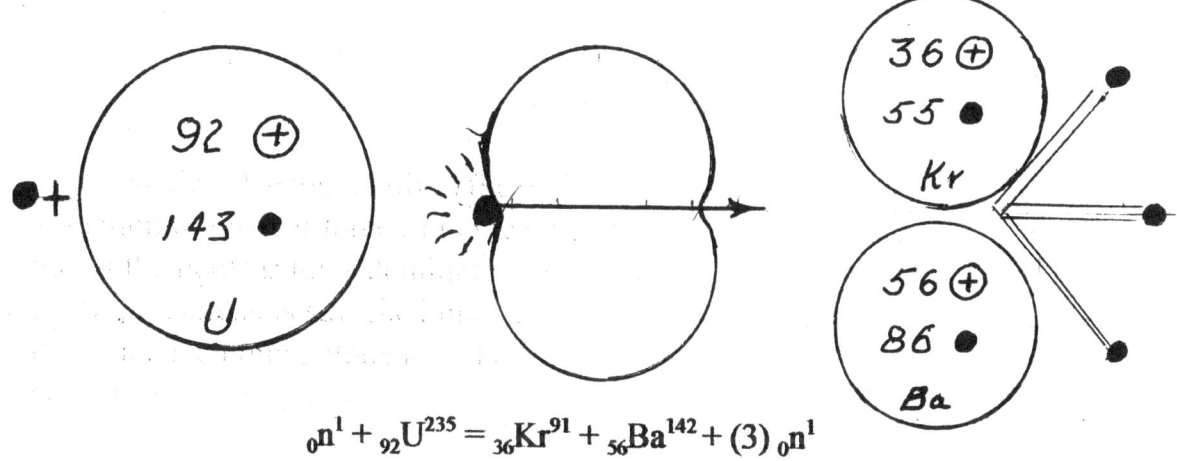

$$_0n^1 + {}_{92}U^{235} = {}_{36}Kr^{91} + {}_{56}Ba^{142} + (3)\ _0n^1$$

This reaction has an energy of two hundred million electron volts (200 MeV). The missing mass is converted into energy by $E = MC^2$. The release of the two or three neutrons can in turn cause the fissioning of two or three other atoms, releasing more energy and a total of four to nine more neutrons. This is the start of a *chain reaction* that can proceed at an ever accelerating rate. The minimum size of the material that will sustain a chain reaction is called the *critical size*; the mass being called the *critical mass*. If the mass of a fissionable material is greater than the critical mass, an explosion of enormous magnitude occurs.

An atomic bomb is constructed by having a minimum of two units, each smaller than the critical size and separated by a short distance. Because of the relatively large surface area of each unit, neutrons escape rather easily and a chain reaction cannot develop. When the pieces are suddenly driven together, the relative surface area is decreased. If the combination of these units exceed the critical mass, a violent explosion takes place. This is a nuclear bomb. The uranium used in the Hiroshima blast consisted of U235 about the size of a baseball.

The heat generated by the nuclear fission explosion is four to five times hotter than the interior of the Sun. This is the reason a fission bomb is used as the trigger for the fusion (hydrogen) bomb. This amount of heat is required from the fission process to initiate the fusion process for greater energy.

A controlled chain reaction of critical size may occur by breaking the uranium into smaller quantities and separating them by a material that slows down neutrons. This is the basis of a nuclear reactor - a nuclear fission power plant. The reactor generates heat and, by running water through pipes in the reactor, causes the water to be converted into steam, which drives a steam turbine. This turbine is connected to an electric generator that converts this mechanical energy into electrical energy. This energy is now sent to a transformer, to power lines and eventually to homes and factories. Energy has been converted from nuclear into steam power, into mechanical power which produced the electrical power.

FUSION

Fusion - the process of the Sun's existence - is merely the combining of four hydrogen atoms to form helium atoms through the proton - proton cycle. Thermonuclear reactions are analogous to ordinary chemical combustion. In both chemical and nuclear burning, high temperature starts the reaction, and the release of energy by the reaction maintains a high enough temperature to spread the fire. The net result of the chemical reaction is the combination of atoms into more tightly bound molecules. In nuclear burning, the high temperature starts a reaction or series of reactions with the net result of producing more tightly bound nuclei. The difference between chemical and nuclear burning is essentially one of scale, or magnitude.

The critical mass of a fissionable material limits the size of a fission (atomic) bomb; no such limit is imposed on a fusion (thermonuclear, or hydrogen) bomb. There is no theoretical limit to the size of a fusion bomb and this fusion bomb cannot be less energetic than its "trigger", an atomic bomb.

Controlling fusion for commercial use is undergoing extensive research. The high temperatures required are the basis of the present problem - how do you create and sustain these high temperatures and what do you use to contain them. It is as hard a question to answer as, "What do you use to contain a universal solvent?"

$$E = MC^2 \qquad \text{Joules} = kg(m/sec)^2 \qquad 1 \text{ kg} = 2.1875 \text{ pounds}$$

1 kg of matter is $1 \times (3 \times 10^8)^2 = 9 \times 10^{16}$ Joules

Energy produced in the fusion process:
    Conversion of four protons to form the helium nucleus:

Mass of four protons		$= 6.6943 \times 10^{-27}$ kg
Mass of helium	(less)	$= 6.6466 \times 10^{-27}$ kg
Difference in mass		$= 0.0477 \times 10^{-27}$ kg

$E = MC^2 = $ mass *in kg* times the speed of light *in meters per second*, squared
$E = 0.0477 \times 10^{-27} (3 \times 10^8)^2$
   $= 0.0477 \times 10^{-27} (9 \times 10^{16})$
   $= 0.4293 \times 10^{-11}$
   $= 4.2930 \times 10^{-12}$ Joules

This is the energy created by the forming of *each* helium atom. Multiply this by the number of helium atoms formed per kilogram to determine magnitude.

INTERESTING THINGS

When an apple falls from a tree, the *gravity* exerted by Earth has overcome the *electromagnetic force* of the atoms in the stem of the apple.

The *electron* was identified in 1897.
The *proton* was identified in 1919.
The *neutron* was identified in 1932.

Four forces:	Strong	Four *fundamental* forces:	Up quark
	Weak		Down quark
	Electromagnetic		Electron
	Gravitational		Neutrino

Electricity and a magnetic field; mass and gravity; North magnetic pole and South magnetic pole - none of these can exist individually - they require each other.

Mass and gravity - attractive.   Antiparticle and antigravity - repulsive?

We notice advancements in some field each day. There is almost no limit to the expectations we anticipate in any area we discuss. There is, however, one area of knowledge that would stop everything in its tracks and cause the world to revert immediately to the eighteenth century if we knew *nothing* about this subject. We would continue to progress because of what we know of water power, natural gas, crude oil and the steam engine. It would be an active, growing world progressing in directions we could not imagine today.

What is this one area of unknown knowledge that could cause the world to head in such a direction? It is something that every person on Earth uses hundreds of times a day. This has caused mankind to advance to the state we enjoy today. It has allowed us to go to the Moon and return. The list would appear to be endless! What would the state of Earth be if we knew nothing about ELECTRICITY! Make a list of everything that does not have any connection to electricity from the beginning of its existence until it is consumed. Remember, batteries and solar cells exist due to the knowledge of electricity.

INTRODUCTION TO LIGHT

Interest in the nature of light must go back to prehistory. It is undoubtedly as old as any problem in science and probably one of the most rewarding scientific problems ever tackled. Interwoven with the whole history of modern physics, from the seventeenth century to the present, is the study of light. This has been closely tied to the development of the theories of electromagnetism, relativity and quantum mechanics, as well as to the practical science of optics, to numerous discoveries in mathematics and, in the modern period, to such things as radar and lasers. From Olaus Romer's measurement of the speed of light, based on observations of the times of the appearance of the moons of Jupiter in 1675, to modern measurements of quasi-stellar red shifts and high energy photon reactions, the history of the study of light has closely paralleled the overall history of physics. In those three hundred years the frontiers of physics have diverged from the solar system upward to the Universe at large and downward to the world of elementary particles.

According to the brilliant prediction of James Clerk Maxwell in 1865, light is a propagating and oscillating combination of electric and magnetic fields - an electromagnetic wave. Although this insight was the greatest milestone in the history of light, it was by no means the end of the road. In 1905 Albert Einstein introduced the photon concept - the energy is not spread evenly along the wave, it appears in bundles - photons.

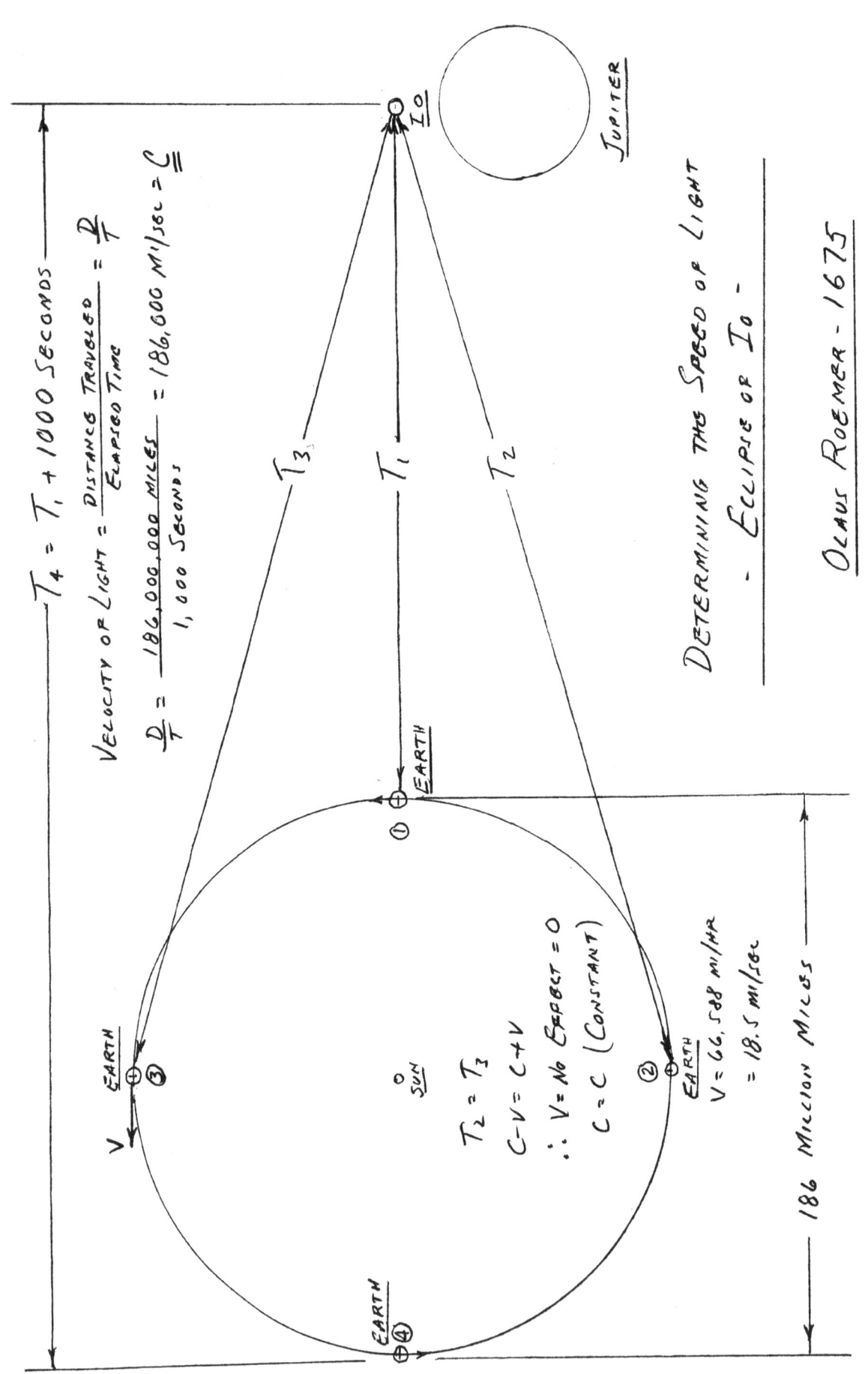

$T_4 = T_1 + 1000$ SECONDS

VELOCITY OF LIGHT = $\dfrac{\text{DISTANCE TRAVELED}}{\text{ELAPSED TIME}} = \dfrac{D}{T}$

$\dfrac{D}{T} = \dfrac{186,000,000 \text{ MILES}}{1,000 \text{ SECONDS}} = 186,000 \text{ MI/SEC} = C$

JUPITER

Io

T_3

T_1

T_2

EARTH

①

EARTH
③
V

SUN

$T_2 = T_3$

$C - V = C + V$

$\therefore V = $ NO EFFECT $= 0$

$C = C$ (CONSTANT)

EARTH
②
$V = 66,588 \text{ MI/HR}$
$= 18.5 \text{ MI/SEC}$

EARTH
④

DETERMINING THE SPEED OF LIGHT
- ECLIPSE OF Io -

OLAUS ROEMER - 1675

186 MILLION MILES

FREQUENCY

The definition of a wavelength is the distance, measured in the direction of the progression of a wave, from any given point to the next point in the same phase. Frequency is best explained as the number of wavelengths that pass a given point in a certain period of time. In physics we use cycles per second (cps) or the notation Hz, an abbreviation for the man who discovered that these types of waves existed - Heinrich Hertz. Cps or Hz each mean the number of *wavelengths* passing a given point in one second.

There is a formula that states that the speed of light is equal to a wavelength multiplied by its frequency ($c = \lambda f$). Since the speed of light is known, knowing one of the two unknowns of either wavelength or frequency, we can easily solve the equation to find the other. Using this formula we can determine that the wavelength being used for the AM radio program listed at 1000 kc (1000 kHz) frequency, or one million cycles per second, is about one thousand feet long. Watching channel 4 on television is made possible by twelve foot long wavelengths. Listening to a FM radio station broadcasting on the 100 mc (100 MHZ) frequency, or one hundred million cycles per second, would be received on a wavelength of about ten feet. Switching to TV channel 13, the wavelength would be five feet. The higher you progress in your TV channel selections, the length decreases to almost one foot. A microwave wavelength can be almost twelve inches and this is the reason small window openings are safely placed in the doors of the microwave ovens for our viewing convenience. As we have reduced the wavelength, the frequency must be increased proportionally to maintain the constant - the speed of light.

As we reduce the wavelengths further, we go to exponentials or use the Ångstrom unit (Å to denote the diameter of an atom - 10^{-8} cm). The infrared wavelength would equal 8,500 atoms lined up side by side, or 8,500 Å; the color red wavelength would 7,000 atoms side by side, or 7,000 Å; violet color's wavelength is 3,500 atoms, or 3,500 Å; ultraviolet waves have 100 atoms side by side, 100 Å; x-rays are 10 atoms sitting side by side, or 10 Å; and gamma rays are one ten-thousandths of an atom's diameter, 10^{-6} Å. We see that the shorter the wavelength results in higher frequencies with correspondingly higher energy levels. Low frequencies such as radio, television, microwave, infrared and visible light do not bother you. High frequencies of ultraviolet, x-rays and gamma rays can even kill you. This is the reason you must obey any warning on frequencies - they are there to protect you.

Electrons emit radiation with a frequency equal to their own frequency of oscillation. The outermost, or valence, electrons of atoms are described as being responsible for the emission of visible and ultraviolet due to their frequencies. Electrons closer to the nucleus, those in the inner shells, experience a higher frequency and are responsible for the higher frequency x-rays. Molecules have a more complex range of possible emission frequencies. Molecules contain the electron frequencies of the different atoms in its makeup plus the bulk motions and vibration of the molecule itself. Since the atoms composing a molecule are much more massive than electrons, they move slower and emit lower frequency radiation than electrons. Molecular rotation and vibration are important sources of infrared radiation. Some quantum transitions are so small in certain atoms and molecules with their energy changes that the emitted radiation is in the weaker microwave and radio region.

Within a nucleus, a proton, almost two thousand times more massive than an electron, experiences such a strong nuclear force that its oscillation frequency is much greater than the oscillation frequencies of its atomic electrons. The oscillation frequency of the proton is about one million times greater than its valence electrons. A typical nuclear gamma ray photon, the result of the oscillation of nuclear protons, has an energy of about one MeV (Million electron Volts), while a photon of visible light has an energy of about one eV (electron Volt), one million times less.

Molecules exhibit three methods of emitting photons:
1. An electron of one of the elements in the molecule moves from the excited state to the ground state.
2. The rotation of the elements in the molecule moves to a slower rate.
3. The vibrations of the elements in the molecule move to a slower rate.

A good emitter of radiation is a good absorber of the same radiation frequency. Molecules readily absorb infrared radiation, atoms absorb visible and ultraviolet light, with nuclei preferentially absorbing gamma rays. X-rays are emitted by an inner electron of a particular atom and readily absorbed by the same kind of atom. However, the very high frequency x-rays emitted by the heavier atoms are not easily absorbed by lighter atoms, for in the lighter atom, not even the innermost electron has a sufficiently high frequency to match the x-ray frequency. Because of this, x-rays easily penetrate organic matter composed of hydrogen, carbon, nitrogen and oxygen (atomic numbers 1, 6, 7 and 8) but penetrate calcium rich bone (atomic number 20) with difficulty, and heavy metals scarcely at all.

The properties of electromagnetic waves are polarization, superposition, reflection, refraction, diffraction and interference.

$$C = \lambda f \qquad \lambda = C/f \qquad f = C/\lambda$$

λ = Wavelength = Distance

f = Frequency = No. Wavelengths/Second

C = Speed of Light in a Vacuum

3×10^5 km/sec

3×10^8 m/sec

3×10^{10} cm/sec

Visible Percentage:

$$\frac{\text{Visible Range}}{\text{Total Range}} = \frac{0.4285 \times 10^{15} \text{ Hz}}{3 \times 10^{24} \text{ Hz}}$$

$$= 14.28 \text{ Billionth of a Percent}$$

EMITTANCE	f sec^{-1}	λ metric	λ English	E (eV)
VLF	3 kHz - 30 kHz	100 km - 10 km	62.1 mi - 6.21 mi	12×10^{-12} - 12×10^{-11}
LF	30 kHz - 300 kHz	10 km - 1 km	6.21 mi - 3280 ft	12×10^{-11} - 12×10^{-10}
MF	300 kHz - 3000 kHz	1000 m - 100 m	3280 ft - 328 ft	12×10^{-10} - 12×10^{-9}
AM	535 kHz - 1605 kHz	560 m - 193.1 m	1836.8 ft - 633.36 ft	2.14×10^{-9} - 6.42×10^{-9}
HF	3 MHz - 30 MHz	100 m - 10 m	328 ft - 32.8 ft	12×10^{-9} - 12×10^{-8}
VHF	30 MHz - 300 MHz	10 m - 1 m	32.8 ft - 3.28 ft	12×10^{-8} - 12×10^{-7}
TV 2,3,4	54 MHz - 72 MHz	5.55 m - 4.14 m	18.2 ft - 13.68 ft	2.16×10^{-7} - 2.88×10^{-7}
TV 5,6	70 MHz - 88 MHz	3.95 m - 3.4 m	12.96 ft - 11.18 ft	2.8×10^{-7} - 3.52×10^{-7}
FM	88 MHz - 216 MHz	3.4 m - 1.38 m	11.18 ft - 4.53 ft	3.52×10^{-7} - 8.64×10^{-7}
TV 7-13	174 MHz - 216 MHz	1.73 m - 1.38 m	5.58 ft - 4.53 ft	6.96×10^{-7} - 8.64×10^{-7}
UHF	300 MHz - 3,000 MHz	100 cm - 10 cm	3.28 ft - 3.937 in	12×10^{-7} - 12×10^{-6}
TV 14-83	470 MHz - 890 MHz	63.98 cm - 33.5 cm	2.1 ft - 1.1 ft	1.88×10^{-6} - 3.56×10^{-6}
SHF	3,000 MHz - 30,000 MHz	10 cm - 1 cm	3.937 in - 0.3937 in	12×10^{-6} - 12×10^{-5}
EHF	30,000 MHz - 300,000 MHz	1 cm - 0.1 cm	0.3937 in - 0.03937 in	12×10^{-5} - 12×10^{-4}
Infrared	3×10^{11} Hz - 0.4285×10^{15} Hz	0.1 cm - 7×10^{-5} cm	10^7 Å - 7000 Å	12×10^{-4} - 1.714
Visible	0.4285×10^{15} Hz - 0.857×10^{15} Hz	7×10^{-5} cm - 3.5×10^{-5} cm	7000 Å - 3500 Å	1.714 - 3.428
UltraViolet	0.857×10^{15} Hz - 3×10^{16} Hz	3.5×10^{-5} cm - 10^{-6} cm	3500 Å - 100 Å	3.428 - 120
X-Ray	3×10^{16} Hz - 3×10^{18} Hz	10^{-6} cm - 10^{-8} cm	100 Å - 1 Å	120 - 12×10^3
Gamma Ray	3×10^{18} Hz - 3×10^{24} Hz	10^{-8} cm - 10^{-14} cm	1 Å - 10^{-6} Å	12×10^3 - 12×10^9

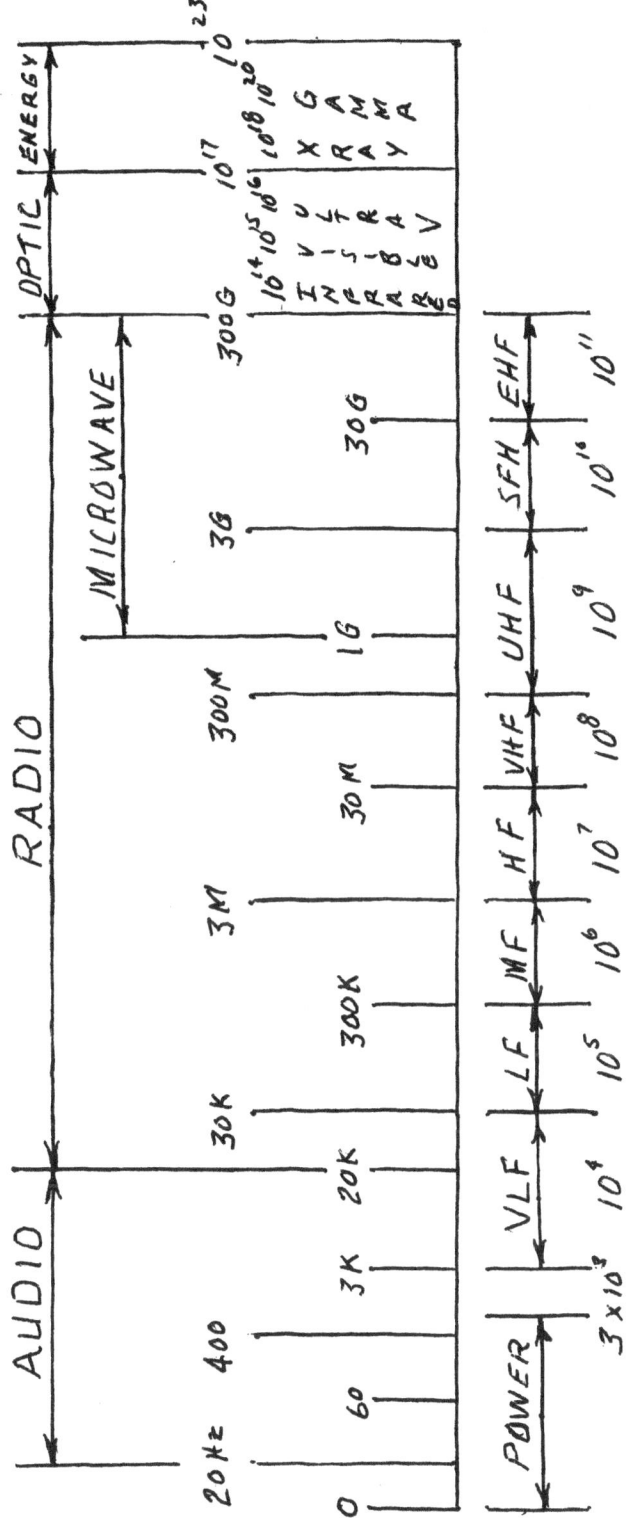

Frequency Spectrum

LIGHT

The speed of light in a vacuum is 299,792.458 kilometers per second, or 186,282.399 miles per second. One meter is the distance light travels in the short span of 1/299,792,458ths of a second.

One of the characteristics of light is that it is a wave. As such, it is related to things like surf coming into a beach or a vibrating guitar string. All waves share one characteristic: The motion of a wave is not the same as the motion of the medium (water, air, vacuum) in which it travels. When the surf comes into a beach the waves come in straight, but the water on which the wave moves goes up and down. As the wave goes by, the motion one experiences is being lifted up and let down; you are not moved toward the beach with the wave.

This distinction between the wave and the medium on which it moves is important, for it tells us that the wave is *different* and *distinct* from the medium. A light wave is the ultimate realization of that distinction, for it is a wave that requires no medium at all - light travels through the vacuum of space from the stars to Earth.

The internal structure of a light (electromagnetic radiation) wave consists of electric and magnetic fields that are perpendicular to each other and to the direction in which the wave moves. These fields are a description of the forces that the light wave exerts on objects it passes, just as the wave exerts on an object in the ocean. These electric and magnetic fields are *always* together - one cannot exist without the other. When an electromagnetic wave encounters an object with an electrical charge, the outer electron of an atom, it will cause the electron to move up and down in the energy levels of its shell.

How is light created? To understand the origin of light, we are going to have to understand another nature of the electrons of the atom. Using our classical view of an atom of the nucleus centered, with several spheres of electrons surrounding it, we can imagine these as steps for us to walk from the nucleus outward to the valence electron shell. Now that you have pictured this, forget it and transform these steps as being available in *each* electron's shell. Now that you are thoroughly confused, I think it is time for you to take another break.

These steps we now visualize in each electron's shell are *available*. Let's pretend we can place ourselves in a shell of an electron and walk up these steps or walk down these steps as we desire. You can move up or down these steps, but can never stand between the steps - you must be on one step or the other, there is no middle ground - the same requirements an electron must observe. You must *expend* energy to move up the stairs but you *gain* energy as you descend these stairs. In the same way, energy must be added to an electron field (shell) for the electron to "climb the stairs" and move to a higher stair, a *higher energy level* within its orbital. An electron that has been forced to this higher energy level will quickly "roll down the stairs" to a lower level releasing energy. These "climbing the steps" are *energy levels* that an electron is forced into when a photon of the same frequency encounters this electron. It adds energy to the electron, forcing it to a higher energy level. The photon that caused this increase in energy to the electron is used up during this process and disappears forever. Now, with nothing keeping the electron forced into this higher energy level, the electron immediately falls back to its original energy level (it "rolls down the stairs"), its *ground state*, and in doing so, causes the release of energy, a *photon* of the same frequency of the one that caused it to move up the stairs in the first place. This photon is now free to move at the speed of light until it encounters another electron, causing a repeat of this process. Energy must be added to cause the electron to move to a higher energy level by a photon slamming into it and causing this action. The return to its original level, its ground state, releases a photon which releases energy. The electron is now stable, or in its ground state again.

Quantum mechanics tells us that when an electron in an atom changes its energy level from one step to another, it does so in a *quantum leap*. It actually disappears from one orbit and appears in the other without traversing the space in between. These quantum leaps occur in both "climbing the stairs" and "rolling down the stairs." You already know that it takes a photon of an equal frequency to move the electron to a higher energy level, and, when the electron returns to its ground state, it releases a photon of its frequency (equal to the frequency of the photon that caused all this trouble in the first place), so we won't mention it again.

The speed of light varies depending on the density of the medium through which it passes - vacuum, air, glass or water. The speed of light in an absolute vacuum, denoted as (c), is the maximum attainable. This is 186,282.399 miles per second, almost three hundred thousand kilometers per second. This constant, (c), has been referred to as "Einstein's constant." The reason the speed of light varies with the density of the medium it is passing through is due to the density for atom to atom transfer of the photons (how close are the atoms to each other). The speed of light through water is only 73 percent of (c), 143,437 miles per second. Light can be slowed to a crawl depending on the medium encountered. In laboratory experiments, using three lasers and a microscopic cloud of sodium atoms, the photon can be captured for an instant and then released to resume its original speed. Therefore, in this laboratory, you would be able to walk faster than the speed of THIS light. We can now state that it is possible for many objects to move faster than the speed of light, depending on the situation. Nothing, not even information, can exceed (c), the speed of light in a vacuum.

The most striking property of light is that it is divided into the various colors we see in rainbows. All light has the ability to be divided into these colors, but making them show their *colors* requires a little help from a prism or water droplets. The light we see is not coming to us in a single wavelength - it is composed of wavelengths of (using our atom count for wavelength) between 7000 atoms all the way down to 3500 atoms in length. The color red has a wavelength of 7000 atoms, or 7000 Å, through orange, yellow, green, blue and finally to violet with the wavelength of 3500 Å. It is obvious that length of the violet wavelength is only one-half of the length of the red wavelength. Returning to our formula of the *speed of light is equal to the wavelength multiplied by frequency* we can see a doubling of the frequency of the violet wavelength in relation to the red is required to maintain the speed of light. This doubling of frequency is also a doubling of energy inherent in this wavelength - as the wavelengths get shorter, the energy of each rises in proportion. For this reason light cannot perform the services that the x-rays carry out very easily - light wavelengths are just too long!

Light exists between the lower energy level infrared and the high energy ultraviolet range. Infrared is mostly associated with radiation that is emitted from items that are warmer than the surrounding area - night vision goggles are one example of utilizing this frequency to our advantage. The ultraviolet energy range, just above the color violet, is very harmful. You cannot see either the infrared or ultraviolet frequencies - infrared frequencies are too low and ultraviolet frequencies are too high. These higher frequencies - those higher than the color violet - are what are harmful to living creatures. We know the damage ultraviolet causes to us when we sunburn or get cancer - x-rays and gamma rays are worse because they can kill you faster. We are thankful for the ozone layer because without it, no life could exist on the surface of the Earth because ultraviolet frequencies in these unimpeded doses does not allow life to exist in their presence at these levels. The ozone layer permits only enough ultraviolet frequencies through which allows life to exist on the surface of Earth - but you must be careful about excessive Sun exposure.

To expand on the formation of photons: Electron shells contain electrons in pairs that have opposite spin to conform to the Pauli Exclusion Principle. When a photon encounters the outer valence electron, it imparts its energy to this electron. This transfer of energy causes the spin of the electron to reverse. The normal form of opposite spin of these two electrons explains the closeness of their orbits - opposites attract. When the spin of the outer electron is reversed, it has absorbed the photon's energy. It is now spinning in the same direction as its partner and is pushed away, pushed "up" - like actions, or charges, repel. When the electron releases a photon, returning to its ground state, its spin is reversed to its normal condition. It can then return to its closer position with its partner - the ground state.

How does light come through the picture window in our living room to make it bright on a sunny day? We know that the Sun emitted a photon (due to many thermonuclear reactions continuously occurring) that took eight minutes and twenty seconds to traverse the ninety-three million miles to our window. This carrier of light, the photon, consist of all the colors within its frequency range. This photon slams into our plate glass window at the speed of light but does not pass through the glass! As a matter of fact, it doesn't even get past the first atom in the glass. It hits the outer valence electron, adding energy to it. This kicks the electron "up the stairs" for an instant, and then the electron returns "down the step" and the energy the electron gave up in returning "down the step" resulted in a release of energy - a photon. Now we have a photon loose inside the glass of the picture window and since the atoms next to it are the closest, this photon jumps in the nearest atom's outer valence electron kicking it "up the stairs" for its process of coming "down the stairs" and release of its photon. This process continues until the photon succession has no more atoms of glass to excite and suddenly the last photon emitted finds itself inside the room with the same properties that the original photon possessed, except for the time it took for the billions of photons to act in penetrating this dense material for the last one to appear in a less dense medium - atmosphere. Here it can speed up again because the atoms are further apart and fewer atom to atom transfer of photons is required. From this you can deduce the reason for the speed of light in a vacuum being the ultimate - no atom to atom transfer of photons. That little sucker doesn't waste time with any valence electron encounters.

Let us assume you have a stained glass border around your picture window. The colors of red, blue and green appear in the sunlight. In the absence of all light, these cut pieces of glass would be dark - not colored. Light (frequencies in the visible range) is required to make them appear colored. We will assume our picture window is clear which allows the photon exiting the glass in the living room to duplicate the photon from the Sun hitting the glass. The glass that appears blue with the presence of light occurs in this manner. From an atomic view, electrons in the pigment molecules are set into oscillation ("up the stairs") by the frequencies of blue light (the higher part of the energy side of the clear light), and they reradiate this energy from molecule to molecule in the glass. The other frequencies (red, orange, yellow and green) are absorbed by the molecules as a whole and are not reradiated. The kinetic energy of the molecules is increased by the absorbed frequencies and the glass is warmed - transform of frequencies as radiation energy. Therefore, only the blue light frequency photon is reradiated atom to atom through the glass whose color we identify as blue - this is the only frequency accepted here. The glass that appears red (the lower part of the energy side of the clear light) absorbs the frequencies of all the other colors while only transmitting the low frequency of red. The red glass will become warmer than the blue glass in its absorption because it is retaining the higher energy frequency of the blue and emitting the lower energy red frequency.

When you view these various colors on your wall in the living room you are still not seeing the light that entered your living room through the glass. Those photons hit the outer valence electrons of the air and went atom to atom until they hit the painted wall and caused - "up the stairs" and "down the stairs" on the molecules of paint covering the wall which were transmitted back through the atmosphere of the room to your eyes - boy, what a trip these photons had to make just for me to see the wall. Wait, that is not all of it - this photon hits your retina's outer valence electron (here we go again) - then a neuron (little computers) which travels to your brain electrically at 220 miles per hour so that you recognize the color and say, "Isn't that pretty." The darker the color of the paint, the warmer the room will become due to the absorption of most of the frequencies of visible light. The lighter the color, reradiation occurs with minimum absorption of these energies, the result being a cooler room.

COLOR	WAVELENGTH (METERS)	WAVELENGTH (ÅNGSTROM)	FREQUENCY (HZ)
Violet	$350\text{-}424 \times 10^{-9}$	3500-4240 Å	$0.85714\text{-}0.70755 \times 10^{15}$
Blue	$424\text{-}491 \times 10^{-9}$	4240-4912 Å	$0.70755\text{-}0.61075 \times 10^{15}$
Green	$491\text{-}556 \times 10^{-9}$	4912-5560 Å	$0.61075\text{-}0.53975 \times 10^{15}$
Yellow	$556\text{-}585 \times 10^{-9}$	5560-5850 Å	$0.53975\text{-}0.51282 \times 10^{15}$
Orange	$585\text{-}647 \times 10^{-9}$	5850-6470 Å	$0.51282\text{-}0.46368 \times 10^{15}$
Red	$647\text{-}700 \times 10^{-9}$	6470-7000 Å	$0.46368\text{-}0.42857 \times 10^{15}$

Maximum visibility is realized in the yellow 5560 - 5850 Å range.

256

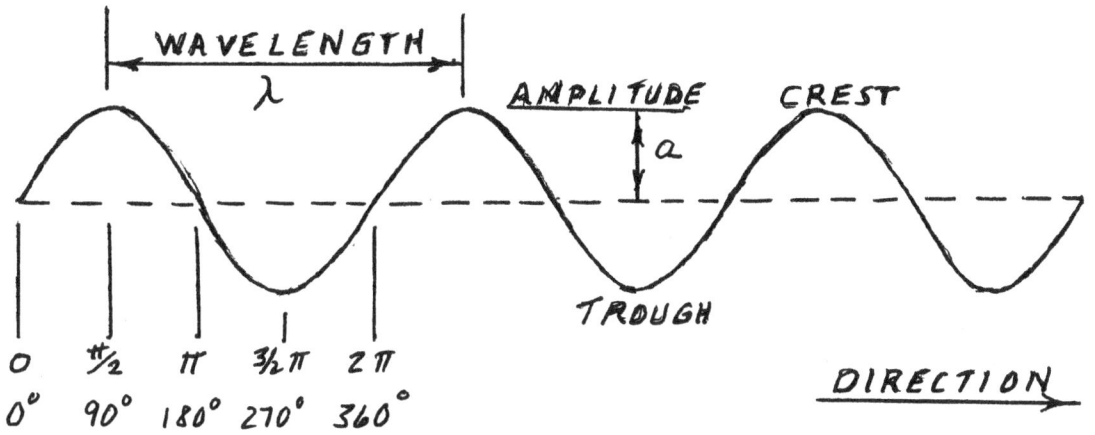

A WAVE IS A WAY IN WHICH ENERGY CAN BE
TRANSFERRED FROM PLACE TO PLACE WITHOUT
PHYSICAL MOVEMENT OF MATERIAL FROM ONE
LOCATION TO ANOTHER. ENERGY IS CARRIED
BY A DISTURBANCE OF SOME SORT.

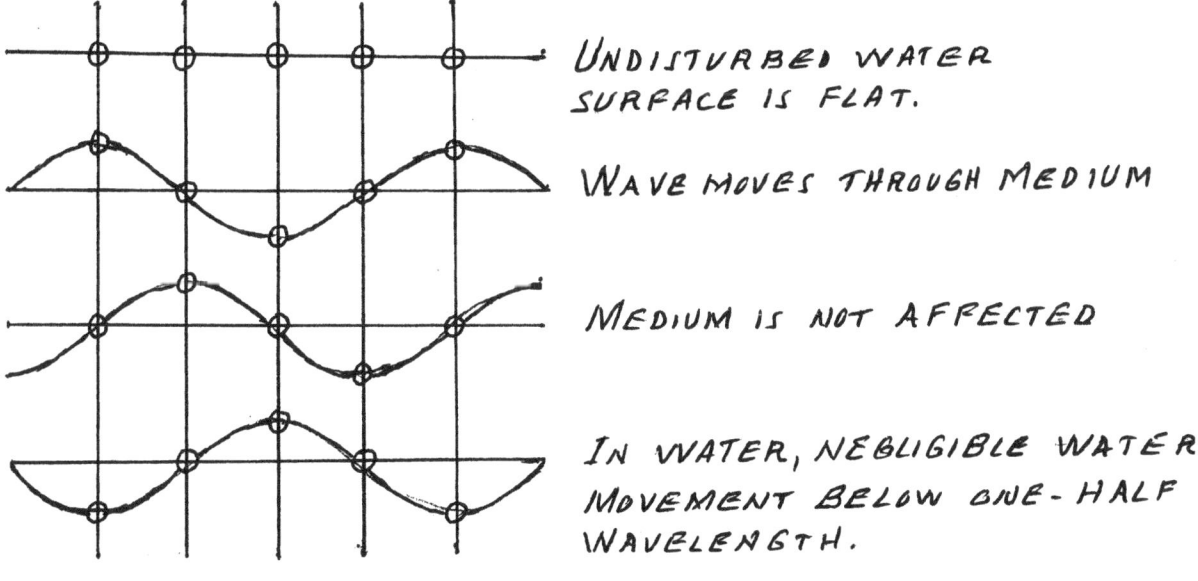

UNDISTURBED WATER
SURFACE IS FLAT.

WAVE MOVES THROUGH MEDIUM

MEDIUM IS NOT AFFECTED

IN WATER, NEGLIGIBLE WATER
MOVEMENT BELOW ONE-HALF
WAVELENGTH.

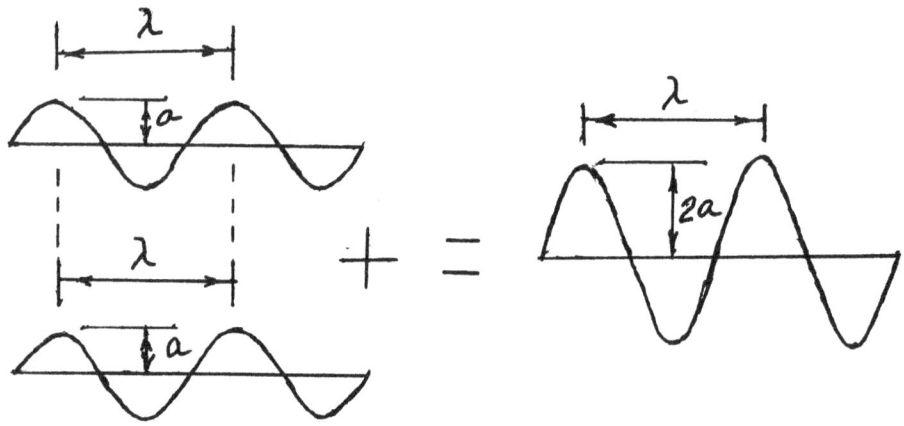

CRESTS AND TROUGHS REINFORCE.
THEY ARE IN PHASE.

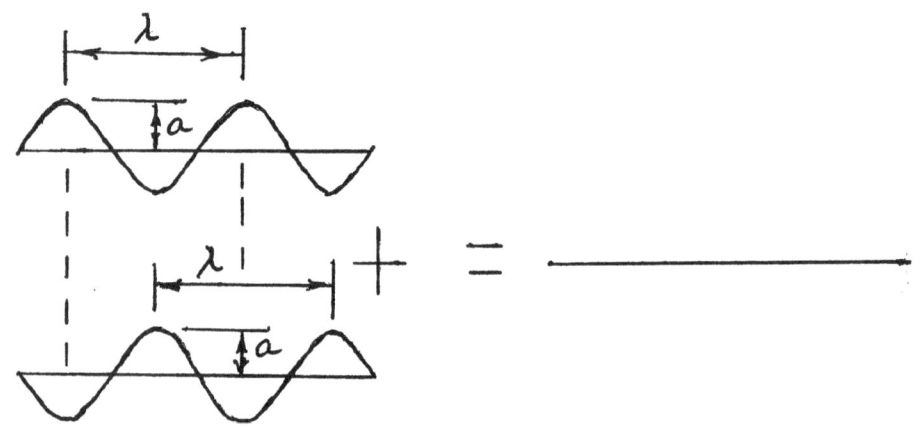

CRESTS AND TROUGHS CANCEL.
THEY ARE 180° OUT OF PHASE.

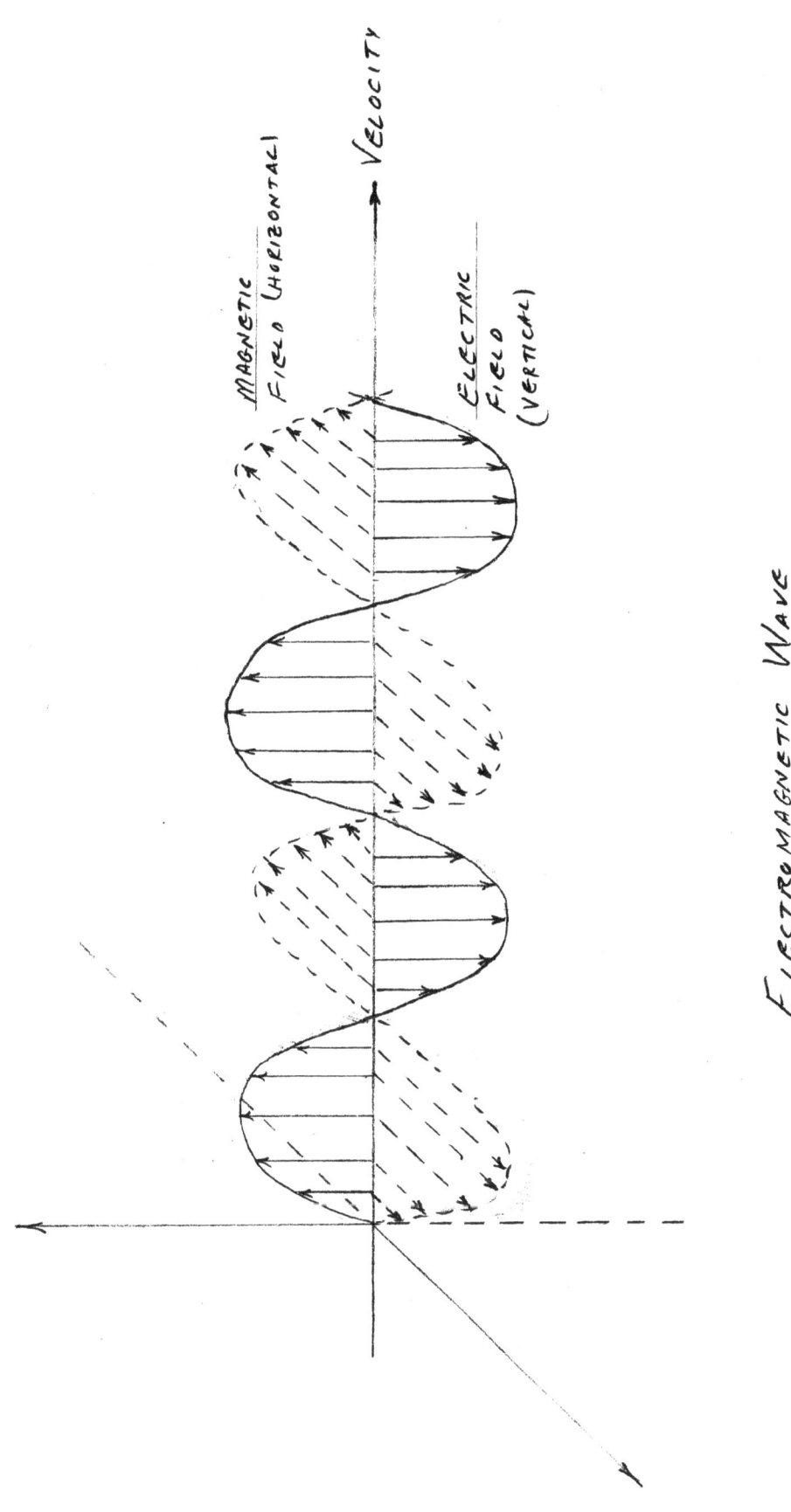

ELECTROMAGNETIC WAVE

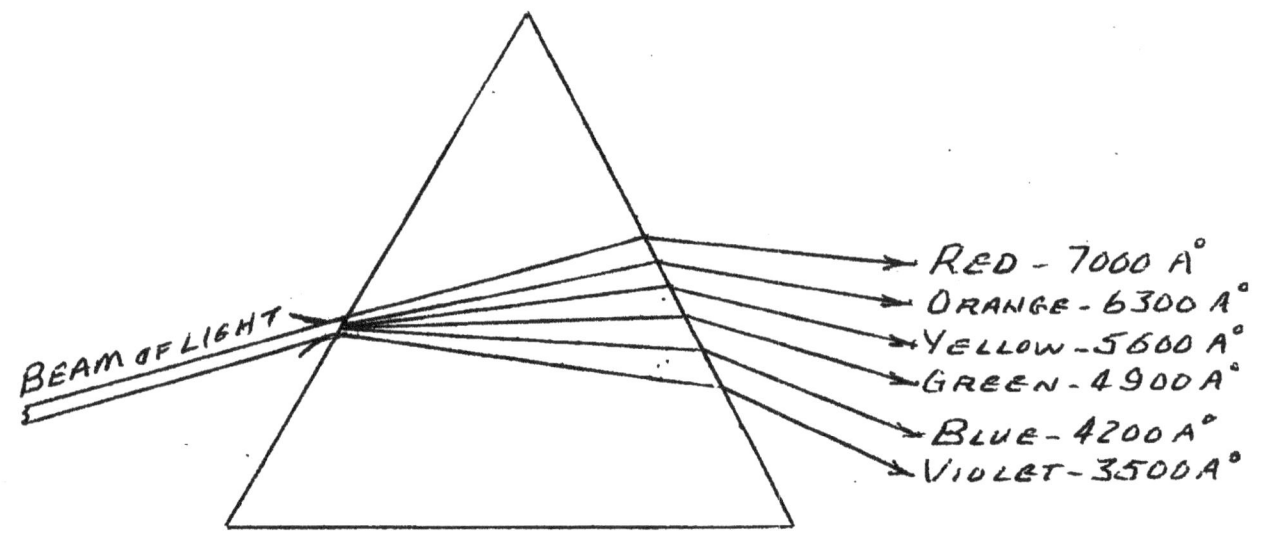

RED - 7000 Å°
ORANGE - 6300 Å°
YELLOW - 5600 Å°
GREEN - 4900 Å°
BLUE - 4200 Å°
VIOLET - 3500 Å°

BEAM OF LIGHT

GLASS PRISM

$$1 \text{ Å}° = 10^{-8} \text{ cm} = 10^{-10} \text{ m}$$

RED $\lambda = 7 \times 10^{-5}$ cm $f = 0.429 \times 10^{15}$ Hz

VIOLET $\lambda = 3.5 \times 10^{-5}$ cm $f = 0.857 \times 10^{15}$ Hz

SUN EMITS MOST OF ITS ENERGY AROUND 5500 Å°

REASON FOR VISIBLE LIGHT:

PERIOD OF ELECTRON'S ORBIT AROUND NUCLEUS - 10^{-15} sec
FREQUENCY OF VISIBLE LIGHT $\approx 10^{15}$ CYCLES PER SECOND
(ELECTRON ORBITS 1,000,000,000,000,000 TIMES EACH SECOND)

MIRROR

When you look at something in a mirror, it is a reradiating process of light (electromagnetic radiation) in the frequencies of 7000 - 3500 Å. These frequencies come in, excite the outer electrons which reemit these frequencies because no absorption of any magnitude occurs. The angle of incidence is equal to the angle of reflection (reradiation). This occurs only on reflective surfaces. A virtual object is the result in a reflective surface and appears equidistant in a flat mirror from the object itself. Convex or concave mirrors distort in the vertical and horizontal directions for special effects.

INCANDESCENCE

Incandescence is light produced as a result of a high temperature. Electrons undergo transitions between energy levels quite unaffected by the presence of neighboring atoms. When the atoms are closely packed, as in a solid tungsten filament, electrons of the outer orbits make transitions not only within the energy levels of the "parent atoms" but also between the levels of neighboring atoms. These energy level transitions are no longer well defined but are altered by interactions between neighboring atoms, resulting in an infinite variety of transitions (an infinite number of visible radiation frequencies).

As the solid is heated further, wider energy level transitions take place and higher frequencies of radiation are emitted. The most predominant frequency of emitted radiation - the peak frequency - is directly proportional to the absorbed temperature of the emitter: $f \sim T$. The electromagnetic waves of violet light have twice the frequency of red light waves. A violet-hot star therefore has twice the temperature of the red-hot star. The time rate at which an object radiates energy is proportional to the fourth power of its Kelvin temperature, $f_r \sim T_K^4$. This doubling of temperature corresponds to a doubling of the frequency of radiation but a sixteen fold increase in the rate of radiation - sixteen times as many photons per second would be emitted and be sixteen times brighter. The temperature of incandescent bodies, including stars, can be determined by measuring the peak frequency (color) or radiation they emit.

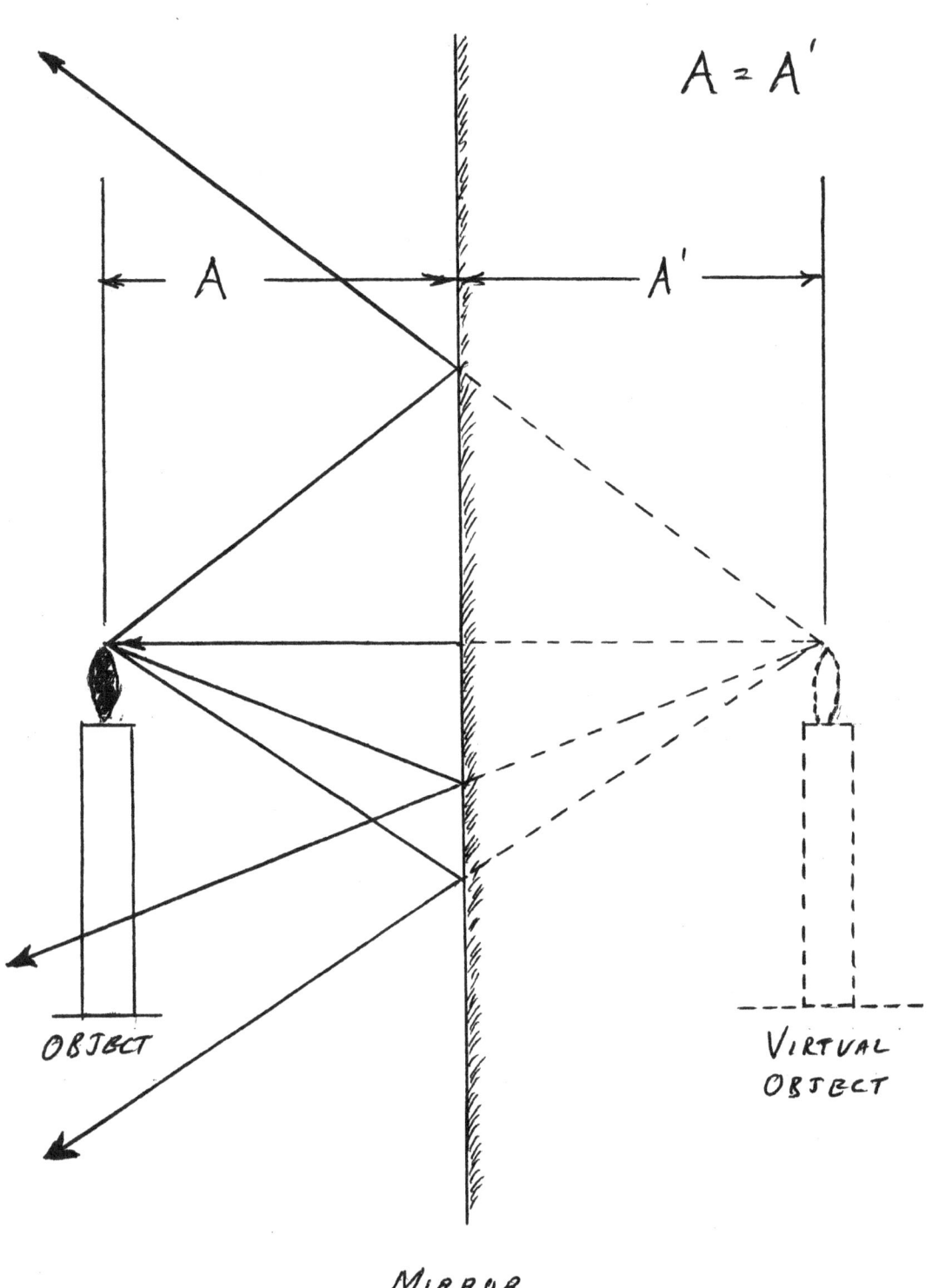

$A = A'$

A

A'

OBJECT

VIRTUAL
OBJECT

MIRROR

LIGHT - FADING

You have heard it all your life - IT'S FADED! Your grandmother, your mother, your friends and even you have complained of how the curtains around your windows fade as time passes. The bright days are so nice to enjoy with all the sunshine. The reason the days are so bright is due to the sunlight - lots of photons. Summer days are longer, brighter and warmer this time of year in the Northern Hemisphere due to the tilt of Earth's axis.

Curtains are, by design, located where much prolonged sunlight passes throughout the years. Curtains, and in most cases the couches sitting in front of the window, are colored. We know that when light strikes an object, it is not reflected, it is *absorbed* and *reradiated*. The photons of light entering room through the clear window contain the frequencies of visible light, including some infrared and ultraviolet frequencies. Since most material colors are dark, the photons of light hitting the outer valence electrons of the first atoms they encounter on the curtain or couch are absorbed with only the corresponding frequency of color being reradiated. This continual absorption of all the other frequencies while only reradiating a single frequency makes everything warmer - the electrons and molecules are more active.

This continual bombardment over time causes the loss of an outer (valence) electron or so. The atom is ionized (loses electrons) by this action making the atom's charge positive because there are now more protons than electrons. Atoms are basically neutral - the same quantity of protons (positive charge) and electrons (negative charge). Since each electron has its own set of energy levels in a neutral atom, any change (ionization) will rearrange some of these energy levels and the absorption levels are changed, reducing them which allows a reradiation of more frequencies. The more frequencies that are reradiated reduces the intensity of the original color - it appears to fade. Over a prolonged period, the outer electrons of the atoms that are subjected to this photon bombardment are shifted enough in their energy levels so that all absorbed photon frequencies are reradiated - it has faded to almost white. The creases are still the same color as the original material because they have been folded together allowing no photon activity to take place.

It is very easy to eliminate fading - NO LIGHT! This would not be practical at all because the colors would theoretically last forever - there would be no photons to cause any change. Impractical because colored items in a darkened room could never be enjoyed by anyone. Color can only be seen in the presence of light, otherwise there is no color. Photons must be *active*. This is the reason the cave paintings of thousands of years ago and objects of the Egyptian pyramids remain bright - they have not been subject to photon bombardment - no light has entered.

So , to reduce fading, place objects so they will not receive direct sunlight and it is obvious that white, or light colored, objects would survive longer because these reradiate almost all the frequencies. Then there is another problem associated with long exposure to continuous sunlight - fabric disintegration. Ask your chemist friend to help you figure this one out.

RAINBOWS

It requires both the Sun and rain to make a rainbow. The Sun must not be too high in the sky and there must be some rain falling past you and the Sun. The rainbow is simply the addition of millions of drops of rain that each act like a prism, breaking the light into different colors. Raindrops are tear-shaped, causing millions of prisms to fall past you and the Sun. If the Sun is too high, you see the reflection of the rain. If the Sun is at an angle of 40-42 degrees above the horizon, you see the rainbow due to the angle of the Sun causing the raindrops to become prisms.

Since the Sun is round, the rainbow is round. It is like looking at an open end of a giant ice cream cone as the sunlight passes through the big end of the cone. When you are standing on the ground, you cannot see the bottom half of the cone - you only see the upper half of the cone. From an airplane you can see the complete circle of the end of the cone. The angle of observation is almost the same as when you view it from the ground except in the airplane you do not have the ground interfering with the bottom half of the circle.

So, there may be a pot of gold at the end of the rainbow, but we will never know because one cannot reach the end of the rainbow to find out - you are always between the Sun and the rainbow. As you move toward the rainbow, it continues to move away from you at the same pace due to the angle of the Sun until you run out of rain - then the rainbow disappears because there are no more prisms to create the separation of the frequencies of the sunlight into the spectrum you were witnessing.

WHY DOESN'T AIR HAVE COLOR?

This is a subject that I am sure has not crossed any normal person's mind. Leave it to the physicist to ask this one. Everything has a cause and effect, an action and reaction, a reason.

The reason air appears transparent (clear) is determined by the elements that make it up - nitrogen, oxygen, hydrogen, etc. Lucky for us, the frequencies the photons make in these elements are in the visible range - the emission of all the colors of the rainbow. Oxygen atoms within the air are always bound into pairs - O_2. That paired structure has levels with a good-sized gap around the energies of the visible colors, resulting in it being transparent. If there were many level spacings within the ordinary color range, the matching light would be absorbed, and air would have color. If sulfur were in abundance in our air, our air would appear yellow. Every atom with its corresponding electron shells is subject to the effect of quantum mechanics, the photon movement caused by elevating the electron from the ground state to an excited state and its return.

WHY IS THE SKY BLUE?

Is the sky blue because it reflects the color of the ocean, or is the ocean blue because it reflects the color of the surrounding sky? The answer is the second due to the process in which light is scattered by air molecules. Scattering means the process by which radiation is absorbed and then reradiated by the material it has encountered - the atom to atom transfer of photons.

As sunlight passes through our atmosphere, it is scattered by the many gas molecules in the air. Blue light is more scattered than red light because the wavelength of blue light (3500 Å) is closer to the size of air molecules than is the wavelength of red light (7000 Å). When the Sun is at a reasonably high elevation, the blue component of the incoming sunlight will scatter much more than any other color. Some blue light is removed from the line of sight between us and the Sun, and scatters many times in the atmosphere before entering our eyes. Red or yellow light is scattered relatively little, and arrives at our eyes predominately along the line of sight to the Sun. The net effect of this is that the Sun is "reddened" slightly, due to the absence of blue light, while the sky away from the Sun appears blue. In outer space, where there is no atmosphere, there can be no scattering of sunlight causing space to be black.

When the sunlight hits the atoms of air in the high atmosphere, some blue light is scattered out to color the sky and the remaining light, minus the blue component, contains only the lower frequencies giving the resultant yellow appearance. This is the reason the Sun appears yellow to us. It also appears almost red at sunrise and sunset because the light is passing through a much greater depth of atmosphere. This causes the scattering of all the higher frequencies from blue all the way down to the orange, leaving only the lowest frequency, red, to be seen. If the atmosphere is too dense, we don't even get to see the red sphere of the Sun - only a scattering of light - a smoggy or cloudy day.

The apparent size of the Sun is increased from the noontime size due to the vastly increased distance the sunlight travels through the atmosphere at sunrise and sunset. This magnification effect is also apparent on the Moon as it rises and sets. This is due to the refraction of light - change in the direction of light as it passes from one medium to another - from the emptiness of space through the various densities of our atmosphere in the atom to atom transfer of photons.

We do not see blue sky on a cloudy day due to the clouds. To reach the ground on Earth the photons from the Sun enter the ozone layer and are transferred atom to atom through the atmosphere normally until they encounter the molecules of the clouds. These molecules transfer the frequencies associated with their chemical makeup and reradiate through the cloud to us on the ground - very slowly compared to the transfer speed in the upper atmosphere. These frequencies are passed to us identifying the densities of the clouds - fluffy white or gray menacing dark thunderheads. It will become a dark, gray day.

Smog - smoky-fog - is seen due to the impurities air contains from the accumulation of automobile exhaust. These elements emit only part of the visible spectrum resulting in this dark colored haze. If life could have evolved on Venus with its atmosphere - how would they have been able to see anything?

REFRACTION

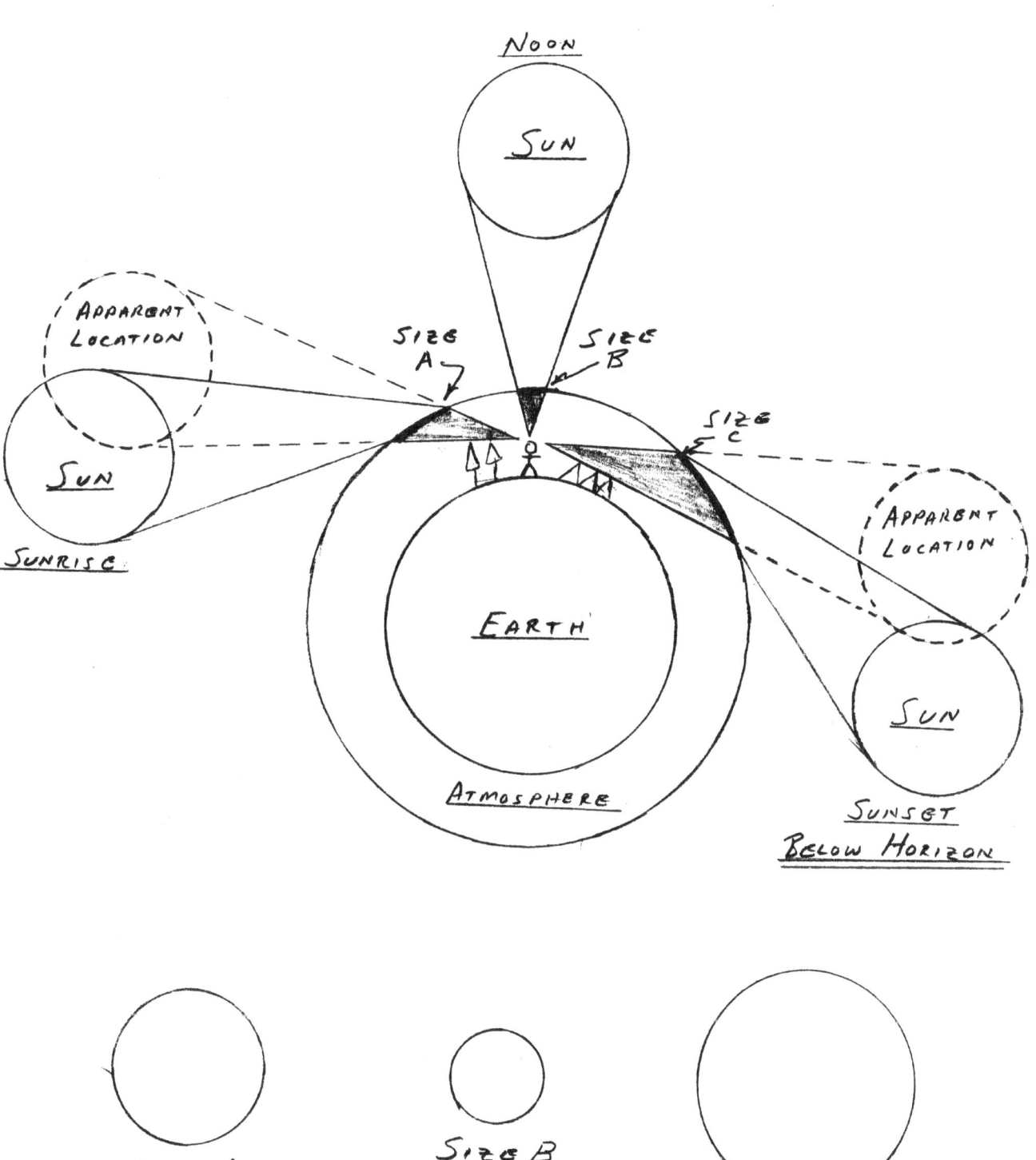

NOON

SUN

APPARENT
LOCATION

SIZE
A

SIZE
B

SIZE
C

SUN

SUNRISE

EARTH

ATMOSPHERE

APPARENT
LOCATION

SUN

SUNSET
BELOW HORIZON

SIZE A

SIZE B

SIZE C

ASTRONOMY - SPACE

Our telescopes show us the billions of stars in the Milky Way and the billions of galaxies in our view as they were when the light they emitted left them. If a galaxy is five billion light years away, it took five billion years for the light from this galaxy to reach Earth. Consequently, our telescopes show the galaxy not as it is today, but as it was five billion years ago when the light we are now receiving had just left that galaxy on its way to Earth. A telescope is a time machine - it carries us back into the past. Telescopes in space are able see much further out into the Universe than any telescope on the ground - this means that it will also see further back into the *past*.

The telescopes that have been sent into space or will be going into space are fairly modest in size, the largest to date is barely half as large as the two hundred inch telescope on Mount Palomar. Why do these relatively small telescopes in space see further than the two hundred inch telescope on the ground? The answer is connected with the fact that rays of light (photons) from stars and galaxies have to pass through Earth's atmosphere; atom to atom transfer of photons, the absorption and reemitting of photons all the way to the ground telescope. As soon as the photon emitted from the star or galaxy reaches the outer edges of our atmosphere, it will encounter an atom of oxygen in the ozone layer and be absorbed. This starts the process of atom to atom transfer of the photons through the ozone layer into the stratosphere. The process occurs through the stratosphere to eventually encounter the glass in the lens of the telescope. The travel of photons vary as the atoms in the atmosphere vary from time to time. Therefore, we are not receiving in the ground based telescope an exact copy (frequency) of the photon that left the star or galaxy years ago - we are receiving only *gossip*! This travel through the atmosphere is also the reason stars "twinkle" at times. Observatories are placed on mountain tops in remote areas of the world to reduce this atmospheric interference. The height of the observatory reduces the density of the atmosphere while the remoteness eliminates a high percentage of man-made light pollution - cars, cities, industries, etc.

The atmospheric blurring of the telescope's images limits the distance and clarity of the celestial objects it can detect. Consider a very distant object, such as a galaxy. This distant galaxy produces a faint image in the telescope because the intensity of the light from any object becomes less concentrated with increasing distance. If a faint image is spread out and diffused by the blurring effect of the atmosphere, it becomes fainter still. In fact, at times it will be submerged in the background illumination of the night sky and cannot be seen at all.

With the telescope in space, the blurring effect of the atmosphere is totally eliminated, the images formed by the telescope are much sharper as they also stand out better against the dark background of space - the photons from the source are received directly into the lens of the telescope. Telescopes in space have other advantages. The information we receive about the cosmos comes to us in the form of electromagnetic radiation of various wavelengths which normally have to pass through Earth's atmosphere. Many of these frequencies are absorbed by the atmosphere before they even get started. Most of the wavelengths of infrared and ultraviolet are blocked out, but the frequencies of x-rays and gamma rays are totally stopped. The only frequencies to penetrate are those in the narrow band whose wavelengths we know as "light" (visible radiation), some infrared, some ultraviolet and the lower radio wave frequencies.

There are many objects under study that emit the radiation frequencies of infrared, ultraviolet, x-rays and gamma rays that cannot penetrate Earth's atmosphere. Black holes and quasars emit gamma rays and x-rays while distant stars that may have planets forming around them emit infrared radiation. All these waves and their information were locked out to astronomers until the advent of space astronomy. When NASA put gamma ray, x-ray and infrared telescopes in satellites orbiting above Earth's atmosphere, new windows to the Universe were opened to astronomers.

The Gamma Ray Observatory will be in some ways the most interesting astronomical satellite of all. Gamma rays have tremendous penetrating power in space and can travel great distances through the Universe. When we view a gamma ray that has come from a distant origin, we see a view of the Universe as it was at a very early time, perhaps not long after the big bang.

X-ray telescopes circling the Earth in orbit revealed that many stars and galaxies are also powerful sources of energy. One intense shower of x-rays seemed to be coming from a bright star. Careful observation showed that the x-rays were not coming from the star itself, but from an invisible object very close to it. It was determined that the invisible object was a black hole circling around the bright visible star. According to calculations the gravitational pull of the black hole was so powerful that it tears streamers of gaseous matter off the surface of the bright star. As the gas approaches the black hole, it picks up speed and the gas becomes quite hot as the atoms collide with each other. Near the surface of the black hole, the hot gas reaches temperatures of millions of degrees and, at that temperature, radiation of intense streams of x-rays occurs. We cannot "see" a black hole, but we can identify its location by knowing the source of these x-ray streams.

Through the use of these space telescopes and satellites, we have discovered that we are able to use these instruments to thoroughly study the most obvious planet of all - Earth. The view from space using advanced optics and instruments has enabled our planet to be mapped, surveyed, analyzed, studied and investigated in all the fields that are available. Most important, specialists began to appreciate how Earth works as a dynamic, adaptable system, greater than the sum of its parts. The earliest applications of this new learning were directed toward crop predictions. Specialized Earth-observation satellites harvested new, detailed data with their large number of optical and infrared sensors. These detect disease and insect infestations in key crops. Additional applications for these detailed, digitally processed pictures followed. Surveys of third-world nations rendered pictures of massive clear-cut areas. Radar pictures from Earth Resources Survey spacecraft even penetrated beneath the surface of the Sahara to reveal a prehistoric riverbed where archaeologists hope to find human and prehistoric relics. These and other satellites are also used in the study of climate, its manner of keeping its climatic balance, man's true impact on the planet, and ways in which environmental disasters can be assessed, predicted and averted. New ideas for data are continually being studied and implemented to assist mankind in his habitation of this planet.

HUBBLE SPACE TELESCOPE

The telescope's overall dimensions are 13 meters (43 feet) long, 12 meters (39 feet) across with solar arrays extended and weighing 11 thousand kilograms (12½ tons). The heart of the Hubble Space Telescope (HST) is a 2.4 meter (94.5 inch) diameter mirror designed to capture infrared, optical and ultraviolet radiation before these can reach Earth's atmosphere. The HST also contains x-ray and gamma ray instruments which capture these frequencies for our analysis and study.

The telescope reflects light from its large mirror to a smaller, 0.3 meter (12 inch) secondary mirror, which in turn sends the light through a hole in the doughnut-shaped main mirror and into the rear bay of the spacecraft. There, any of five major scientific instruments wait to analyze the incoming radiation. Most of the instruments are contained in individual modules about the size of a telephone booth for easy maintenance, replacement or updating. This allows for improved or new equipment to be replaced quickly by space shuttle crews.

HST is designed to allow astronomers to probe the Universe with at least ten times finer resolution and with some thirty times greater sensitivity to incoming radiation than existing Earth-based devices. HST orbits the Earth about once every 95 minutes at an altitude of about 600 kilometers (380 miles). It is operated remotely from the ground.

TELESCOPES

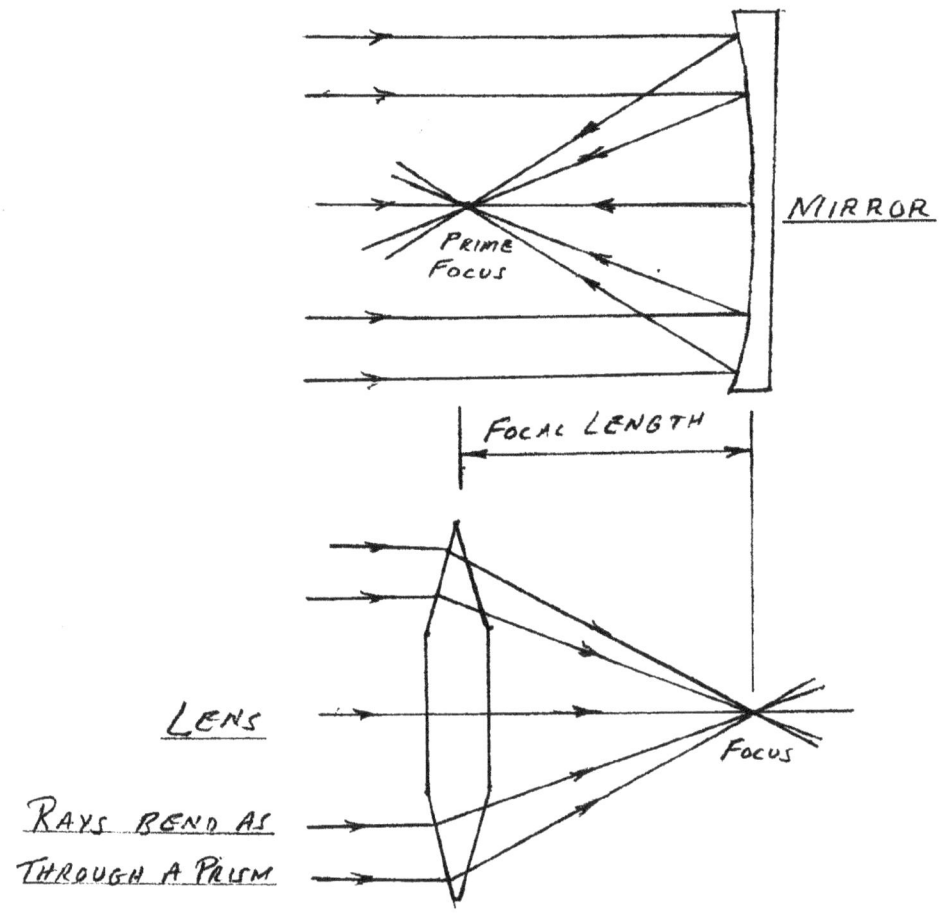

MIRROR

Prime
Focus

FOCAL LENGTH

LENS

FOCUS

RAYS BEND AS
THROUGH A PRISM

ASTRONOMICAL TELESCOPE

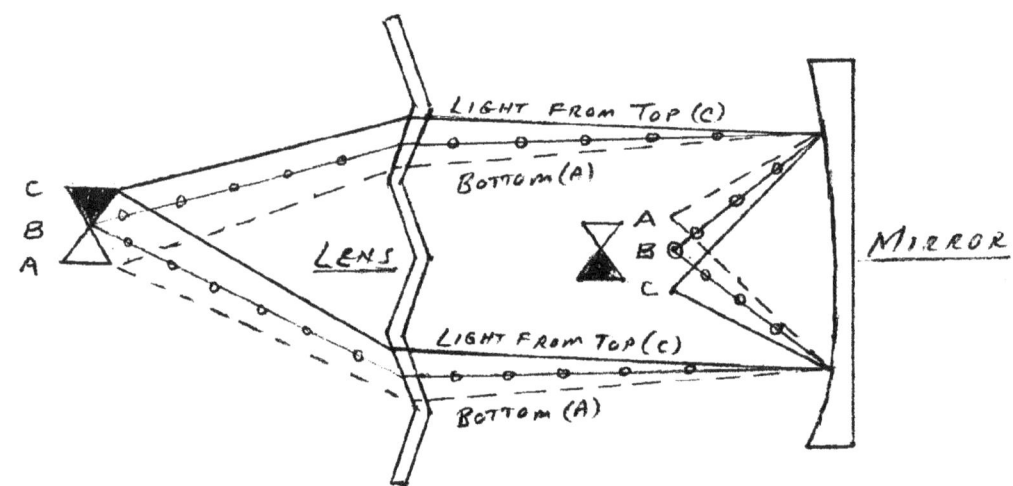

LIGHT FROM TOP (C)

BOTTOM (A)

C
B
A

LENS

A
B
C

LIGHT FROM TOP (C)

BOTTOM (A)

MIRROR

IMAGE WILL BE INVERTED   A - IMAGE OF BOTTOM

B - IMAGE OF CENTER

C - IMAGE OF TOP

A
B
C

SPECTROSCOPE

Since an ORANGE we eat is called an ORANGE, why isn't a LEMON called a YELLOW?

White light is a mixture of many different colors - red, orange, yellow, green, blue and violet - the divisions you see when a beam of white light is passed through a prism. This red to violet range of colors is called a *spectrum*. This range of colors - this *spectrum* - is seen each time you look at a rainbow. Human eyes are insensitive to those wavelengths of radiation shorter than the color violet or longer than the color red. Radiations outside these limits are invisible to the human eye, but they do exist as radio waves, infrared, ultraviolet, x-rays and gamma rays.

The *spectroscope* is an instrument that can perform studies and analysis of a *spectrum*. The *spectroscope* is probably the greatest scientific invention of all time. Since it is light that is emitted from the stars, astronomers use this instrument to perform many functions. The spectrum of this visible light is shown as red vertical lines next to orange vertical lines through the group of colors until the last color of violet is shown. Each color is vivid at its center and seems to blend to the adjacent colors - there are no discernable divisions between the colors because the spacing of frequencies is so close with an almost infinite quantity available.

If you were seated on the Hubble Space Telescope and aimed a spectroscope at a nearby star, you would expect to see a perfect spectrum since there appears to be nothing between you and the star. You will be surprised at what your spectroscope shows and how much information you will be able to determine about this star. You are above the atmosphere of Earth and in the vacuum of space with this star. Space is cold, it is -270^0 Celsius (three Kelvin), and also a vacuum. You would not expect anything to be between you and this star - but there *may* be.

Gases, when heated, give off light of only part of the spectrum. Different gases give off different color combinations and the spectra shown of each is not continuous - the result being thin lines of different colors that are spaced apart in a pattern. These patterns are the fingerprints of each gas - all being different from each other. A catalogue of each element is available showing the spectral lines of its identification. When a gas is heated, these spectral frequencies are emitted. These are called the *emission lines* of the element. Molecules present more lines than an element because they consist of multiple elements with a variety of other motions. All these identifications were made using a spectroscope. The action to create these frequencies of emission of each was the quantum action created by photon action of the valence electrons in elements or one of several methods in molecules. So what does all of this have to do with me as I sit here getting cold on this big telescope? This has nothing to do with you yet, but you do get involved pretty soon.

A cold gas does not emit spectra. Shine a light through a cold gas and the photon action will cause the gas to activate its spectrum *in all directions*. It has absorbed and reemitted in its spectrum range. This reemittance reduces the amount of light in these particular frequencies to continue in one direction at full strength, these frequencies, or colors, now head out of the electron in all directions. This will cause the spectrum of this light to be darker in these frequency lines. These lines that are not shown in the spectrum of the light which passed through this cold gas are called *absorption lines*. These lines are identical to the *emission lines* of the heated gas and instead of being the only colored lines shown, they are the dark, or the missing, lines of the spectrum of light. This also identifies the gas.

Now back to your seat on the Hubble Space Telescope with your handy spectroscope in your cold hands. You look at the spectrum of the star you are observing and notice that some thin lines of color are missing! Atoms of hydrogen and helium are in space and the missing lines were absorbed and reemitted by these atoms which caused a reduction of intensity of these frequencies and eventual darkness on the spectrum. You have identified two elements that were absorbed between you and the star - one of the many uses of the spectroscope in astronomy.

Helium was first identified as an element in 1868. In that year an expedition had traveled to India to study the Sun during an eclipse. A spectroscopic analysis of the Sun's chromosphere yielded a yellow line. In 1871 Sir Joseph Norman Lockyer and Pierre Jules Cesar Janssen conclusively proved that the line was not due to any element thus far found on Earth. The source of its discovery was commemorated in the name of the new element, the word "helium" being derived from *helios*, the Greek name for the Sun. In 1895, almost a quarter of a century later, helium was found to occur naturally on Earth.

The German physicist Gustav Kirchoff summarized the relationships between the three types of spectra - continuous, emission line and absorption line - in 1859. He listed three spectroscopic rules, now known as Kirchoff's laws, governing the formation of spectra:

1. A luminous solid or liquid, or a sufficiently dense gas, emits light of all wavelengths and so produces a *continuous spectrum* of radiation.

2. A low-density, hot gas emits light whose spectrum consists of a series of bright *emission lines*. These lines are characteristic of the chemical composition of the gas.

3. A cool, thin gas absorbs certain wavelengths from a continuous spectrum, leaving dark *absorption lines* in their place, superimposed on the continuous spectrum. These occur at precisely the same wavelength as the *emission lines* produced by that gas at higher temperatures.

The incoming beam of radiation from a star carries a wealth of information in its spectra. Typically the spectra of many elements are superimposed on one another, but their individual "fingerprints" can identify each to unravel the information supplied.

Observed Spectral Characteristics	Information Provided
Peak frequency or wavelength (Continuous spectra only)	Temperature (Wein's law)
Lines present	Composition
Line intensities	Composition, temperature
Line width	Temperature, turbulence, rotation speed, density, magnetic field
Doppler shift	Line-of-sight velocity

ASTRONOMY AT MANY WAVELENGTHS

λ / f Range	General Considerations	Common Applications
Radio	*Radio radiation can penetrate dusty regions of interstellar space. *Earth's atmosphere is largely transparent to radio wavelengths.	Radar studies of planets. Planetary magnetic fields. Interstellar gas clouds. Center of the Milky Way.

λ / f Range	General Considerations	Common Applications
Radio (Cont.)	*Radio emissions can be detected both day and night. *High resolution at long wavelengths requires very large telescopes.	Galactic structure. Active galaxies. Cosmic background radiation.
Infrared	*IR radiation can penetrate dusty regions of interstellar space. *Earth's atmosphere is only partially transparent to IR radiation - some observations must be made from space.	Star formation. Cool stars. Center of the Milky Way. Active galaxies. Large-scale structure in the Universe.
Visible	*Earth's atmosphere is transparent to visible light.	Planets. Stars and stellar evolution. Galactic structure. Large-scale structure in the Universe.
Ultra-violet	*Earth's atmosphere is opaque to UV radiation - must be observed from space.	Interstellar medium. Hot stars.
X-ray	*Earth's atmosphere is opaque to x-rays - must be observed from space. *Special mirror configurations are needed to form images	Stellar atmospheres. Neutron stars and black holes. Hot gas in galaxy clusters. Active galactic nuclei.
Gamma rays	*Earth's atmosphere is opaque to gamma rays - must be observed from space. *Cannot form images.	Neutron stars. Active galactic nuclei.

LASER

Light Amplification by Stimulated Emission of Radiation.

The laser is not a *source* of energy - it is a *converter* of energy. It takes advantage of stimulated emission to concentrate a certain fraction of its energy into radiation of a single frequency moving in a single direction. Because of this concentration, a laser beam with less total power than an ordinary light bulb can burn a hole through a metal plate or send a message over hundreds of miles.

In a ruby laser a flash of light from a xenon discharge tube excites atoms in a ruby crystal. *When we say "excites atoms" we are referring to the absorption of the photon by the valence electron and the "up the stairs, down the stairs" action.* These atoms are first elevated to a high energy level, releasing their extra energy in the form of a photon of light. This photon is "tuned" to a specific frequency, so that when it hits another excited atom, it stimulates that atom to emit a photon of its own. The second photon travels in the same direction as the one that triggered it placing the two photons in phase. The photons continue their passage down the ruby crystal, colliding with other excited atoms, stimulating them to emit photons. The laser is so constructed that the light is reflected back and forth between mirrors at each end of the ruby crystal causing the photons of light to effectively travel a long distance through the ruby rod. This causes interactions with many atoms, creating a growing quantity of photons. When the intensity is at the desired level, laser light emerges through one of the mirrors, which has a semi-silvered surface. The result is *coherent* laser light - light in which the electromagnetic waves maintain a fixed phase relationship. Different energies are emitted by each laser to fulfill the industrial or medical service desired - to burn holes through metals or perform eye surgery.

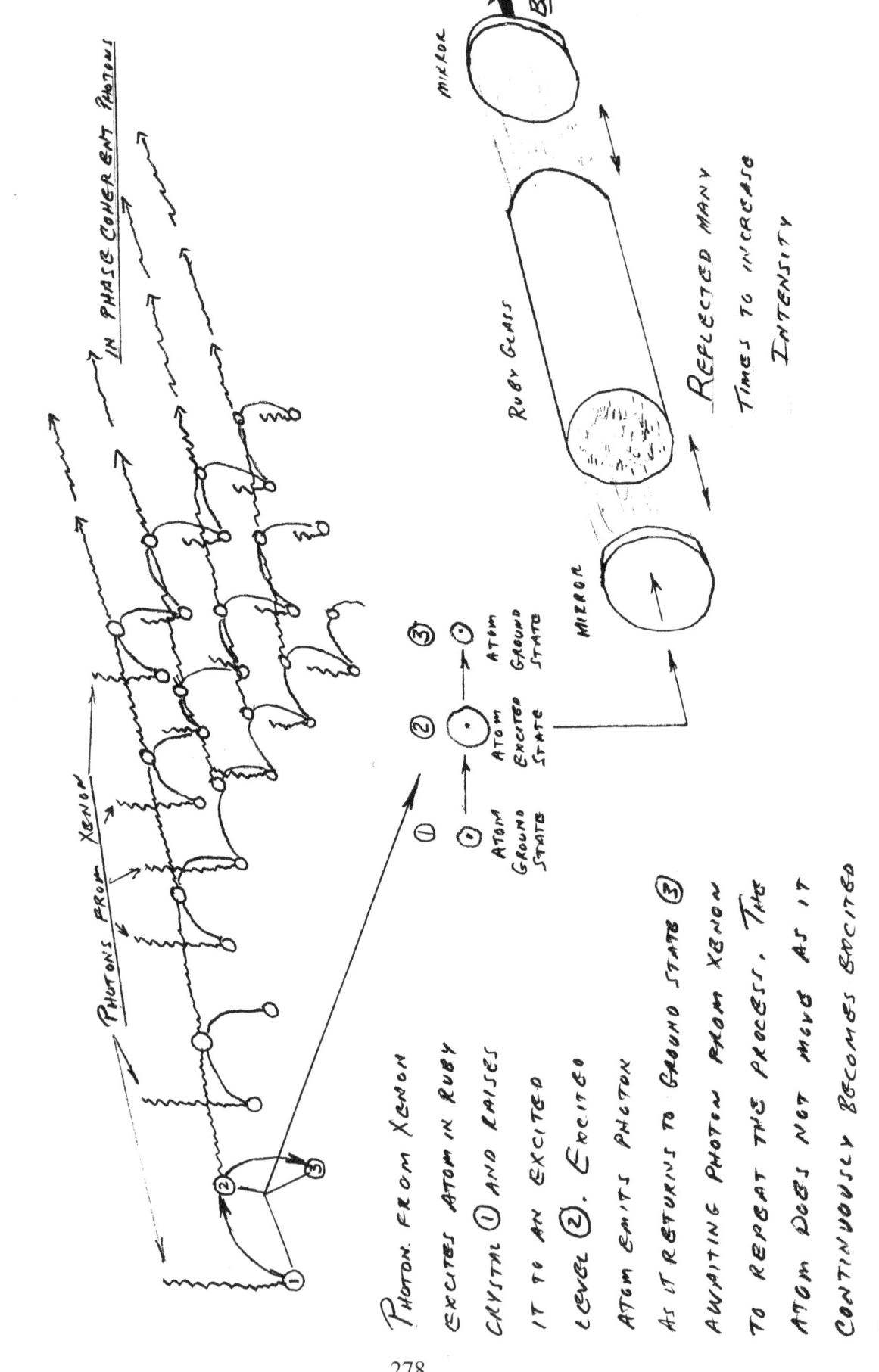

RUBY CRYSTAL LASER
7000 A°

IN PHASE COHERENT PHOTONS

PHOTONS FROM XENON

① ② ③
·O   O·   ·O
ATOM   ATOM   ATOM
GROUND EXCITED GROUND
STATE   STATE   STATE

MIRROR

RUBY GLASS

REFLECTED MANY
TIMES TO INCREASE
INTENSITY

MIRROR

BEAM

PHOTON FROM XENON
EXCITES ATOM IN RUBY
CRYSTAL ① AND RAISES
IT TO AN EXCITED
LEVEL ②. EXCITED
ATOM EMITS PHOTON
AS IT RETURNS TO GROUND STATE ③
AWAITING PHOTON FROM XENON
TO REPEAT THE PROCESS. THE
ATOM DOES NOT MOVE AS IT
CONTINUOUSLY BECOMES EXCITED
EMITTING PHOTONS.

278

SOUND

The speed of a sound wave is governed by the nature of the vibrating medium. The speed of light is a constant associated with the coupling between electric and magnetic fields. The speed of sound is determined by the interaction between atoms and molecules. In a gas, the speed of sound depends on the temperature and the molecular weight, the quantities that affect molecular speed. In air at normal temperature, sound waves travel at 0.343 kilometers per second, about one millionth the speed of light. In a solid, the closely packed atoms conduct sound more rapidly. In a vacuum (space, on the Moon, etc.) there is no sound because there is no medium for sound to vibrate, no atmosphere. There would be sound on most of the planets because of their various atmospheres. The sound on each would vary according to the composition and densities of the gases present.

Sound waves travel faster in warm air than in cold air. The frequency range of sound is 15-20,000 cycles per second (ultrasonic). Low frequency sound waves tend to spread in every direction while ultrasonic waves tend to travel in beams, like light. Sound waves transmit mechanical energy: Speech - 60 decibels, a boiler factory would be 130 decibels.

MATERIAL	SPEED OF SOUND AT ROOM TEMPERATURE	
	km/sec	mi/hr
Air	0.343	767.7
Hydrogen	1.32	2,952.7
Water	1.49	3,333
Glass - Pyrex	5.6	12,527
Iron	6.0	13,422
Aluminum	6.4	14,316

RELATIVE MOTION

Bearing - *constant*! Range - *closing*! These four words to anyone who served aboard a ship or who flew above these ships in the Navy had an instant response to them - *COLLISION*! A constant bearing with the distance decreasing between ships, planes, cars, trucks or people will result in these objects running into each other.

A simple example of this would be your spotting a friend in the park across the way with each of you running to the corner of the park to meet. You have to look a little to your left to see your friend as he starts running to the corner. As you run to meet him and never change the direction you are looking at him, you *will* meet at the corner. If neither of you slow down, this meeting will result in a *collision*! This never changing direction of your looking at him as you travel to the corner is the same as the bearing being constant. You knew the distance was decreasing because your friend was getting larger the closer he came.

The relative motion between the two of you would be a line from a point from you to your friend at the start connected to a point where you would meet. This line drawn between these two points is called a *vector* - a relative motion line the two of you created. If you observe these conditions when driving and see a train approaching, you had better slow down or stop because if you continue, the train will win - it is much larger and heavier!

When you are flying above the clouds, you notice a small object in the distance, hopefully above or below your altitude. As time passes, the object becomes larger and larger as it seems to be headed toward you - at a lower or higher altitude. This is a crossing, not a collision, because of the difference in the altitudes of the planes. As the other plane passes below, you notice that the nose is not pointed in the same direction as the path the plane is flying. Can planes fly sideways? Of course not, you are seeing the resultant relative motion between the two planes.

We know where we are going - is there any way we can determine the other planes destination? We can easily find out using vector analysis. Our direction and speed are known and we have a line showing the time we first spotted the plane until it passed under us. From this we can determine the other plane's speed and direction. These vectors work whether you are a submarine in the ocean, a car on land, a plane in the air or a spacecraft in space. Relative motion is vector analysis, not *relativity*. We will talk about relativity later in the book.

In our example we are flying north-northeast at 450 miles per hour (Point A to Point B). We determine that the other plane is flying at 525 miles per hour in a southeastern direction (Point A_1 to Point B_1). The relative speed between our two planes is 975 miles per hour in a south-southeasterly direction. Notice the planes are pointed in the right directions on their actual paths. The *relative* line shows the other plane pointed in one direction while tracking in another. This is the view we see when observing the other plane pass underneath us.

Relative motion is very important to us even in our everyday lives. We take this into consideration when passing a car on a two lane road as we pull back in front of the car we passed. If we do not do this properly, an accident will occur. Relative motion is very important in space because the objects are traveling at 17,000 miles per hour at three hundred miles above the surface of Earth. They are generally all headed in the same direction at these speeds. This means that the relative speeds between these objects is very low and in some cases it is zero. Once in space you basically drive yourself around to find the satellite you want and slowly coast near to it, slowing until a docking is completed. Automobile race car drivers run an entire race at hundreds of miles per hour - spectators see this - but these drivers live with a relative speed to other drivers from zero up to five miles per hour as they race around the track in mass.

DIRECTIONS TO DRAWING THE GRAPH

The diagram at the top shows the direction and distance actually traveled by each plane. A to B is the track of your plane, A_1 to B_1 is the track of the other plane. A is your position when you see the other plane at A_1 and B is your position when you see the other plane at B_1.

1. Plot your direction and distance, A to B, on the graph.
2. Plot your initial observed direction and distance to the other plane, A to A_1, on the graph.
3. Plot your final observed direction and distance to the other plane, B to B_1, on the graph.
4. Connect A_1 to B_1. This is the relative motion line.
5. Transfer the relative motion line, A_1 to B_1, to the center of the graph, placing B_1 at the center.
6. Connect A_1 to B to produce the line A_1 to B. This is the direction and speed of the other plane.

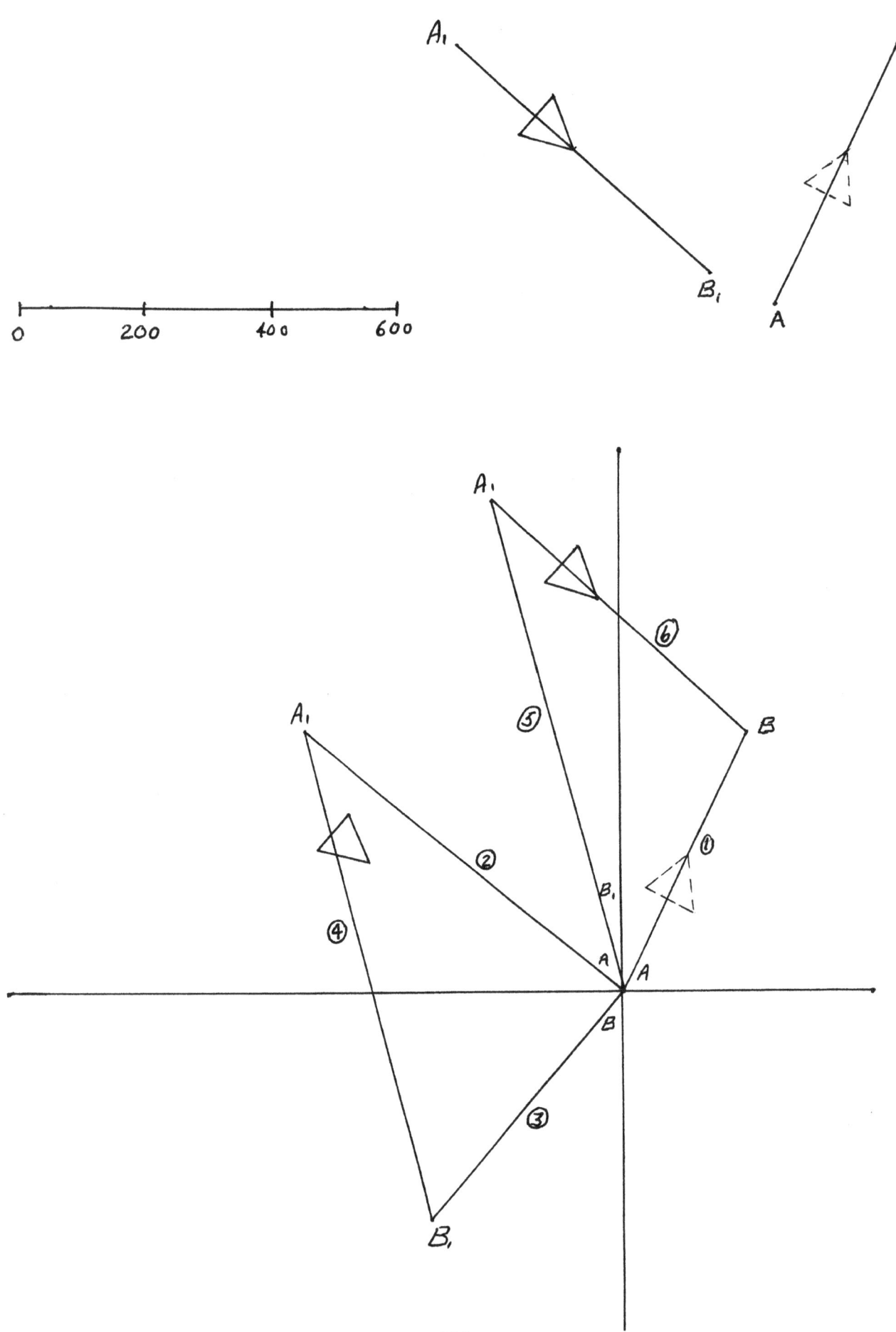

DNA

Deoxyribonucleic acid - the molecular structure and chemical process of life.

Billions of years ago, the primordial atmosphere of Earth had no oxygen. *Primordial* just means that something existing at or from the very beginning. The early atmosphere was subject to lightning, intense solar ultraviolet radiation, cosmic rays and volcanic heat. The incoming energy stimulated the molecules of the existing atmospheric gases to form five new compounds. The compounds became condensing droplets of matter that fell into the oceans, which were a primordial soup. The compounds that fell were the chemical letters of the genetic alphabet: adenine, thymine, guanine and cytosine. Together, they form deoxyribonucleic acid, otherwise known as DNA. The fifth letter of the genetic alphabet formed in the primordial atmosphere of Earth was uracil, which is a component of RNA, the "yes man" of the DNA molecule, the genetic messenger that carries out the orders.

The production of the five basic genetic ingredients from atmospheric gases surrounding a primordial soup has been accomplished in the laboratory. The five bases have also been found aboard a meteorite that dates back to the formation of the solar system about four and a half billion years ago. The same building blocks of life occur in fifty-seven different molecules discovered in the Universe. The conclusion is that life is a relatively common occurrence, if the conditions are reasonable enough.

The essentials of life - hydrogen, carbon, nitrogen, oxygen, phosphorous, sulfur and calcium - were formed following the big bang and have progressed through galaxy formation, star formation and explosion, new star formation (our Sun) and planet formation down to us. We are made from these and will use them during our stay on Earth - returning them to be used over and over throughout eternity upon our death. *We are from the stars!*

From these elements the four molecules of life are formed - nucleic acids, proteins, carbohydrates and lipids. The term "genetic code" refers to the link that leads from these base pairs on a DNA molecule to a particular amino acid holding a certain position along a protein chain. Everything in your body is manufactured, moved, modified and used by proteins that are built from scratch from a sequence of amino acids precisely dictated by the DNA sequences - the reason DNA holds the secret of life.

To determine where the cell obtained the blueprint that told it to form, move into the nucleus and notice the lanky contours of the DNA macromodules secreted within its genes. Each holds a wealth of genetic information accumulated over the course of these four billion years of evolution. Stored in a nucleotide alphabet of four "letters" - made of sugar and phosphate molecules and filled with punctuation marks, reiterations to guard against error and superfluities accumulated in blind alleys of evolutionary history - its message spells out just how to make a human being, from skin and bones to brain cells.

One important example of how quantum phenomena can influence the world of our experience in a drastic way concerns the effects of radiation on genetic material. DNA is a double helix of atoms arranged in a complex pattern and if the pattern is altered in any way, the genetic code is changed and the DNA will not correctly reproduce. If the altered DNA is an egg or a sperm cell, the offspring will be a mutant. DNA can become damaged in many ways, but a Universal threat is that of cosmic radiation - high energy subatomic particles which bombard Earth from outer space. The impact of an energetic particle with a DNA molecule can result in a mutilated genetic code. Mutations are vital to evolution because they supply a variety of alternative life-forms for nature to select or destroy according to their efficiency. By going back far in time, very small changes that occurred then have produced large differences now.

MOLECULAR BIOLOGY

Biology is the study of living things. A simple example is showing how "pain" occurs. We have known that the nerves in our body have an electrical association in transferring information from where the "pain" occurred to our brain for instructions for reactions. Electricity in the body travels at two hundred twenty miles per hour. Biology tells us how all this fits together.

When you hit something sharp with your arm you cause the compression of nerves. These nerves are very fine tubes with a complex wall which is very thin. The cell pumps ions (atoms with a missing electron) through this wall to create positive ions (atoms with an extra electron) on the inside - the same as a capacitor. Visualize this as the addition of a battery to a circuit which creates a current flow to the brain. The brain processes this information and relays a message along the nerve. At the end of the nerve which received the blow to your arm, the nerve branches out into fine little fibers, connected to a structure near a muscle, called an endplate. When the impulse reaches the end of the nerve, little packets of a chemical called acetylcholine are shot at a rate of five to ten molecules at a time affecting the muscle fibre and making it contract. Therefore, your arm reacts and you pull away from the object creating this pressure.

Biology is such an enormously large field. Imagine the other problems your computer (brain) may have also been involved: seeing the object, yelling, jumping away, thinking about your next action, moving your arm, hearing help arriving, smelling fire, . . .

SINGLE ATOM MEDICINE

When the nucleus of a boron atom is hit by a neutron from a nuclear reactor, it emits an alpha particle. If this alpha particle (the nucleus of a helium atom - two protons and two neutrons) were released within living tissue, it would travel no further than the edge of the cell it happened to occupy before it expended its energy and came to a stop. The penetration of an alpha particle is limited to the thickness of a sheet of paper. Then it would attract two stray electrons and turn into a harmless helium atom. But the energy it imparted to the cell in the process of slowing down would kill the cell, making neutron irradiation of boron atoms a promising candidate for cancer therapy. This kind of microscopic control over organic processes is the dream of the medicine of the future.

MICROMEDICINE

In the future it may be possible to build tiny robots, the size of bacteria, that could be injected into the body. There, they would directly destroy unwelcome organisms and fat deposits lining arteries. This type of research is being done with the idea that these robots will be going on "search and destroy" missions for certain failures in the body's chemical system. The health of the human race is always going to improve due to these advances in medical technologies - prevention of and early eradication of any possible health problems.

WAY OUT THINGS

We find ourselves in a bewildering world. We want to make sense of what we see around us and to ask: What is the nature of the Universe? What is our place in it and where did we come from? Why is it the way it is?

Stephen W. Hawking
A Brief History of Time

Strictly speaking, the relevant formula relating mass and energy turned out not to be $E = MC^2$, but $E^2 = M^2E^4$. Taking the square root of both sides result in: $(E^2)^{1/2} = (M^2C^4)^{1/2} = (+/-) E = MC^2$. $E^2 = (+E)(+E) = (-E)(-E)^*$.
*Remember that $(-1)(-1) = (+1)$.

The speed of light is constant for observers in different states of motion.

Space and time are joined together in a single entity - *space-time*.

Why is there something rather than nothing?

$$e = ma \qquad e = mb^2 \qquad e = mc^2 \text{ - - - - yeah!}$$

287

QUANTUM MECHANICS
INTRODUCTION

The quantum nature of atoms dictates: the density of ordinary matter; the behavior of solids, liquids and gases when heated; the transparency, opacity and color of materials; the existence of *physical* change such as freezing and boiling; and all aspects of *chemical* change. Because of the quantum structure of atomic nuclei, nuclear burning lights the Sun and the stars, and nourishes life on Earth. The facts that metals are conductors, that chemical cells generate electricity, that a crystal can be doubly refracting, that light travels more slowly in matter than in empty space - all find their explanation in the *quantum theory*.

ORIGIN AND DEVELOPMENT

1900	quantum units - Max Planck
1912	quantum principle - Niels Bohr
1924	wave nature of matter predicted - de Broglie
1925	electron spin - George Uhlenbeck and Samuel Goudsmit
	exclusion principle - Wolfgang Pauli
	mathematical theory of quantum mechanics - Werner Heisenberg
	wave nature of matter verified - George P. Thompson
	Clinton Davisson
	Lester Germer
1926	wave theory of quantum mechanics - Erwin Schrodinger
	probability interpretation - Max Born
1927	uncertainty principle - Werner Heisenberg
1928	relativistic quantum theory of the electron - Paul Dirac

The normal sequence of revising the outlook we have on the present state of affairs usually appears in a sudden manner when people realize something momentous has happened or is about to happen. These are usually natural catastrophes, medical discoveries or scientific revolutions. These are well publicized and documented for all to see and understand. Information is disseminated and impacts on present thinking and practices may become subject to change because these new facts replace unknowns, suspicions and fears.

The Earth was the center of the Universe and everything revolved around it until Copernicus arrived on the scene. He started the flow of facts verifying that Earth was not at the center of the Universe. This simple statement of fact began the destruction of religious dogma that resulted in the change of European history. It started people thinking and changes happened. Next appeared Charles Darwin with information that we humans were not that special after all - we were but one of many species that appear on the scene and later change or disappear. A great uproar was heard. Edwin Hubble shocked everyone in stating that our Milky Way galaxy was not the entire Universe - it was but one of billions of galaxies in an apparent infinite Universe that was *expanding*! Einstein's special and general relativity theories were widely publicized and discussed by many - but understood by only a few. The immensity of what these theories would mean was almost beyond comprehension.

Einstein's theories were accepted as the ultimate and must, when finally understood, result in the answers to all that was not known. While this was the center of interest, the greatest scientific revolution of all time was going on unnoticed by the general public during these same years, between 1900 and 1930, because comprehension of these facts was almost unbelievable - the *quantum theory*.

The quantum theory was necessary to explain the internal workings of the newly accepted atom. This initiated sub-atomic physics. How could anything be smaller than an atom - ridiculous! This was the beginning of the greatest scientific explosion the world was to witness even though it was only understood by a small group of scientists. It has grown to incorporate modern microphysics, from elementary particles to lasers. Because of quantum mechanics the transistor was possible - making computers and spacecraft ordinary items. Cell phones and the internet became necessities as micro-miniaturization compresses more and more due to the understanding of quantum mechanics.

Much of what the ancients knew about nature, and a great part of what is now known as the classical physics of the seventeenth, eighteenth and nineteenth centuries, rests ultimately on the quantum mechanics of the submicroscopic world. When quantum mechanics was properly developed in the 1920's, it turned science upside down. This was due to its astonishing success in explaining a wide range of physical phenomena. As with the theory of relativity, quantum mechanics swept away many deeply entrenched assumptions about the nature of reality, and demanded a more abstract vision of the world.

QUANTUM MECHANICS

The quantum theory was born in 1900 when Max Planck realized that the only method to explain the spectrum of energy generated by a perfectly radiating object was to abandon the classical assumption that energy is emitted continuously and replace it with the unprecedented hypothesis that energy comes in discrete units - quanta. *Quantum* is defined as the smallest elemental unit of a quantity. *Mechanics* is the analysis of the action of forces on matter or material systems.

Radiant energy is composed of many quanta - each identified as a photon. Quantum mechanics provides a completely deterministic view of the evolution of physical states. In quantum mechanics, if you know the state at one instant, and knowing the machinery of the system, you can calculate what the state is in any following instant.

Particles have a quantum state which is a combination of *position* and *velocity*. As we just stated, calculating where each photon may be and when is easy, but just try to *measure* these properties. This requires us to measure the *location* and *speed* of one of the smallest quantum in the Universe - a single photon.

Let's go back to the workshop where we built the nucleus and later added the electron shells. Comparing the proton we built to the electron we placed in orbit about the nucleus, the weight of the proton is 1836 times greater than the weight of the electron and that the electron is a long way from the nucleus of the atom, relatively speaking. We saw in the section on LIGHT how energy (photon) is absorbed and reemitted by causing the outer electron of the atom to become excited (up the stairs) and return to a ground state (down the stairs) emitting a photon to continue the process of transmitting this particular frequency of light.

Everything we see, everything a microscope sees and everything a x-ray machine sees is the result of electromagnetic radiation - light we see or "rays" that are converted to pictures for us to view. We saw how light is transmitted through glass - atom to atom by the photons being absorbed and reradiated. This is how we see *what* and *where* the object is. There is one thing in common on all the above objects - they are much larger than the measuring devices - photons.

Let's take our photon and make it appear the same size as a ping-pong ball. We know photons are wavelengths, but we still talk about them as if they were solid *things* because, in general, waves and solids act similarly on each other, for our purposes. We know that electromagnetic radiation (light, x-rays, gamma rays, etc.) consists of a continuous string of photons - our ping-pong balls. We can see an apple because photons are "hitting" and "bouncing" off the apple. To maintain the proper perspective with our ping-pong balls, we have to enlarge the apple to equal the size of Earth. Now imagine our ping-pong balls "hitting" and "bouncing" off the Earth in the same manner photons "bounced" off the apple. Both the apple and Earth are standing still relative to the speed of the photon or ping-pong ball, each traveling at the speed of light. These photons or ping-pong balls have no effect on either the apple or Earth because of the relative size of each to the other. Neither the apple or Earth even knows it has been "looked at." Now let us try to take a "look" at another photon or ping-pong ball.

Here is where the trouble begins. Let's try to measure *one* photon or *one* ping-pong ball. So how do we measure this single ping-pong ball that is *moving at the speed of light?* We will use the same method that we used on our Earth-sized apple with our ping-pong balls. All we have to do is hit any ping-pong ball moving very fast with another ping-pong ball that is also moving very fast. When our measuring ping-pong ball intercepts and hits the ping-pong ball we want to measure we see the obvious - a collision occurs and both ping-pong balls ricochet to kingdom come! At least we found out one thing, where IT was - *position.* What about *velocity?* We must repeat the process on another ping-pong ball because we only solved one of the two unknowns. We tried to measure an object with an object of similar size and velocity head on.

What this crude example tried to simplify was Heisenberg's Uncertainty Principle: "In order to predict the future position and velocity of a particle, one has to be able to measure its present position and velocity accurately. One needs to use light of a short wavelength in order to measure the position of the particle precisely. One quantum must be used. This quantum will disturb the particle and change its velocity in a way that cannot be predicted. The more accurately one measures the position, the shorter the wavelength of light that is required - the higher the energy of a single quantum. The velocity of the particle will be disturbed by a larger amount. Therefore, the more accurately you try to measure the position of the particle, the less accurately you can measure its speed. The uncertainty in the position of the particle times the uncertainty in its velocity times the mass of the particle can never by smaller than a certain quantity - Planck's Constant of **h** - the quantum unit of angular momentum in nature."

Quantum mechanics introduces an unavoidable element of predictability or randomness into science due to the duality of waves and particles. Since the structure of molecules and their reactions with each other underlie all of chemistry and biology, quantum mechanics allows us in principle to predict nearly everything we see around us, within the limits set by the Uncertainty Principle.

The strong force is exceedingly strong; it actually becomes stronger with increasing distance. This is why free single quarks are never found under natural conditions in the present Universe. If any two quarks were somehow pulled apart, the force between them would increase until the energy in the strong field would create a new pair of quarks. Compare this to trying to divide a magnet; when a bar magnet is cut in half, the result is two smaller magnets - not two distinct poles. At extremely high energies and densities, however, the strong force becomes negligible, and the quarks are able to behave as if they were completely free particles. There was no reason for the quarks to form particles until the temperature had dropped far enough after the big bang.

Today, the truly fundamental material entities are no longer considered to be particles, but *fields*. Particles are regarded as disturbances in the fields, and so have been reduced to a derivative status. In practice, the physicist usually uses second-order differential equations of calculus. It has been stated by Leon Lederman, director of the Fermi National Accelerator Laboratory near Chicago, "We hope to explain the entire Universe in a single, simple formula that you can wear on your T-shirt." His reference was to a *Lagrangian* - a mathematical expression named after the French physicist Joseph Lagrange.

These *fields* could be described as wavelength. Wavelength denotes frequency and frequency denotes energy. The shorter the wavelength, the higher the frequency and consequently the higher the energy. Energy is an equivalent of mass, $E = MC^2$, which could explain the wave/particle duality of a quantum.

David Bohm revived a theory Louis de Broglie had originated years earlier - the guiding-wave. The guiding-wave theory solves the problem of wave/particle duality by meeting it head on: It allows an entity to be a particle whenever it is caught and a wave whenever it passes through two slits (interference). This is called the Bohm-de Broglie theory. Specifically Bohm imagined the entity as a particle riding on its wave like a leaf in a stream. This wave, known as the guiding-wave, is a real, but invisible, part of every particle. A quantum entity travels as a wave but arrives a particle. "A collapse of the wave function" - the wave becomes a particle when being measured or when interacting with something, then becomes a wave again.

After taking my car in for a checkup, the mechanic asked me, "What is quantum mechanics?" I answered him the only way I could, "It should be easy for you to understand - it is how the smallest motor works."

RELATIVITY - SPECIAL

Special relativity - *relativity* because it declares that all motion is relative, and there is no absolute "frame of reference" in which the laws of physics are uniquely true; *special* because it deals only with constant velocities - steady speed in a straight line. Special relativity is a description of flat space-time.

When traveling with a clock, time does not appear to slow down or speed up; but, to an outside observer, time will have slowed or changed. Time is relative from one observer to another who exist in different frames of motion. An astronaut traveling in the $t = 10t_0$ frame making a trip of one year on his clock would notice that people on Earth had aged ten years to his one year - they are now nine years older. If this astronaut were to be gone for ten of his years, according to his clock, one hundred years would have passed on Earth and none of his generation would still be alive. This would continue as long as he remained in his frame of reference. Imagine that this astronaut could travel into the future in fifty Earth-year jumps as he was only aging five years as he remained in his frame of reference. If he had left Earth at age thirty in the year 2000 and returned in the year 2050, he would only be thirty-five years old and all his friends that had remained on Earth would have aged fifty years. Doing this on a basis of fifty Earth-years for each trip (aging five years on each trip) he could travel to the year 2300 on Earth as he returned on his sixtieth birthday - three hundred years later at the Earth-time age of three hundred thirty.

In objects with length, an object ten feet long at rest would be five feet long at 87 percent the speed of light, one foot long at 99.5 percent the speed of light and have zero length at the speed of light. The object does not contract - only when visualized from another reference position. Mass, on the other hand, at the speed of light would be infinite - requiring infinite energy to maintain momentum of the object which has zero length.

If it were possible to observe actions of others on a planet in a distant galaxy, or even on a planet in our own galaxy, watching their actions would be identical to their observations of us - each would be different from the other in time. To them, due to relativity, we would be moving in their time frame - probably very slowly - and if they could see one of our clocks, it may seem an eternity to them to watch the second hand on our clock make one revolution. This would be the result if our gravity was much greater than their own gravity. This is what we would observe their actions to be if their gravity was much greater than ours. Each system has its own reference points that are only applicable to it.

We can look back in time but we cannot travel back in time because nothing may travel faster than the speed of light. If something could travel as fast as the speed of light and one could travel that fast relative to Earth, one would remain suspended in that particular second of Earth-time. To go back one more second, you would have to travel at the speed of light over your present speed of light - twice the speed of light. Each second gained in going back in time would require another acceleration of another speed of light. You would have to be going at three times the speed of light to traverse three seconds back in time and you would be suspended in that time frame unless you continued your journey. Only in science fiction can this be done - that is why it is called *fiction.*

You measure the mass of a system to determine its energy. When observing a moving object, is the object moving past the observer, or is the observer moving past the object?

An astronaut in a rocket traveling at 99.9 percent of the speed of light would notice that the Sun as being only four million miles from Earth. His trip time for this distance at this speed would be twenty-two seconds according to his clock. To the observer on Earth, the distance to the Sun is ninety-three million miles and the time for the astronaut in the rocket traveling near the speed of light to reach the Sun would be well over eight minutes as measured by the clock on Earth. Both clocks are right *relative* to their positions of observation. This example may clarify the twin paradox so widely known about special relativity. The astronaut is aging slower in his frame of reference *relative* to the observer on Earth. However, when the astronaut returns to Earth and leaves his rocket, he will have aged almost the same as the observer that had remained on Earth. He had left and has returned to his original reference point. All his actions are normal to all the observers who were witnesses to his trip. Let's now change our reference point to the rocket. As the astronaut continued his journey into space, he would look back at Earth and notice that events were hardly moving at all - as though time had slowed almost to a stop. At this speed, outside occurrences will be slower to him in the relative amount that his high speed may appear to other observers in different frames of reference. Each person is aging properly relative to their location.

Another example of two observers in different frames of reference would be when an observer on Earth watches a rocket pass overhead traveling near the speed of light. The rocket is designed so that the observer on Earth can see the interior of the rocket as it passes over his position. A person in the rocket is standing in the center of the rocket pointing a flashlight toward the front of the rocket and another flashlight toward the rear of his rocket. He turns the two flashlights on at the same time. To the person in the rocket, the beams of light hit the front and rear of the rocket at the same time because he is equidistant from each end. This is what would actually happen in the rocket.

To the observer on Earth seeing the lights come on at the same time, he would see the rear-pointing beam of light hit the rear of the rocket some time before the front pointing beam of light hit the front of the rocket. This observer on Earth sees the rear of the rocket approaching (near the speed of light) the rear-pointed beam of light and making contact with it before the forward pointed beam of light reaches the front of the rocket because the front is traveling in the same direction as the beam of light - it will reach the front of the rocket later. This is what the Earth observer would see.

This is special relativity - constant velocities are involved. Both observations are correct. The concept of simultaneity, the same moment in two different places, has no universal meaning. What is judged as to be "now" by one observer can be in the past or in the future as determined by another.

Imagine being on a merry-go-round in space spinning almost at the speed of light. You can see Earth as you come around the circle toward it and watch it disappear from your view as the merry-go-round completes its circle of movement. You decide to shine a beam of light toward Earth as your movement is headed away from Earth. You are in motion away from Earth at nearly the speed of light in your rotation as you shine your beam of light toward Earth. This beam of light leaves the merry-go-round headed for Earth at the speed of light. As your merry-go-round continues its rotation, you are now approaching Earth at almost the speed of light. You take aim at Earth and let loose another beam of light. You may think this second beam will catch up and *pass* the first beam of light, reaching Earth first. Not so, the speed of light, or any electromagnetic radiation, is constant to its reference point. There is no such thing as fast-moving pulses (radiation) overtaking slow-moving pulses. They all arrive at the same speed, in a regular spacing according to when they were sent, or left their source.

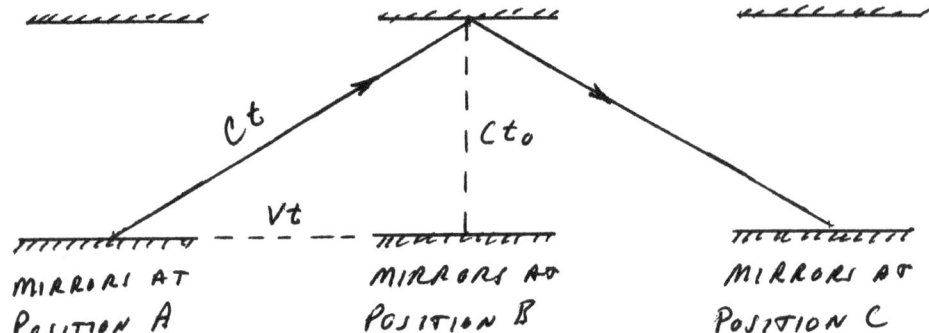

MIRRORS AT POSITION A   MIRRORS AT POSITION B   MIRRORS AT POSITION C

$$c^2 t^2 = c^2 t_0^2 + v^2 t^2$$

$$c^2 t^2 - v^2 t^2 = c^2 t_0^2$$

$$t_0^2 = t^2 \left(1 - \frac{v^2}{c^2}\right)$$

$$t^2 = \frac{t_0^2}{1 - \frac{v^2}{c^2}}$$

$$t = \frac{t_0}{\sqrt{1 - \frac{v^2}{c^2}}}$$

C – SPEED OF LIGHT

V – VELOCITY

t_0 – TIME FIXED

t – TIME MOVING

$t = 1$ @ $V = 0$ $t = t_0$

$t = 1.15$ @ $V = 0.5c$ $t = 1.15 t_0$

$t = 2$ @ $V = 0.87c$ $t = 2 t_0$

$t = 10$ @ $V = 0.995c$ $t = 10 t_0$

$t = \infty$ @ $V = c$ $t = \infty t_0$

$$L = L_0 \sqrt{1 - \frac{v^2}{c^2}}$$

$L = L_0$ @ $V = 0$

$L = 0.5 L_0$ @ $V = 0.87c$

$L = 0.1 L_0$ @ $V = 0.995c$

$L = 0 L_0$ @ $V = c$

L_0 – LENGTH AT REST

L – LENGTH MOVING

$$M = \frac{M_0}{\sqrt{1 - \frac{v^2}{c^2}}}$$

M_0 – MASS AT REST

M – MASS MOVING

$M = M_0$ @ $V = 0$

$M = 1.15 M_0$ @ $V = 0.5c$

$M = 2 M_0$ @ $V = 0.87c$

$M = 10 M_0$ @ $V = 0.995c$

$M = \infty M_0$ @ $V = c$

RELATIVITY - GENERAL

GRAVITY

All atoms emit light at specific frequency characteristics of the vibrational rate of electrons within the atom. Every atom is therefore a "clock" and a slowing down of atomic vibration indicates the slowing down of such clocks. An atom on the Sun, where gravitation is strong, should emit light of a lower frequency (slower vibration) than light emitted by the same kind of atom on Earth. The effect is called the *gravitational red shift*. Since red light is at the low frequency range of the visible spectrum, the lowering of frequency shifts the color toward the red. The gravitational red shift is observed in light from the Sun, but various disturbing influences prevent accurate measurements of this tiny effect. It wasn't until 1960 that an entirely new technique using high frequency gamma rays from radioactive atoms permitted incredibly precise and confirming measurements of the gravitational slowing of time between the top and bottom floors of a laboratory building.

Measurements of time depend not only on relative motion in travel through space but also on the relative gravitational field intensities of the regions in which the events are taking place and being measured. Just as time dilation in special relativity is relative to the differences in motion between the observed frame of reference and the frame of reference from which observations are being made, the gravitational red shift of general relativity is relative to the differences in gravitational field intensities at the location of the event and the location of the observer of the event. Where gravitation is seen to be stronger, events are seen to proceed more slowly into the future. As viewed from Earth, a clock will be measured to tick slower on the surface of a star than on Earth. If the star shrinks, the resulting increase in gravitation at its surface will be seen to be accompanied by a corresponding slowing of time, and we would measure longer intervals between the ticks of the star clock. But if we made our measurement of the star from the star itself, we would notice nothing unusual about the clock ticking.

If a person stood on the surface of a giant star that began collapsing, we, as outside observers, will notice a progressive slowing of time on the clock of the person on the star as the gravitational field increases due to the density of the star increasing as the collapse occurs. He does not notice any difference in his own time because he is viewing events within his own frame of reference, and notices nothing unusual at all about his time. As the collapsing star proceeds toward becoming a black hole with time still proceeding normally to him, to us on the outside, his time approaches a complete stop; we would see him frozen in time with an infinite duration between the ticks of his clock. From our view, his time stops completely. Here, the gravitational red shift, instead of having a tiny effect, is dominating.

It is important to note the relativistic nature of time in both special relativity and general relativity. In both theories, there is no way that you can extend the duration of your own experience. Others moving at different speeds or in different gravitational fields may attribute a great longevity to you, but your longevity is seen from their frame of reference - never your own. Changes in time are always attributed to the person being observed, not you.

Since gravity is strongest on the surface of an object, such as Earth, it decreases both as you go up in a plane or descend in a mineshaft (gravity is zero at the center of the Earth) the person on the ground will age slower than either of his two friends due to the gravitational red shift. The difference would only be a few millionths of a second per decade in this case. It would be much longer in a greater difference of the gravitational forces.

Our present theory of gravity has a built-in restriction that the strength of gravity between two standard masses at a given distance apart is the same wherever in space they are located and whenever in time they exert their force. In the case of Earth, it will pull on an apple with the same force whether Earth is located in the Milky Way or the Andromeda galaxy. Likewise, it will pull the apple equally hard today as it would have done a billion years ago.

Einstein's general relativity forecast the discovery of objects and phenomena that were unimaginable at the time the theory was formulated - space-time, black holes and the expansion of the Universe. These actual discoveries came years later, verifying again the formulas of general relativity. Bent space-time was verified in 1919, the expanding Universe in 1922 and black holes were verified in the 1980's through the study of quasars. All are common knowledge today.

Acceleration is equivalent to gravity; so general relativity, a theory of gravity, also must be a theory of acceleration.

RELATIVITY - GENERAL

GRAVITY WAVES

General relativity requires a new geometry: A geometry not only of curved space but of curved time as well - a geometry of curved four-dimensional *space-time*. Gravity is evidence of space-time geometry; a gravitational field is a geometrical warping of space-time. The presence of mass results in the curvature, or warping, of space-time; and, by the same token, a curvature of space-time identifies a presence of mass. Instead of visualizing gravitational forces between masses, we discontinue the concept of force and now think of masses responding in their motion to the curvature, or warping, of the space-time they inhabit. It is the bumps, depressions and warping of geometrical space-time that are the phenomena of gravity.

We cannot visualize the four-dimensional bumps and depressions in space-time because we are three-dimensional beings. We can get a glimpse of this warping by considering a simplified analogy in three dimensions, a heavy ball resting in the middle of a waterbed. The more massive the ball, the greater it distorts, or sinks into, the two dimensional surface. A marble rolled across the bed, but away from the indentation of the ball, will roll in a relatively straight line path. A marble rolled near the heavy ball's indentation will curve as it rolls across the indented surface. If the curve closes upon itself, its shape is an ellipse. The planets that orbit the Sun also travel along four-dimensional geodesics in the warped space-time about the Sun. Photons of electromagnetic radiation - light, x-rays, gamma rays - have no mass and will pass these space warps, following the curvature in a straight line.

Every object has mass and therefore makes a bump or depression in the surrounding space-time. When an object moves, the surrounding warp of space and time moves to readjust to the new position. These readjustments produce disturbances in the overall geometry of space-time. This is similar to moving the heavy ball that rests on the surface of the waterbed. Upon moving the heavy ball, a disturbance moves across the waterbed surface as a wave. If we move a more massive ball, then we get a greater disturbance and the production of even stronger waves. The waves travel outward from the gravitational sources at the speed of light and are called *gravitational waves*. Shake your hand back and forth - you have just produced a gravitational wave. It is not very strong, but it exists.

One of the more exciting results of Einstein's theory of gravity - the general theory of relativity - is the possibility of gravity waves. The force of gravity is in some respects like the force of electricity between oppositely charged particles, or the attraction between magnets, but with mass playing the role of charge. When electric charges are violently disturbed, such as in a radio transmitter, electromagnetic waves are generated. The reason for this can readily be visualized. If an electric charge is pictured as surrounded by a field, then when the charge is moved the field must also adjust itself to the new position. However, it cannot do this *instantaneously*; the theory of relativity forbids information to travel faster than the speed of light, so the outlying regions of the field do not know that the charge has moved until at least the light travel time from the charge. It follows that the field becomes buckled, or disturbed, because when the charge first moves, the remote regions of the field do not change whereas the field in the proximity of the charge is quick to respond. The effect is to send a kink of electric and magnetic force traveling outward through the field at the speed of light. This electromagnetic radiation transports energy away from the charge into the surrounding space. If the charge is moved back and forth in a systematic way, the field distortion moves likewise, and the spreading kink takes on the feature of a wave. Electromagnetic waves of this sort are experienced by as visible light, radio waves, heat radiation, x-rays and so on, according to their individual wavelengths.

An analogy to the production of electromagnetic waves: We might expect the disturbance of massive bodies to set up kinks in the surrounding gravitational field, which will also spread outwards in the form of gravity waves. In this case, though, the waves are kinks in space itself, because in Einstein's theory, gravity is evidence of distorted space-time. Gravity waves can be visualized as undulations of space, radiating away from the source of disturbance. Gravity waves in the direction of movement would be blue-shifted whereas the following gravity waves would appear to be red-shifted.

With special relativity, Einstein had established that nothing could travel faster than light. A conflict existed due to the fact that the gravitational force appeared to be transmitted instantaneously across space between any pair of objects. The apple falling from its branch to the ground and the Earth reacting to the Sun appear to be instantaneous reactions - faster than the speed of light. This is impossible because time is required to supply the information of the movement to the outer edges of the objects space-time. Whatever is encountered in an objects journey through space is in contact with the object's space-time - the part of the depression in the waterbed that a marble rolled across the bed would encounter. Without the understanding of general relativity, gravity and its effects would continue to be a mystery. Wavelengths are required for the transmission of electromagnetic radiation. Gravitons are the "photons" of gravity waves. Due to mass, gravity exists automatically - one cannot exist without the other.

Diagrams and waterbeds do not project space-time properly because these are simple two and three-dimensional examples from only one point of view. You must imagine an object in space with the curvature drawn toward the object from any and all positions in which you view the object.

Many astrophysical systems, such as orbiting binary stars, merging neutron stars and colliding black holes, emit powerful gravitational waves. According to the theory of general relativity, the waves are generated by any physical system with internal motions that are not spherically symmetric. A pair of stars orbiting each other will produce the waves, but a single star will not. These waves join the gravitational waves produced by quantum processes during the inflationary epoch following the big bang.

Detectors have been built to attempt to capture and measure these faint waves, but final discovery may have to wait until these detection devices can be put in space. This detector would consist of three identical spacecraft flying in a triangular formation and firing five million kilometer long laser beams at one another.

Cosmologists are still asking the same questions that the first stargazers asked as they looked outward:

Where did the Universe come from?

What, if anything, preceded it?

How did the Universe arrive at its present state?

What will be the future of the Universe?

RELATIVITY - GENERAL

CURVATURE OF SPACE

The oldest known maps were clay tablets made in Babylonia sometime around 2500 BC. Egyptian maps are dated around 1300 BC.

A geodesic line is defined as the shortest distance between two points. If you were to draw a straight line between Seattle and Tokyo on a world map you would assume this to be the shortest distance between these two points. We have always been told that the shortest distance between two points is a *straight line.*

Get a globe and some string. Place one end of the string on Seattle and run the string across the surface of the globe and place the other end of the string on Tokyo. Adjust the string back and forth until you know it is tight. Cut the string at Tokyo - this length of string gives you the *distance* from Seattle to Tokyo, the shortest distance. Now place the string back on the globe and draw it tight between Seattle and Tokyo. Draw a line on the globe, following the string. You have now have the *direction* you must travel. This length of string and the line you drew on the globe represent a *geodesic* - a *great circle* route - the shortest distance between these two points.

Now go back to the world map you had earlier with the straight line you had drawn between Seattle and Tokyo. Let's see if the straight line is the correct path we would be flying over the Pacific Ocean. Return to your globe and pick out five or six spots equidistant from each other and list each spot's latitude and longitude. Take this data back to your world map and plot these positions on it. Now connect these points you have just plotted and take a good look at them. They are not on the straight line you had drawn on the map! The new line we have drawn is *curved*! Not only is it curved but it is curved upwards toward the North Pole! And it is LONGER! Before you get too excited, I want you to know the curved line you have just drawn is really the shorter of the two lines and represents a *geodesic* - the *great circle* route. To ease your fears about the curved line being longer and my telling you that this is the shortest route between Seattle and Tokyo, let's explain something about the difference between maps and globes.

Maps are drawn with the latitude lines (the ones going from left to right on the map) and the longitude lines (the ones going from the top to the bottom of the map) shown as being vertical and horizontal, making rectangular boxes all over the map. Look at your globe and tell me where any RIGHT ANGLED RECTANGLES

ARE! The longitude lines, the vertical lines, all begin at the North or South Pole and spread out as they head toward the equator - they widen out as you move closer to the equator. Look at Greenland on your map and take a glance at Greenland on the globe - it is much smaller on the globe in relation to the United States. On the map, Greenland is as big as the United States. What is wrong? If you were to take the vertical lines on the map and pull them closer at the top than at the bottom, you would now be approximating what the globe is trying to tell you - it is the right picture. The map is a cylindrical projection which is distorted as distance from the equator is increased. Look at Greenland on each again to verify this. Maps are made of small areas of the globe and the smaller the area, the easier a rectangle on the map will resemble the actual measurements on the globe - but as you take in more and more area, the distortion begins to show. You just had a map that covered a lot of area where the distortion is greatest. You probably noticed (at least some of you did) that when you pulled the vertical lines on the map closer at the top, the straight line you had drawn on the map curved upward in the middle as the vertical lines at the top got closer to each other. Maps and charts are *projections* of the facts that are shown on globes and as long as the map is covering an area small enough, both latitude and longitudes on each are equal - remember from your integral calculus that a curved line consists of an infinite number of small straight lines. A map uses these small straight lines while a globe automatically gives you a curved line.

I know that in our minds while flying from Seattle to Tokyo, we gaze down and see the surface of either land or the Pacific Ocean and it looks flat but when we look at the horizon, we see the curvature of the Earth. It appears as though we are headed in a straight line (the map) but we are actually traveling through *curved space*. We on Earth do not hold a monopoly on this *curved space* because it happens to be everywhere else also!

The path of a light beam follows a geodesic. If three experimenters located on Earth, Venus and Mars measure the angles of a triangle formed by light beams traveling between these planets as they are positioned equidistant around the Sun - each being 120 degrees apart - they would discover that the light beams bend when passing around the Sun. Remember the ball on the waterbed, the depression caused a surface for the easiest trip around this curve of the depression rather than straight across - there was no *straight across* anyway. An equilateral triangle has three angles of sixty degrees each for a total of one hundred eighty degrees - the geometry of Euclid, *Euclidian* geometry. These experimenters would find that each had angles greater than sixty degrees between each! Euclidian geometry is for flat surfaces; space is curved and a different geometry is required - one included in relativity. In Einsteinian physics, objects move through curved space-time without being influenced by an external force of gravity - they follow geodesics, the paths of least resistance, the "straight lines" of curved space-time. The equations of general relativity describe the behavior of space-time in the presence of mass-energy. What matters, as far as the bending of real space-time is concerned, is the concentration of matter in a small volume. The greater the density of matter, the greater the distortion of space-time. Light, massless photons, travel in straight lines. The straight line near a massive object is curved toward the object which causes the massless photon to appear to curve toward the object. The light the experimenters are viewing travel the outer curved path because they are away from the massive center of the Sun - there is no straight line from planet to planet, only the curved lines are available due to the space-time diagram.

In reality, nothing is flat or straight, it is only a small part of a curved surface. Explanations of these facts are simple when each segment is taken separately and broken down to the basics. Newtonian gravity and general relativity, gravity, are both correct in their applications - Newtonian gravity for maps and charts with Einstein's general relativity for the larger picture of space-time as we become involved in curved space and directions. Both Newtonian gravity and Einstein's general relativity require *things* to be round or become spheres - can't be helped.

CURVATURE OF SPACE IN PRESENCE OF A MASSIVE OBJECT

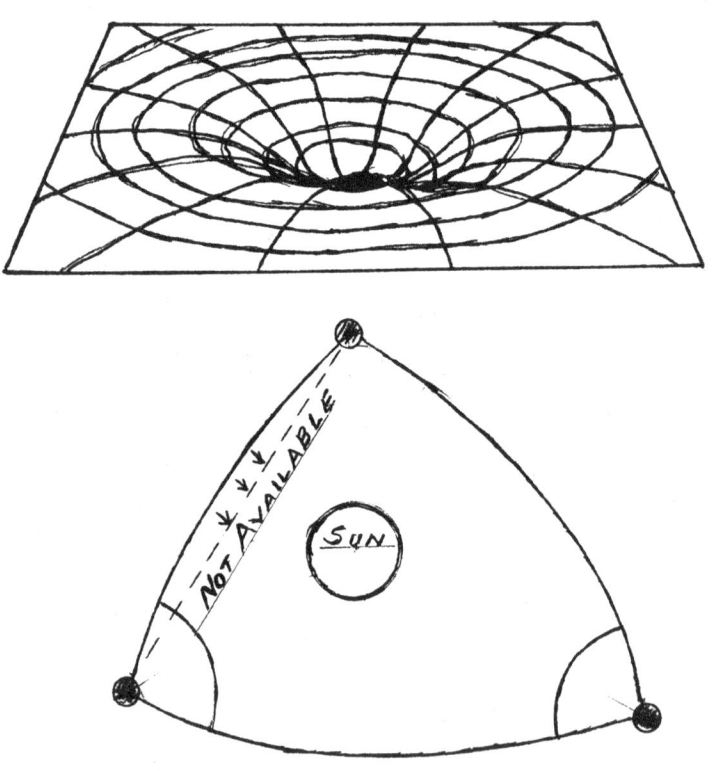

Geodesic - Curved Space

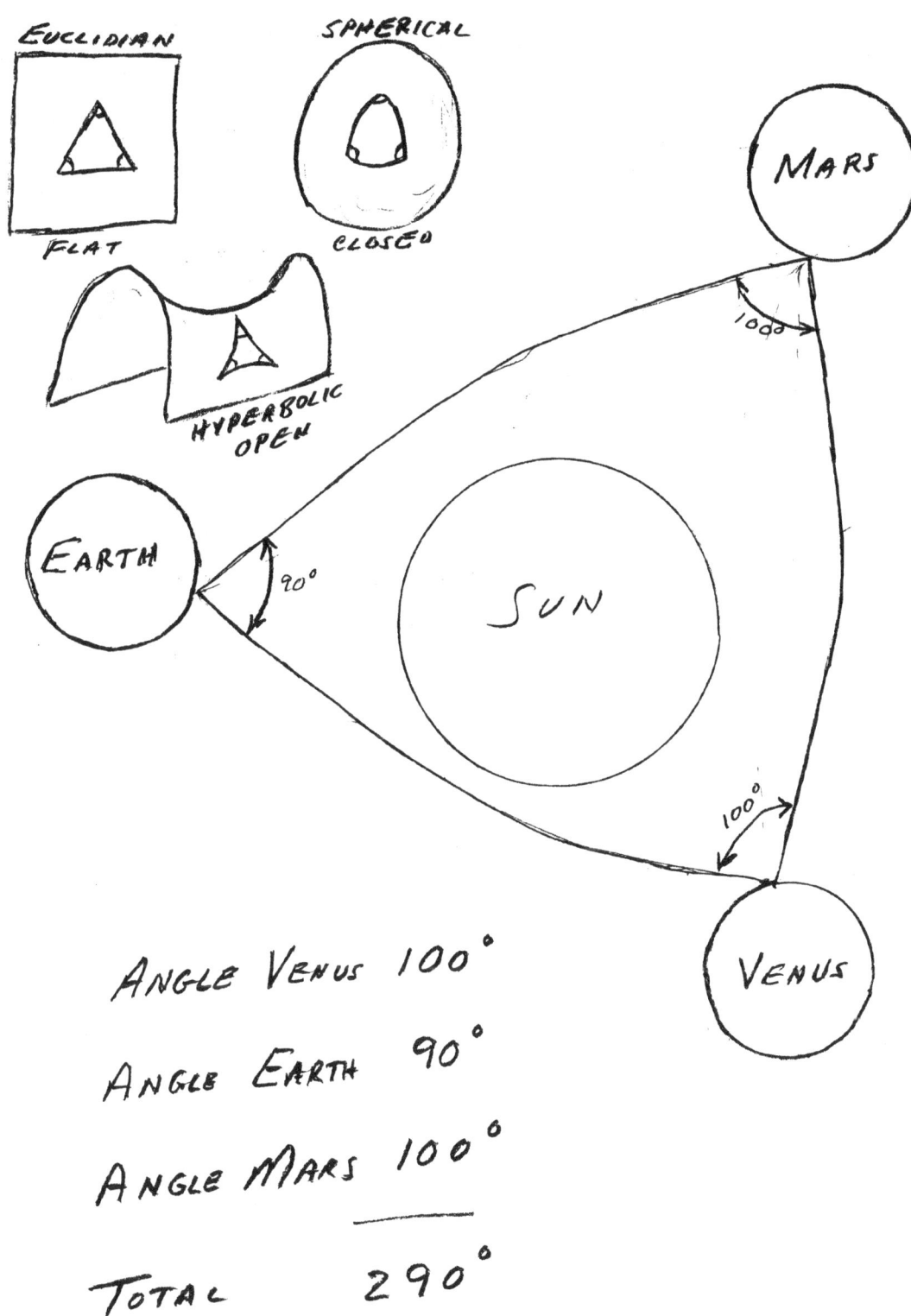

ANGLE VENUS 100°

ANGLE EARTH 90°

ANGLE MARS 100°
_____

TOTAL 290°

RELATIVITY - EXPANSION OF THE UNIVERSE

Space itself was created like matter in the big bang; there was no "outside" into which the explosion occurred. Space is expanding so that the gaps between the galaxies stretch. The red-shift of light - the reducing the frequency of light as we view these distant galaxies - is commonly called the Doppler Effect. A common example of the Doppler Effect is given when standing beside a railroad track. As the train approaches, the whistle becomes louder and higher pitched because the sound waves are being compressed as the train approaches - a higher frequency than is really being sent. After the train passes, the sound of the whistle changes to a lower sound because the source of the whistle is moving away from us causing the wavelengths to become longer and longer with the sound decreasing the further the train speeds from us. Light would have these same properties because of the wave properties of light would react in the same manner. This term "Doppler" was used to explain the recession of all the galaxies from each other and us causing the red-shift of the frequency of light we were receiving from them. A check of relativity revealed that this theory was completely wrong. Galaxies were not racing through space away from each other! The galaxies were actually "hanging out" in their space doing their thing - *space is expanding* and the galaxies are all going along for the ride.

The red-shift we see as we view the galaxies is in no way related to the Doppler Effect. It is the stretching of the wavelength caused by the expanding space. To visualize this action imagine a rubber sheet laid on the floor that is five feet square. Cut out a dozen or so circles six inches in diameter and space them equally around the rubber sheet. Place a pin through the center of each circle so it is attached to the rubber sheet - only attach at the center of each circle because these "galaxies" are just riding on the sheet. Now take a colored marker and make a sinusoidal wave ($\wedge\!\!\!\wedge\!\!\!\wedge$) between each of the circles nearest each other and to the distant ones in its view. Ask three of your friends to help you. Each take hold of a corner and, on a signal, walk away slowly from one another. You will notice that the circles are moving away from each other at a certain speed. You will also notice that the waves you had drawn between the "galaxies" are stretching - getting flatter ($\frown\!\!\!\frown\!\!\!\frown$) which relate to a decrease in frequency - a red-shift - that we observe. The "galaxies" do not change in size because all their action occurs in space (on the sheet) as the pulling of each corner *expands* the sheet. The expansion of space began at the big bang and will continue to create more space in this manner.

You notice that the expansion appears faster at the outer edges of the sheet than in the middle of the sheet. The distances from corner to corner are growing at a faster rate and the waves we drew are becoming flatter and flatter (lower and lower frequencies). Eventually the waves are stretched so far that they cannot be seen - this frequency is below the visible frequencies. This marks the visible horizon and radio telescopes are employed to "see" these lower frequencies. The Universe beyond is still there, but is invisible to our eyes.

If the expansion rate were to slow in the future, galaxies whose frequencies were flat (invisible to us) will become less flat and become visible as though they were slowing down from being "flatter than the frequency of light." The effect of this is that the universal visible horizon grows outward as time passes: The horizon itself, moving outward, appears to expand faster than the Universe, so that as time goes by it reveals more and more galaxies, even though the galaxies continue to retreat from us as space expands at a slower rate.

SPACE-TIME EXPANDS

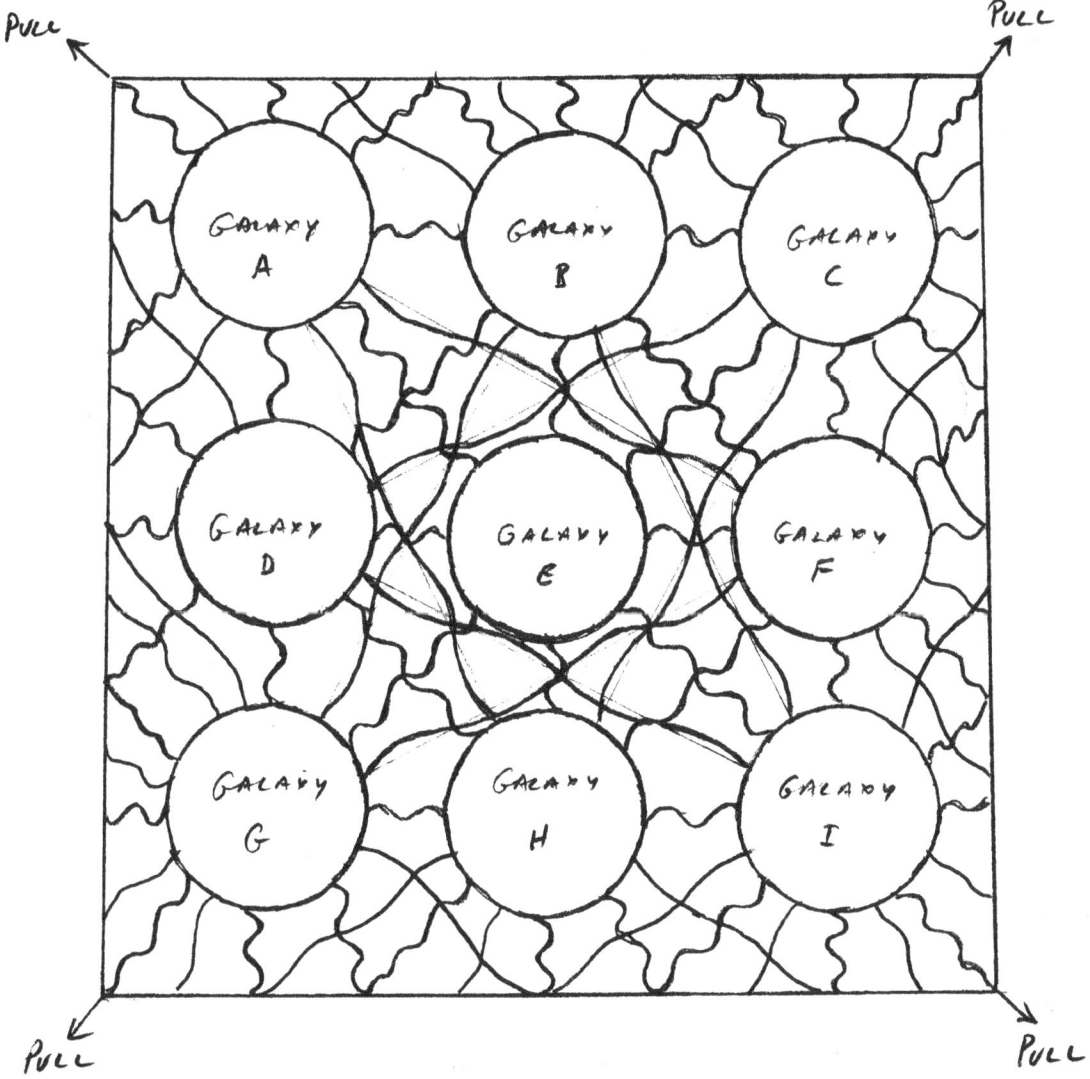

The shape of space has three possible choices - flat, open or closed. Flat space isn't two-dimensional, it just isn't curved. Each shape corresponds to a density of matter denoted by the symbol Omega (Ω). To create a flat universe, matter must reach a so-called critical density, which means Omega equals one (1). In a saddle-shaped universe, Omega is less than one; in a spherical universe, it is more than one.

If Omega is less than one, the universe keeps on expanding forever, but at an ever diminishing rate; that universe has the saddle-shape and is called "open." If Omega is more than one, the universal expansion slows and eventually reverses, collapsing in a cosmic crunch; that universe is spherical and "closed." In a flat universe, where the density of matter is exactly one, the expansion eventually slows very nearly to a stop but never actually reverses. Recent discoveries indicate that we are in a flat universe that is actually accelerating in its expansion.

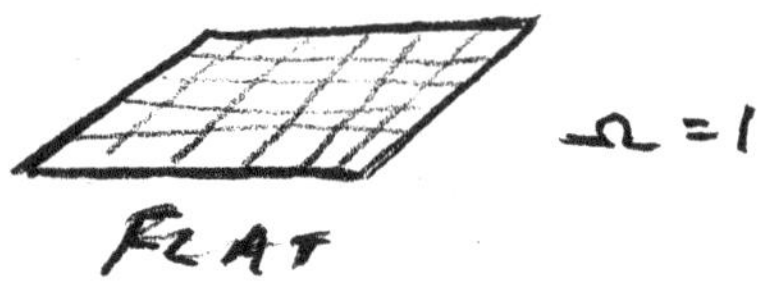

$\Omega = 1$

FLAT

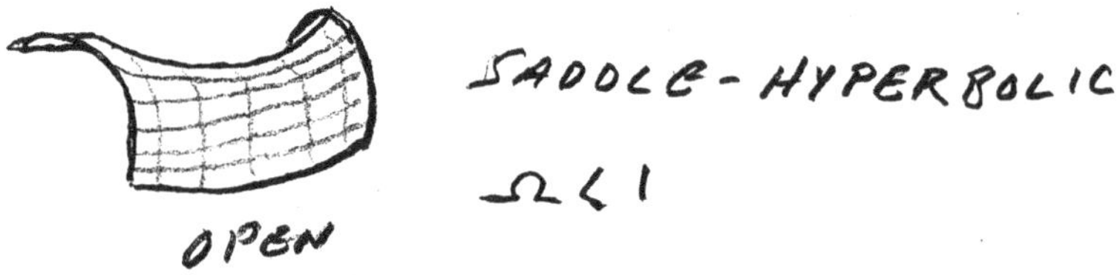

SADDLE-HYPERBOLIC

$\Omega < 1$

OPEN

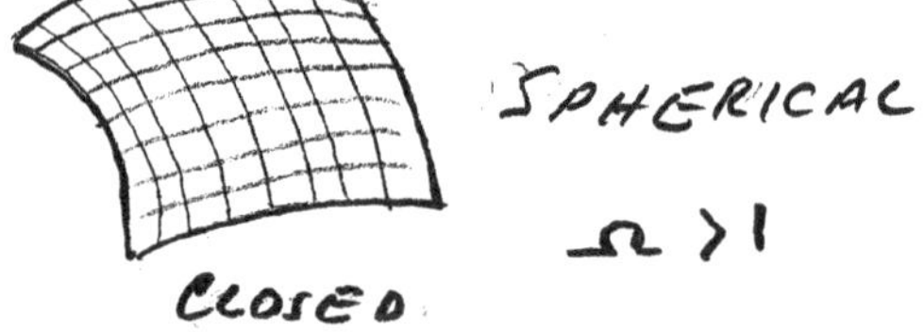

SPHERICAL

$\Omega > 1$

CLOSED

TOP VIEWS IN SPACETIME

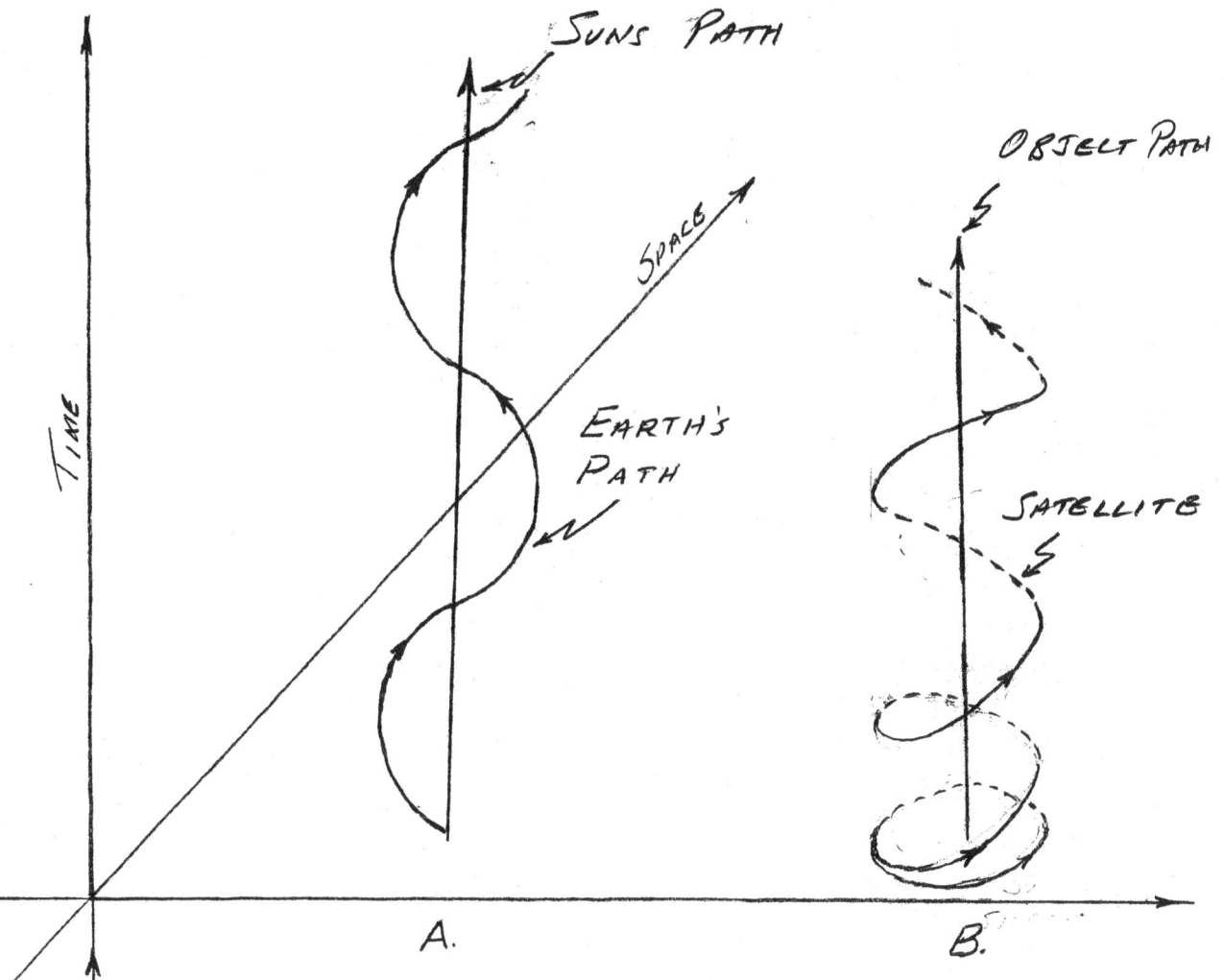

A. EARTH TRACES OUT A SINUSOIDAL PATH
AS IT ORBITS THE SUN IN AN EQUATORIAL ORBIT.

B. UNCOILING OF TWO-DIMENSIONAL ELIPSE
INTO A THREE-DIMENSIONAL HELIX FOR
POLAR ORBITING SATELLITE.

LIGHT CONE

A light cone is developed by creating a picture of distance and time. We know the speed of light and how to calculate space distances. If we were to be able to stand on the Sun and it went "out," we would know that fact instantly because we were there and zero time was involved. What would have happened had we been standing on Mercury when the Sun went "out?" We are now thirty-six million miles from the Sun and it would be three minutes and thirteen seconds later before it got dark here on Mercury - remembering that light takes time to travel from one place to another. A graph of distance and time will create a line that creates a cone - a *light cone* because it would be true no matter which direction is chosen. In the example we see Earth to the right of the Sun at our distance of ninety-three million miles. A line drawn from Earth to the light cone projects to an eight minute twenty second time. From this we can see that Mercury would have been dark for over five minutes before it became dark on Earth had the Sun gone "out." The further from the source, the longer the time. The planets in our solar system are listed in their light-time from the Sun.

PLANET		LIGHT-TIME FROM THE SUN	
Mercury		3 minutes	13 seconds
Venus		6 minutes	
Earth		8 minutes	20 seconds
Mars		12 minutes	42 seconds
Asteroid Belt		22 minutes	22 seconds
Jupiter		43 minutes	13 seconds
Saturn	1 hour	19 minutes	16 seconds
Uranus	2 hours	41 minutes	3 seconds
Neptune	4 hours	10 minutes	31 seconds
Pluto (minimum)	4 hours	10 minutes	31 seconds
(maximum)	6 hours	51 minutes	34 seconds
(mean)	5 hours	31 minutes	2 seconds
Earth to Moon			1.28 seconds
Earth to Mars	(nearest)	4 minutes	23 seconds
	(farthest)	21 minutes	

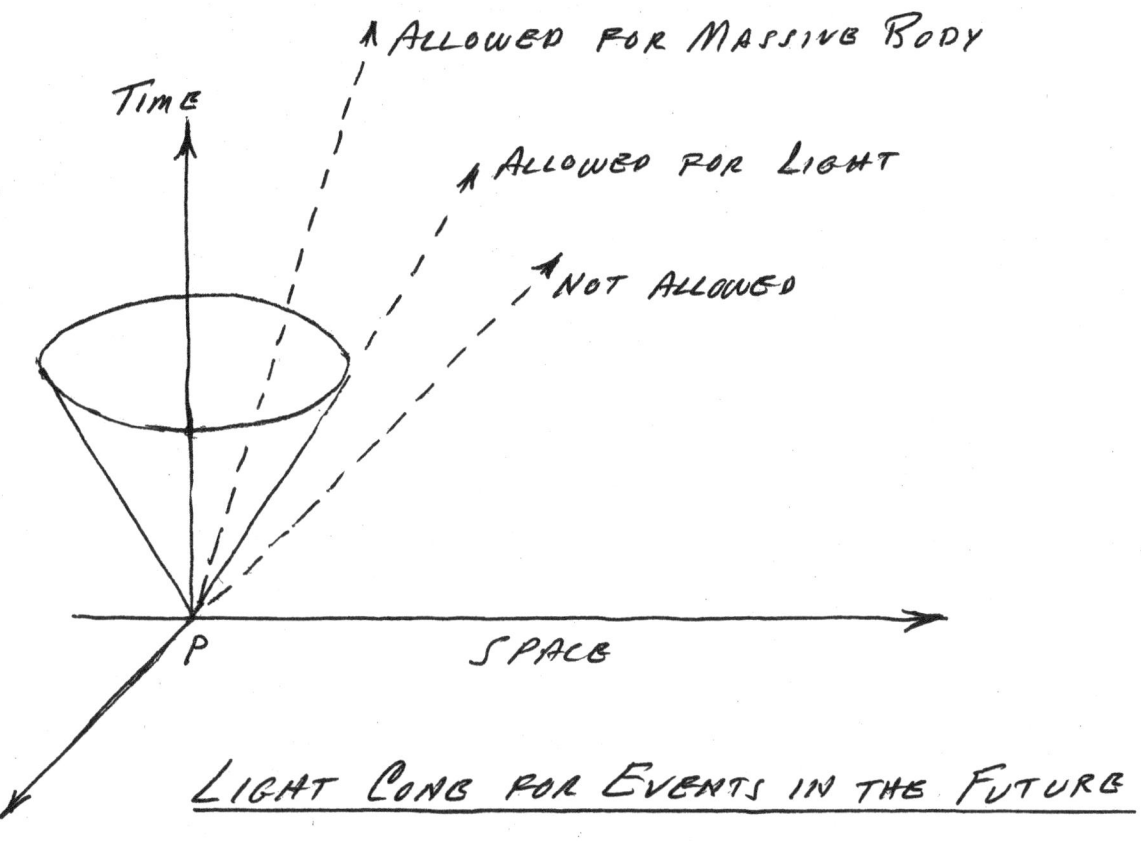

TIME

ALLOWED FOR MASSIVE BODY

ALLOWED FOR LIGHT

NOT ALLOWED

P

SPACE

LIGHT CONE FOR EVENTS IN THE FUTURE

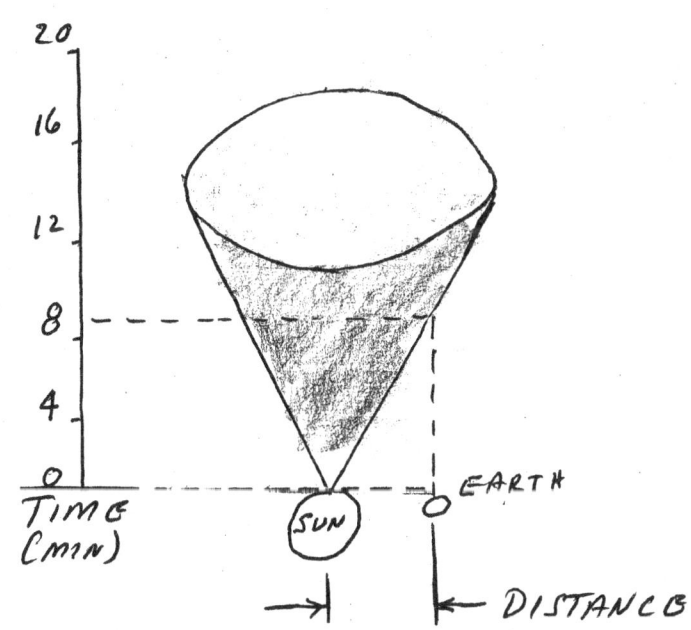

20
16
12
8
4
0

TIME
(MIN)

SUN

EARTH

DISTANCE

IF SUN STOPPED SHINING, IT WOULD BE
EIGHT MINUTES BEFORE EARTH ENTERS
THE DARK LIGHT CONE

BLACK HOLE

According to general relativity, there must be a singularity of infinite density and space-time curvature within a black hole. Eventually, when a star has shrunk to a certain critical radius, the gravitational field at the surface becomes so strong that the light cones are bent inward so much that light can no longer escape - space-time closes in on itself.

J. Robert Oppenheimer, the physicist who had headed the Manhattan Project that developed the atom bomb during World War II, developed a theory in 1939 that is now accepted without question. He calculated that if the mass of a star was 3.2 times more than the mass the Sun, the star would collapse until it was concentrated at a point - later to be called a *black hole*.

A black hole ought to emit particles and radiation as if it were a hot body with a temperature that depends only on the black hole's mass: the higher the mass, the lower the temperature. The particles do not come from within the black hole, but from the "empty" space just outside the black hole's event horizon, according to the quantum theory. A black hole with a mass of three times that of our Sun would have a temperature of only one ten-millionth (10^{-7}) of a degree above absolute zero, which is zero Kelvin (-273^0 Celsius).

If something falls into a black hole, its mass will increase, but eventually the energy equivalent of that extra mass will be returned to the Universe in the form of radiation - a recycling of energy. The fact that gravity is attractive (space-time curved) means that it will tend to draw the matter in the Universe together to form objects like stars and galaxies. However, eventually the heat or the angular momentum will be carried away and the object will begin to shrink. If the mass is less than about one and a half times that of the Sun, the contraction can be stopped by the pressure of electrons on the neutrons. The object will settle down to be a white dwarf. This is called the Chandrasekhas Limit - 1.44 solar masses (SM). Neutron stars fall in the range above this limit to about three solar masses.

If the mass of the collapsing star is greater than three solar masses, there is nothing that can hold it up and stop it continuing to contract. Once it has shrunk to a certain critical size, the gravitational field at its surface will be so strong that the light cones will be bent inward. The outgoing light rays are bent toward each other, causing them to be converging rather than diverging. This means there is a closed trapped surface - a black hole.

One of the weirdest implications of Einstein's general relativity theory is that as a black hole spins, it pulls space-time along for the ride. The spinning of the central black hole of a galaxy inside a huge magnetic field produces power the same as an electric generator. This energy contributes to the bright glow of iron atoms and ultra-hot matter swirling in a region called the corona. Space isn't sitting there stationary outside the black hole, it is always being stretched and pulled into the black hole. Time is also being stretched due to wavelength stretching. To an outside observer, any object approaching a black hole with activity present would show a slowing of this activity the closer it came to the black hole. Think of a black hole not simply as a place where gravity is extremely strong but as a place where the fabric of space-time is being pulled continuously into the hole.

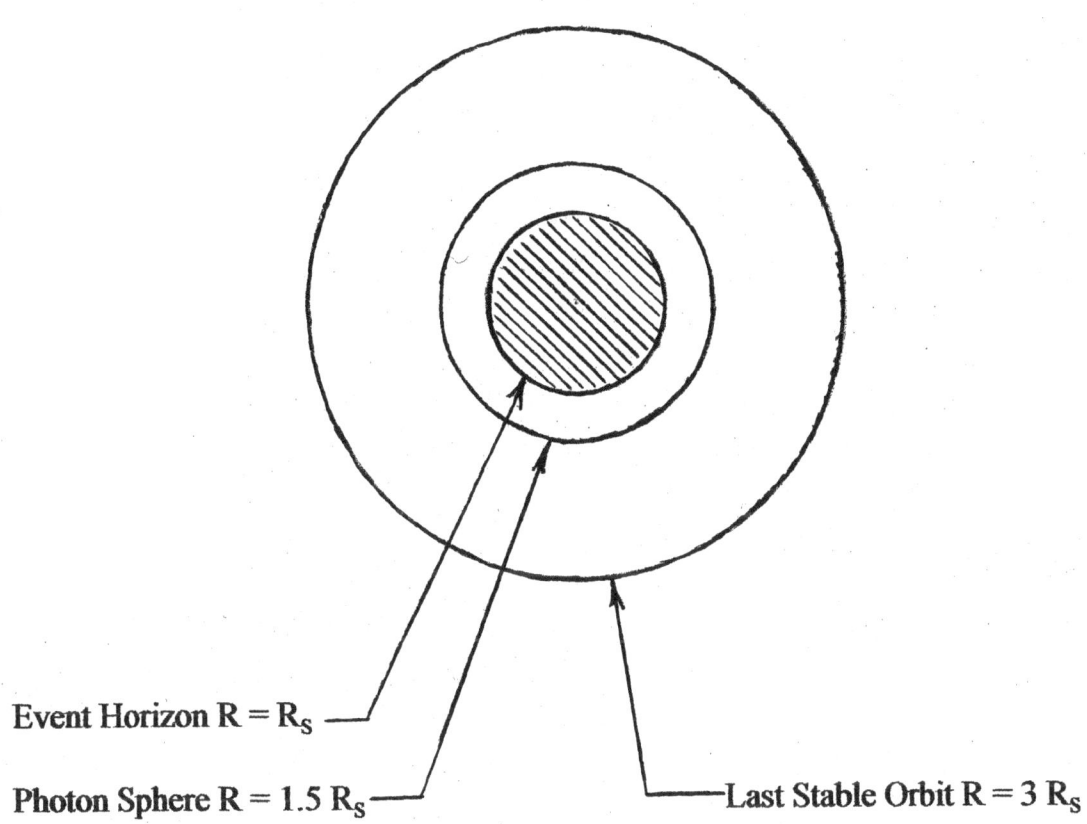

Event Horizon R = R_S

Photon Sphere R = 1.5 R_S

Last Stable Orbit R = 3 R_S

SCHWARZSCHILD BLACK HOLE

1.5 R_S - Light beam turns back and traces out a circular orbit.

3 R_S - No material object can orbit fast enough to resist infall.

A black hole is space-time bent back on itself.

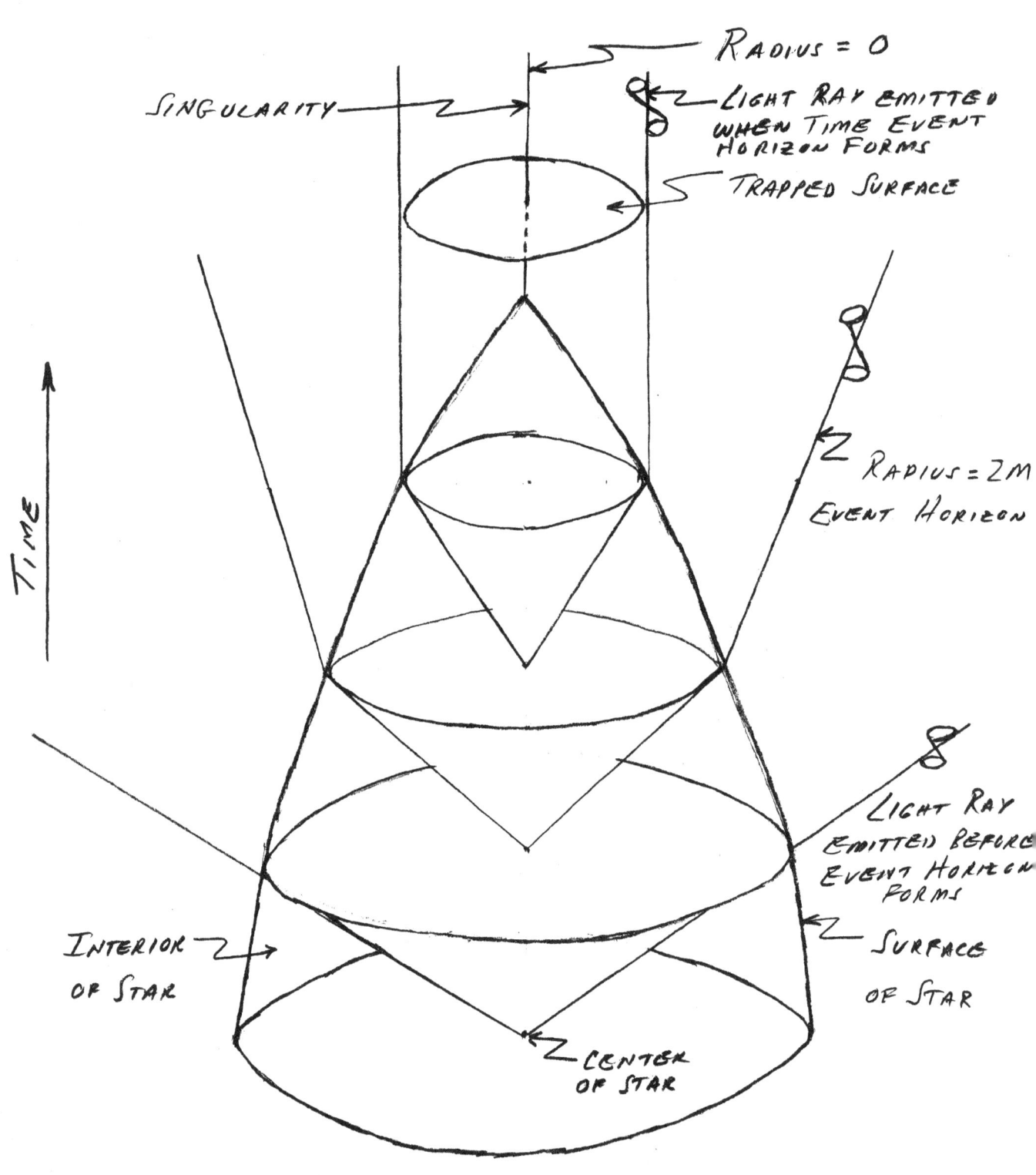

RADIUS = 0

LIGHT RAY EMITTED
WHEN TIME EVENT
HORIZON FORMS

TRAPPED SURFACE

SINGULARITY

RADIUS = 2M
EVENT HORIZON

TIME

LIGHT RAY
EMITTED BEFORE
EVENT HORIZON
FORMS

SURFACE
OF STAR

INTERIOR
OF STAR

CENTER
OF STAR

A SPACE-TIME PICTURE OF THE COLLAPSE
OF A STAR TO FORM A BLACK HOLE. SHOWN
IS THE EVENT HORIZON AND A CLOSED
TRAPPED SURFACE.

BLACK HOLE DIAMETER:

2.9 KM TIMES NUMBER OF SOLAR MASSES.

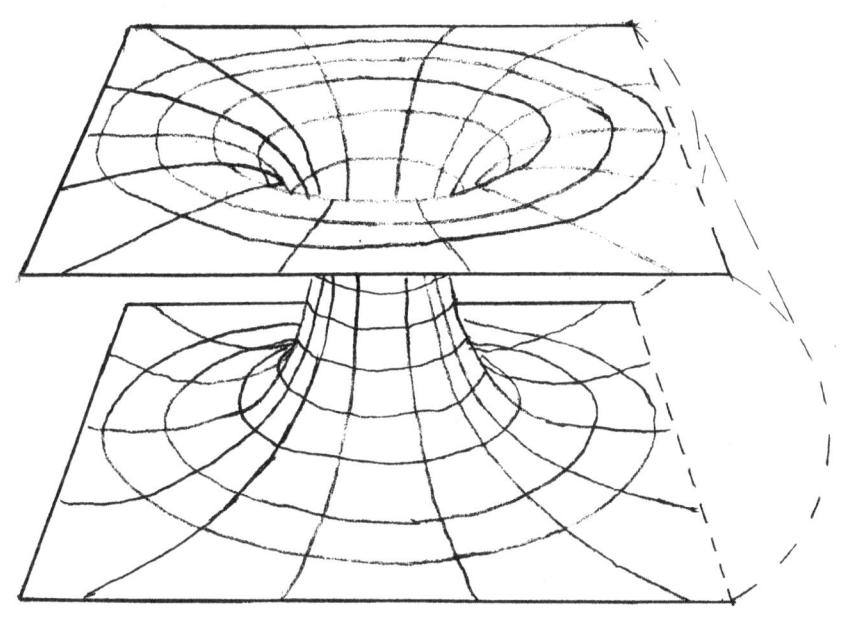

BLACK HOLE ——

WORM HOLE - - - -

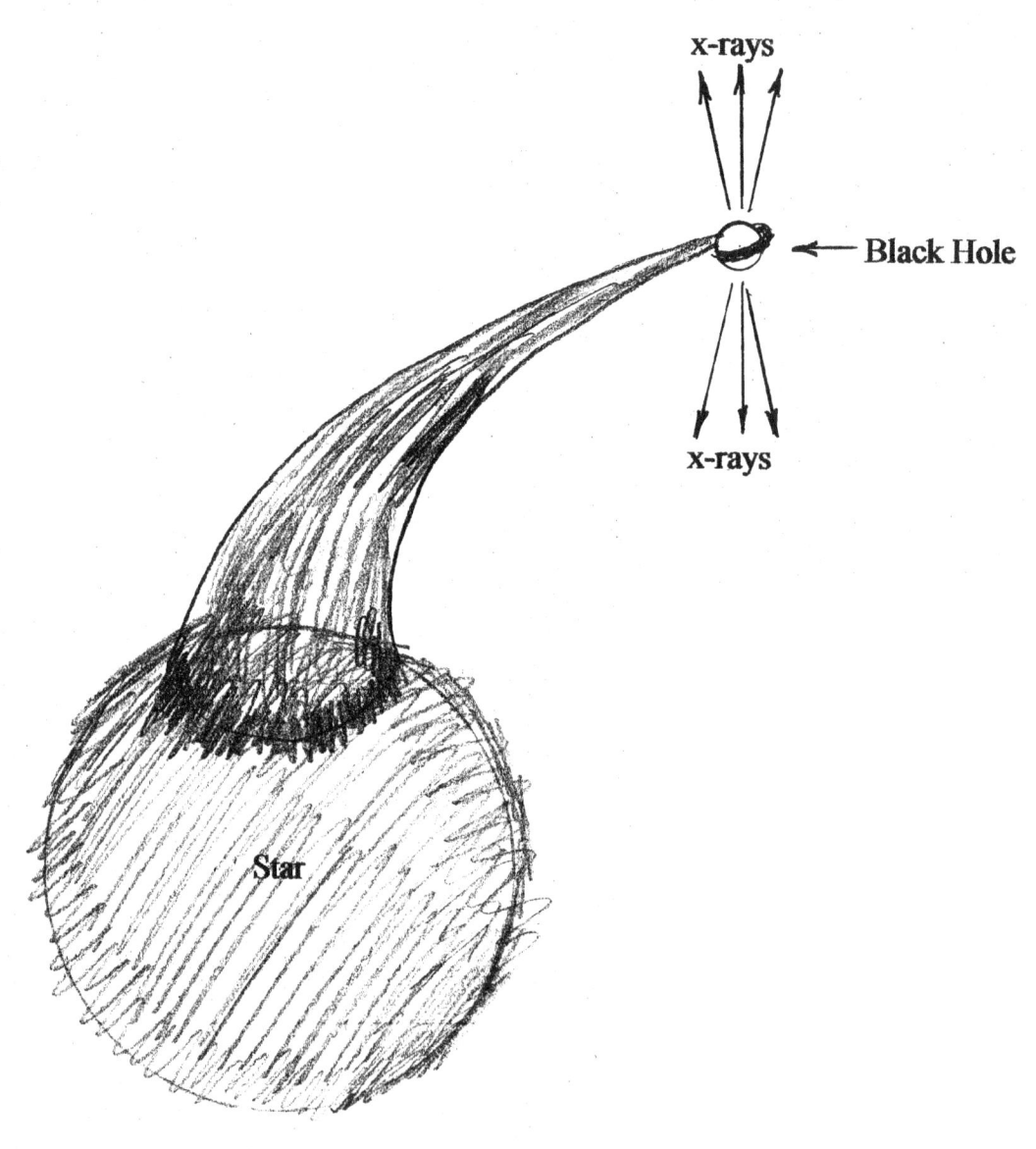

THE FUTURE OF THE UNIVERSE

The prolific radiation of energy by the Sun has to be paid for in nuclear fuel, and eventually the fuel reserves will start to run out - this should occur in the next four to five billion years. Since the age of the Universe is around thirteen and one-half billion years old and the age of the Sun is already four and one-half billion years old, this makes the Sun comfortably middle aged.

As the fuel runs low, the Sun will expand until it becomes a red giant. The core of the Sun will shrink further and further until quantum effects intervene to stabilize it. At this stage the Sun may have become so enlarged that the inner planets of Mercury and Venus will have been engulfed, the Earth's atmosphere stripped away and the solid rocks melted and then vaporized. Thereafter the Sun will embark on a new and erratic career, in which the nuclear burning of hydrogen fuel will be replaced by the less efficient burning of helium and then by heavier and heavier elements.

When all the fuel is exhausted, the Sun will consist of moderately heavy elements such as iron. Any further fusion of nuclei would not result in the release of energy. Iron is the most stable nuclear form, and according to the second law of thermodynamics, all systems seek out their most stable state. During this phase the Sun's central temperature will have risen steadily toward a billion degrees. Since all fuel is spent, the internal pressure will cease and gravity will take command. The Sun will begin to contract under its own weight, crushing the material to increase the density to over eighteen tons per cubic inch. The shrunken, burnt-out Sun will be reduced to the size that Earth used to be. It will then become inert for countless billions of years, slowly fading and cooling to end its career as a black dwarf star. The remaining planets will continue in their orbits - with orbits modified to fit the new system - but still orbiting their dead star. Gravity - space-time - has not disappeared, it has been modified.

The same pattern of instability, enlargement, fuel starvation and collapse will be repeated throughout our galaxy and in every other galaxy. One by one the stars will burn their way through the nuclear cycle until they can no longer hold up their own weight against the collapsing force of gravity. Some stars will die in more spectacular fashion, as supernova, blowing themselves apart as their cores implode catastrophically and release tremendous energy. The remnants of the lighter of these exploding stars will consist of dispersed debris surrounding a chunk of ultra-crushed matter in which the equivalent of one and one-half to three solar masses is compressed into a spherical volume of only a few miles across. So great is the gravity of such an object that a teaspoonful of matter would weigh more than the

317

mass of Earth. The force of gravity is great enough to cause the atoms to be compressed to become neutrons - a neutron star.

As for the dead stars themselves, there is plenty of further activity in a time scale that is greatly increased. Black holes will tend to swallow up any object they encounter and the massive black holes in the center of galaxies will continue to grow. The orbits of the stars will slowly decay due to the emission of gravitational radiation - wavelike ripples of space that reduce the orbital energy of all massive objects. Over a vastness of time, the stellar remnants will tend to drift closer and closer to their galactic center, eventually to be consumed in their enormous central black hole. Some dead stars may escape this fate as a result of a fortunate collision with another dying star which will knock them out of the galaxy altogether, to roam in solitary confinement in the vastness of intergalactic space.

For the stars that do escape, and for any gas or dust that avoids a black hole death, the reprieve is only temporary. The neutrons and protons decay into positrons and electrons, which then start to annihilate each other - opposite charges of the same object, when they meet, eliminate each other to produce a final neutral charge. All solid matter disintegrates and the outcome of this carnage depends on the speed with which the Universe is actually expanding. The electrons and positrons will be swept apart by the expansion faster than they can annihilate each other, so complete annihilation will not occur and there will always be some particles left. Those that do not annihilate produce gamma radiation, which itself slowly weakens with cosmic expansion. In addition there will be neutrinos and heat radiation left over from the big bang - all these components will gradually cool towards absolute zero, but at different rates.

While all this is happening, the gas clouds and nebula we observe will be starting their life cycles that will exist for billions of years. During these times as the expansion continues, particles that cannot be noticed, due to lack of density, may start to accumulate over time to continue forming new gas clouds and nebula to cause this process to be continuous. We do not know the future - only what might occur if present conditions, as we understand them, continue.

SEARCH FOR EXTRATERRESTRIAL INTELLIGENCE

In 1887, Heinrich Hertz demonstrated that the oscillations of electric charge in an ordinary spark emit radio waves. Earth has regularly transmitted radio frequencies since the 1920s, radar since the 1940s and television since the 1950s. We know that it is possible for life to form elsewhere in the Universe if certain conditions exist. To an outside observer, intelligent life on Earth developed after 1920 because this was the time man was able to transmit electromagnetic waves that could travel throughout the Universe. Until this date, man had only communicated with each other through the use of vocal chords - the same as animals. The first scheduled radio program was aired over KDKA in Pittsburgh, Pennsylvania the evening of November 2, 1920, when returns from the Harding - Cox Presidential Election were broadcast to an audience of a few thousand listening in on homemade sets. This event marked the birth of scheduled broadcasting.

We know that the transmitted radio, radar and television signals travel at the speed of light and how far they can travel each year - a light year. Since our first broadcast over eighty years ago, the farthest our initial broadcasts have reached by the year 2000 is eighty light years. A planet seventy light years away, assuming intelligent life existed with the capacity to receive and send radio messages, could study the waves for ten years - 1990 to 2000 (the signals sent from 1920 to 1930) - and then communicate with Earth. The signals sent at the end of their ten year study, the year 2000, would be received on Earth during the year 2070 - seventy years later. To communicate with this planet, a period of one hundred forty years is required between the sending a message and receiving a reply to THAT message.

Receivers on Earth to receive any signals from sources other than our own have only been active since the 1960's. These were initially set up because of the radio frequencies we were able to receive due to our technological advances - the frequencies from space which were first thought to have been originated by *others*. Any messages from any *intelligent life* to Earth in which a reply was expected prior to the 1960's would not have been acknowledged and the sender could have moved his research to other planets in search of intelligent life.

The search for extraterrestrial intelligence - SETI - has been active for many years with no results to date and becomes more intense as new technologies evolve. We know Earth's range since 1920, but if other intelligence evolved hundreds, thousands or millions of years earlier than Earth, there should be plenty of signals for us to receive and study. From the cosmological point of view, intelligence only exists between observers when they are able to communicate. This method, using today's technologies of radio, radar and television - electromagnetic radiation in the form of quantum - is all we now have. It is very obvious that, due to the time required, the person sending a message will *not* be the one receiving the answer.

SPACE TRAVEL

Science fiction has us traveling through space at "warp" speed from planet to planet and galaxy to galaxy. Technology is continually studying means to increase the speed in space. Any travel outside the solar system would involve travel at the speed of light for a minimum of 4.4 light years just to orbit Alpha Centauri - you cannot land on a star - too hot. Travel outside our galaxy - - - ?

If a location to land outside our solar system existed, it would probably be located at least fifty light years away, maybe even thousands. The chart shows the number of years required to travel these various distances at various speeds. To travel to a location fifty light years distant at the speed of one hundred million miles per hour would require three hundred thirty-five and a third years plus the travel time required to accelerate to this speed and then slow down enough to land. The studies in these fields will push the limits of relativity and quantum physics. Do not be surprised to see advances in each of these fields as time passes. A new science, in the field of physics, may be required.

At present the fastest object placed in orbit from Earth is a satellite on a polar orbit around the Sun. It was sent by way of Jupiter to give it an additional push in speed by "slingshotting" it using Jupiter's gravity. Its speed is a little over eighteen miles per second - sixty-seven thousand miles per hour. The speed of light is ten thousand times faster. It is obvious that additional energy on board a satellite is required for increases in speed. These answers are being sought through vigorous research and tests. One detail that must not be overlooked in considering space travel - the *guarantee* of an adequate food and water supply for the journey and sustenance on landing until new food and water resources are established.

SPACE TRAVEL - NUMBER OF YEARS REQUIRED

DISTANCE LIGHT YRS	SPEED - MILES PER HOUR			
	1,000,000	10,000,000	100,000,000	500,000,000
1	670.77	67.077	6.708	1.3415
5	3,353.85	335.385	33.539	6.7075
10	6,707.70	670.770	67.077	13.4150
15	10,061.55	1,006.155	100.616	20.1225
20	13,415.40	1,341.540	134.154	26.8300
25	16,769.25	1,676.925	167.693	33.5275
30	20,123.10	2,012.310	201.231	40.2450
35	23,476.95	2,347.695	234.770	46.9525
40	26,830.80	2,683.080	268.308	53.6600
45	30,184.65	3,018.456	301.846	60.3675
50	33,538.50	3,353.850	335.385	67.0750
100	67,077.00	6,707.700	670.770	134.1500
200	134,154.00	13,415.400	1,341.540	268.3000
300	201,231.00	20,123.100	2,012.310	402.4500
400	268,308.00	26,830.800	2,683.080	536.6000
500	335,385.00	33,538.500	3,353.850	670.7500
1,000	670,770.00	67,077.000	6,707.700	1,341.5000

One light year = 5.99×10^{12} miles - almost six trillion miles.

Speed of light (C) = 186,282.399 miles per second = 670,615,200 miles per hour.

Distance traveled per million miles per hour = 8.766×10^9 miles.

Example: To travel 100 LY @ 10,000,000 mi/hr = 6,707.7 years.

ON THE OTHER HAND - - -

What if visitors from outside our solar system have visited Earth in the past and continue to do so today? Why would they have chosen to visit Earth? If they have been here often, why haven't they left verifiable evidence of their visits? Why have they not attempted to officially communicate with the people of Earth? Where would this intelligent life have originated? We don't know these answers, so let's reverse the questions. When capable, where and why would Earth send explorers outside our solar system?

The following capabilities and knowledge must be known:
1. Possess the ability to travel to other solar systems. Return?
2. Identify a solar system that can supply a planet or planets that may have the basic requirements for life that could support us.
3. The solar systems that developed with ours should have developed within a period of plus or minus one billion years.
4. Upon arrival, expect to be either years ahead or years behind the host's intelligence capabilities.
5. Use extreme caution on all contacts until full capabilities of the place visited is known - diseases, climate, etc.

Why would we even attempt to reach such a place? Exploration is in our blood and we would want to make the journey just because it is "there." After discovering other solar systems that have planets that qualify for the sustenance of our lifestyle, statistical and probability studies would be made to choose a candidate for exploration. Exploration could be the main objective of the trip, but other needs may become more necessary; need for a new source of energy because the Sun has become erratic, or is running out of fuel; food problems projected due to over-population or environmental damage; a projected orbital fatal impact due to an asteroid; and other problems that may be projected to occur.

Since our candidates for exploration could be somewhere between five hundred to a thousand light years away, we must have achieved the ability to travel at very fast speeds. If our capabilities are one hundred million miles per hour, it would take a minimum of 3,354 years to travel to a planet located five hundred light years away, or only 6,708 years to travel to a planet located one thousand light years away. If we had the capability to travel at five hundred million miles an hour (the speed of light is six hundred seventy million miles per hour), it would take only 672 years to reach our destination if it were only five hundred light years away and 1,352 years to reach the destination of one thousand light years distant. Taking relativity into account, the travelers would age at a slower rate in relation to the speed involved, but they would still age.

The travel time denotes that the generation that lifted off from Earth would not be the one arriving at the destination. Life support must be able to be maintained so that the people arriving would appear much like their ancestors who departed on this journey. Artificial gravity must be used along with other amenities to make the journey comfortable. If we sent them without any form of gravity, they would arrive evolved into globs that would not be able to function. Communication with the home planet would always take twice as long as the distance traveled from home. Messages sent five hundred light years distant would reqire one thousand years for the sender to receive the answer - so much for any help such as, "Houston, we have a problem."

Upon arrival at our destination, an orbital exploration and mapping of the planet would be performed. This study would hopefully validate the reason for the choice of this planet. Assuming the information was correct, the next step would be to select a landing place remote from any life forms to assess the situation. It would be a nervous time for the explorers and much caution must be observed. How and when contact with the inhabitants of this planet would be accomplished is best left to the science-fiction novelists because I have no idea. There would be more of them than us in their land and we would act accordingly. Would you like to be the one that disembarked first onto this strange planet?

Before exiting to the surface of this new planet, we would have completed all our sampling tests of the atmosphere and also verified that no bacteria was present that would effect our health. Now would be the time to open the hatch and descend the ladder to our new home. We would send a message stating that we had landed safely and someone would receive this message on Earth by the time our fifth or sixth generation of descendants were now living here. One of the next items on the agenda would be trying to communicate with the intelligent life of the planet we are "visiting." We would want to indicate to them that we certainly come in peace - we do not have conquest on our mind; what would we do with this place if we did? We have to remember that the technology we possess is somewhere between 671 and 6,708 years old, depending on the planet we landed. We may be ahead of our host planet in technology, or behind. We have come for our survival, not our funerals. The sooner we and the other intelligent life can adjust to each other's presence, the better everything will be.

On Earth, some people think "they" have been orbiting and dropping in from time to time sampling over the past centuries. We would not want to waste several generations doing this after such a long trip - what would be the purpose? Most of this is beyond our comprehension at this time, but worth thinking about and planning for just in case either is possible - "them" visiting "us" or "us" visiting "them." If "they" are visiting "us" here in the early part of the twenty-first century, I wish they would land outside my window and invite me in so I could just sit back and listen and learn.

For those who listen to radio talk shows or watch television programs that feature "experts" on UFO's, space travel, communications, black holes and worm holes can believe or not believe what they hear and see according to their own outlook and knowledge of facts versus science-fiction dreams. These "experts" state, "At a black hole I think that as you pass through it, everything can go faster than the speed of light." Also, "Enter a worm hole and you might exit in another Universe - everything would happen faster than the speed of light." We really don't know because we haven't heard from anyone that has succeeded in this endeavor. All this kind of talk is easy for these "experts" to say but there is one thing that they seem to forget, you must get to the black hole or worm hole first! No one living will ever get to any of these because of the distances involved and time for travel - time is not on their side. I have better things to think about and I prefer to stay at home, anyway.

Myths that need to be explained scientifically are: People on the International Space Station, or any other orbiting vehicle, have escaped Earth's gravity. In fact, the astronauts experience a full 89 percent as much gravitational pull in orbit as they do at home on Earth in their living rooms. The sensation of weightlessness arises because any orbiting vehicle is constantly falling along its curved orbit, like a roller coaster taking a steep plunge, making everyone on board feel weightless. In the TV space programs, space engines do not have a *whooshing* sound - there is no atmosphere in space, therefore, no sound. Stars do not streak by these speeding spacecraft - light takes many years to traverse these distances. Pilots of the fighter spacecraft could not dodge enemy laser beams because nothing travels faster than the speed of light, so there would be no way to know that a beam was headed your way until it arrived.

You already knew all this because you have almost finised reading this book and have learned that fact is stranger than fiction and that learning the reason behind everything was a lot of fun. Now you can reason things out for yourself.

WHAT'S NEXT?

There is not anyone who can state in precise terms what the states of science, the arts and society will be one hundred years from today - that is a FACT. There is one statement that will be true - all will be more advanced in ways that we cannot comprehend.

As the 1800's came on the scene with sailing ships and muskets, people had no inclination or foresight into the fact that the 1900's would be entered with the beginning of the automotive world and the airplane - horses were going to be replaced! The steam engine and the cotton gin had been put into use during the nineteenth century - what more could man's ingenuity create? The introduction of the theories of relativity and quantum mechanics in 1905 were too extreme to be considered - God would not permit such things. Electricity was in a continual state of experimentation that might produce some results - no one really knew, so on we went living day to day, but still wanting to be the first to have any "new" item.

Who could envision what actually happened in the twentieth century? No one had the foggiest idea. The only prediction for the new century, which would also end by entering a new millennium, was the continual prediction of the coming of Armageddon and the end of the world. This was standard practice for the turning of the calendar over another hundred years by the religious fanatics that used the Gregorian calendar. These people were not intelligent enough to realize that a calendar is just an arbitrary time keeping mechanism for days and years. Many other calendars were also in use around the world by different religious groups, each with their own set of numbers, and they did not worry because they were in years of completely different numbers anyway.

What is covered in this book is but a pittance, a fraction of the knowledge gained in science - what about the arts and all the other advances realized? Not any individual in 1900 would even believe what has happened unless they had lived through these experiences and advances. We even went to the Moon! And we RETURNED! SAFELY!!! Had limits finally been reached? Hardly.

The people of today think there is no limit in man's ability to advance in any direction chosen. These people realize that the future is unlimited in all respects even though they have no idea of how we will arrive with the answers. Genetic engineering has many great possibilities with final results that we TODAY cannot state with finality what these may be - the results will be positive. It would be hard to surprise today's knowledgeable inhabitants about the future - they may think that you are not thinking futuristic enough when you describe plausible objectives that are being considered! We will still be surprised in 2100. So beware, you who write with such "knowledge" about the future - you will be the laughing stock of the future generations. That is what the future is all about - the UNKNOWN.

Mass ***	Matter	
Compound	Matter	
Molecule	Matter	
Element	Atomic	
Atom	Atomic	
Neutron	Atomic Particle	Nuclear
Proton	Atomic Particle	Nuclear
Electron	Atomic Particle	
Particle	Quantum	
Quark	Quantum	
Field	Quantum	
Wavelength	Quantum	
Frequency	Quantum	
Energy	Quantum	
Mass ***	Quantum	$E = MC^2$

DOES THE UNIVERSE REALLY EXIST?

A little over two hundred million years ago a dinosaur passed a mass of mud and clay that had transformed itself into a solid that we see as a boulder today. This boulder has existed for only one revolution of our galaxy - we know this due to science, astronomy and the knowledge of our galaxy and the Universe. As this same dinosaur continued his journey, he passed other rocks that had formed over two hundred million years earlier - these rocks have now existed for two revolutions of our galaxy. We can be sure that this dinosaur only had one instinct - survival - utilizing his limited intelligence capabilities.

John Wheeler, colleague of Albert Einstein and Niels Bohr, and also the scientist who chose the name "black hole," is concentrating on one question: "How come existence?" William Wooters, one of Wheeler's many students stated: "I don't know if human intelligence is capable of answering that question. We don't expect dogs or ants to be able to figure out everything about the Universe. And in the sweep of evolution, I doubt that we're the last word in intelligence. There might be higher levels later. So why should we think we're at the point where we can understand everything?" Humans have a much higher level of intelligence than our dinosaur friend and the animal world and have evolved intelligence to include *observation* and *analysis.* The Universe exists to us - the intelligent humans.

As we observe the animal world in each of their daily activities, we notice that they continue to have the same primary instinct as the dinosaurs - survival. To the animal world there is no "out there" - no Universe exists to them. The eventual appearance of Homo Sapiens Sapiens caused a change in the level of *observation* with the later evolution of *analysis.* These humans further developed *time, measurement, written languages* and *religion* for their daily use. To them, "out there" did exist and time has brought us to where we are today. If the human race were to disappear tomorrow, would the Universe continue to exist? It exists to us because we know of its existence. Remove the human race and *time, measurement, written languages, religions* and the *Universe* would cease to exist because the remaining life forms have no need, or knowledge, of any of them. To the animals, none of these ever existed - they were not cognizant of their existence, so they would have lost nothing. To the animal world, humans were just another species that had appeared on the scene and disappeared, as many species have.

The scientist may ask, "Did the Universe really exist before we started *looking* at it?" The answer would be, "The Universe *looks* as if it existed before we started *looking* at it." The Universe and the observer exist as a pair - in the absence of observers, our Universe is dead - it does not exist! No device can detect the existence of the Universe unless there is a person present to *observe* the detection and begin the collection of information.

In Russia there is a fairy tale about two frogs trapped in a can of sour cream. The frogs were drowning in the cream - there was nothing solid there, so they could not jump from this can. One of the frogs understood there was no hope, and he stopped beating the sour cream with his legs and just died - he drowned in sour cream. The other one did not want to give up, knowing there was absolutely no way he could change anything, but just kept kicking and kicking and kicking. Then, all of a sudden, the sour cream was churned into butter. Then the frog stood on the butter and jumped out of the can. When you look at the sour cream you think, "There is no way I can do anything with that." But sometimes, unexpected things happen. Intelligence did not save the surviving frog, persistence did. Science can be described as combining intelligence with persistence.

EPILOGUE

One of the most interesting times of my life occurred at the end of my freshman year in college in the spring of 1949. Everything was getting back to normal following World War II even though the Korean "Police Action" was evolving. The United States Congress knew that the future would bring growth and eventual statehood to the Territory of Alaska. One of the first ventures in opening communications and understanding between Alaska and the "lower 48" was to send people from the "lower 48."

The existing Alaska Railroad tracks that had been laid many decades earlier had used the standard smaller rails which ran from Seward, through Anchorage, to Fairbanks. These rails were to be replaced with the latest, larger rail system now in use on a new railroad bed, where possible, considering the topography of the area. Many construction companies were involved due to the several hundred miles of new construction that was required.

The U. S. Congress decided that the "people" to begin this venture of communication would be engineering students - two each from selected Universities throughout the country. Over fifty of us were chosen to assist in the surveying requirements for the summer in the construction of this new railroad route. I was overjoyed at being selected and then realized that I was the only freshman involved and I was preparing for a degree in Electrical Engineering, specializing in the growing field of electronics. Electrical Engineering has no requirements for surveying. As luck would have it, the other student selected was a sophomore Chemical Engineering student - who, by the way, graduated a couple of years later with straight A's throughout his academic career - also had the same problem about surveying, so we had to learn surveying fast because we had to leave for Alaska in two weeks. Most all of the other engineering students selected around the country were Civil Engineering students, so they had two weeks of vacation.

We had these two weeks to learn surveying and since we had just finished our finals, the Professor of Civil Engineering agreed to teach us two separate surveying courses, each a semester long, in the two weeks we had remaining. We surveyed from before daylight until after dark and successfully completed these courses. Bring on the Civil Engineers - we were ready!

The students from the South and the East Coast met in Chicago with the students from the midwestern universities and departed on an Alaska Airlines DC-4 that had been sent from Anchorage. This was my first airplane trip. We stopped in Great Falls, Montana for refueling and breakfast. We learned a lesson about Montana economics in the Airport Cafe as we paid for our breakfast with the $10 and $20 traveler's checks we all carried. Montana did not use "greenbacks" - we all left leaning to one side loaded with either nine or nineteen silver dollars. I still have those silver dollars.

Our next stop was Seattle, Washington for refueling and lunch. The West Coast students joined us there. All students were now aboard as we departed for Anchorage, Alaska to begin our surveying careers for the summer. Arriving at Anchorage, we had dinner and spent the night at the Alaska Railroad barracks. At our breakfast and organizational meeting the next morning we were to be notified of our work assignments on the surveying crews and with which construction company we would be employed. What we heard was somewhat different. The appropriations that Congress had sought for this project were not approved. The construction companies were to adjust accordingly. What about us?

We were instantly adopted by Alaska Railroad. All of us were in excellent physical condition, one of our qualifying requirements, and were made employees of the railroad for the summer. All of our activities that had been scheduled for us as students as a group were approved by the railroad, so we would not miss any of those activities. We were to become "gandy-dancers," or the common term "section hands." The railroad needed additional workers to maintain the existing tracks during the construction of the new railroad bed. Since railroads are divided into sections, each university pair was assigned to a section and dropped off at locations from Seward to Fairbanks. Our section was Rainbow - a few sections south of Anchorage. So much for our expertise in surveying!

A "section hand" is a part of a group of eight workers that make up a "section gang." These eight report to the Assistant Foreman who is responsible to the FOREMAN! The duties of the section gang are to maintain the integrity of the existing infrastructure of the railroad - of course we did not hear these words in this context from our foreman. He had been a Polish refugee that had immigrated to Alaska prior to World War II to work for the railroad. He had worked his way up to this position and was about five feet tall with equal girth - a happy-go-lucky, knowledgeable and fully qualified fellow. His explanation of our "job description" was that we were to pull out the bad crossties, put in the new ones, replace worn rails and level the track by pushing the gravel of the railroad bed under the crossties until his "trained eye decided that the tracks were level for the comfortable passage of certain personages," or something like that. Pushing the gravel under the crossties performing this leveling operation - is called GANDY-DANCING!

Time passed quickly because we were busy. Nothing too fast, you realize, because the other six members of our "gang" were Native Alaskans whose internal clocks were tuned to a pace from past centuries in which projects and schedules were unknown, much less scheduled completion dates. Another of these jovial people's practices occurred every payday - this occurred every two weeks. A schedule when each person could have the weekend off and travel to Anchorage for rest and recreation (R & R) was posted. Each time a Native Alaskan left for his two days of R & R with his paycheck, it was for these two days plus an additional two weeks. These two weeks were always scheduled for their absence because each had a habit of spending two weeks in jail for activities they encountered during their two days of R & R. All sections on the railroad had this situation with their workers and now you know why the Alaska Railroad joyfully rescued our group of sober, well-conditioned young men into their system. None of the students ever complained of our assignment because everything was an eyeopener - the people were terrific and the scenery was beautiful.

As soon as our foreman discovered we were studying engineering and seeing our efficient work results, he put us on a pedestal and always pointed us out to the others. Each time a locomotive would pass, he would point to us and then to the locomotive "engineer" and proudly say to the rest of our section gang, "These boys will be up there some day when they get through studying." We let the issue rest because neither of us had the heart to try to explain Electrical or Chemical Engineering to him - he was having such a good time telling everyone this. We loved talking to him, especially trying to decipher his broken English.

From the above, it is easy to compute that our standard "gang" of eight hardly ever exceeded six workers due to the overlapping of the schedule of two days R&R plus two weeks. Our section gang operated very well adapted to this schedule. The section was about eight miles of track, four miles in a northerly direction from our centrally located section house, and four miles south. This area is located in some of the most beautiful country I had ever seen.

During the summer a special train was sent to collect all of us several times for special events. We went to Seward, to Anchorage several times, and to Fairbanks. The highlight of our trip was to the University of Alaska in Fairbanks for the installation of the second president of the University - he was replacing the retiring original president. It was quite a historic moment for all of us. We all returned home to enter our fall semesters at our Universities, greatly fulfilled with experiences none of us would ever trade.

As the saying goes, "I told you THAT so I could tell you THIS!" When I became an Eagle Scout when I was fifteen, I thought I knew a lot about nature and the stars. This trip in which I saw much more of the United States, the Pacific Northwest, of Canada, and a large portion of the Territory of Alaska made me realize that I had not even seen the tip of the iceberg concerning the makeup of our planet. Flying over the Rocky Mountains, the Cascade Mountain Range, the Olympic Mountain Range, the mountains of Canada and witnessing the Alaskan ranges, including Denali (Mount McKinley) and the other beauty of Alaska, the Matanuska Valley, required research on my part because of the difference between what I was seeing on this trip as opposed to the much older and smaller (worn down) mountain ranges and valleys of which I was familiar - the Ouachita and Ozark Mountains. I wanted to know what had been happening to create such diversity of topography in each area. Later as I saw more of our planet I knew there had to be an explanation.

The German meteorologist, Alfred Wegner, published *The Origins of Continents and Oceans* in 1915. He proposed that at one time the continents of Earth had been joined as one - by squeezing them all together they seemed to fit into one piece. Plate tectonics was later discovered and verified his theories. I was now able to see why things are as they are and how and why they change. It was in the late 1950's when I purchased my copy of this book and it made sense. Physics, astronomy and all the sciences were advancing at an extraordinary pace and I became involved in "keeping up" with these fields. I remember the early 1960's in studying both the "big bang" and the "steady state" theories before the confirmation of the big bang in 1964. What an interesting time to live.

Through the years I have returned to Alaska to see the growth that was anticipated. The era I experienced was a rugged frontier atmosphere in which rail and air were the only modes of transportation. The factors that impress me the most in these last fifty years of growth are the successful efforts to maintain the integrity of nature - the land and the wildlife.

Now back to what I intended to say before getting sidetracked to Alaska. The physics of today can act as a clear signpost for the physics of tomorrow. In fact, we recognize when a new discovery is REQUIRED! When Einstein released his general theory of relativity, it was realized that Newton's theory of gravity is not really wrong, it merely has a limited range of validity. The special theory of relativity is a more useful theory for high-speed systems. As Robert Jastrow stated, "The oysters on the seashore are not aware that they've been overtaken and surpassed" when discussing the intelligence evolution.

Finally, this small presentation of the history of science is as a grain of sand on an endless beach. The most rewarding and exciting activity one can pursue is to read. Read contemporary history, ancient history, the Sumerians, Egyptians, Greeks, Romans, the Asians, the explorations; and explore the sciences. There is so much to enjoy.

At the beginning of this book, explaining what had happened during the last four hundred years in science was the intended purpose. During research and noting the fragile twists history has taken - just imagine:

During the eighteenth century Russia owned Alaska and a settlement north of San Francisco. Japan, having a heavy population in the Hawaiian Islands, lusted these islands and part of California as part of their expansion possibilities. The Spanish, through Mexico, controlled much of the Southwest United States and California. Had the South successfully seceded without a Civil War, the United States would have entered the twentieth century consisting of five nations:

United States of America

Confederate States of America

Mexican America

Russian America

Japanese America

Four language and religious groups would have existed - a duplication of the structure of Europe:

Buddhism - Japanese

Greek Orthodox - Russian

Protestant - the two Americas

Roman Catholic - Mexican

Germany would probably have won the Great War in Europe that began in 1914. It would have been unrealistic to think that these five nations of the North American continent would have had any interest to assist or intervene on either side. These American nations would probably have been fighting their own wars against each other. All the history of these centuries would have been different. See how fragile the string that holds our world together can be.

Those who do not learn from history are obliged to repeat its mistakes.

George Santayana

EPITAPH

As an Eagle Scout I had received the unique opportunity to view in depth nature, wildlife and the paths of the stars while deep in the forests of Southeastern Arkansas. A harmony seemed to exist and I wanted to understand it.

I hope to be listening to George Gershwin's *Rhapsody in Blue* when I die, and I will die as everything that "lives" does. I then want my ashes spread over the Cascade Mountain Range, which include Mount St. Helens and Mount Rainier, or the Olympic Mountain Range, which includes Mount Olympus, both located in the Pacific Northwest of the United States in the State of Washington.

The atoms that constitute my body were formed in the big bang over thirteen and a half billion years ago. These atoms began as part of a gaseous cloud and were later part of a contraction that became a star, which later exploded forming our Sun and its planets. The many atoms that were available went to work on our planet and formed life. By my returning these atoms to nature in form of gas and ash molecules, they will be available for use in the growing of: plants or trees, a bird or a bear, rain or a rock, or even an atom or so in another person or persons. I want the next object in need of certain atoms that I possess to be able to use them as the need arises. This process will continue for countless billions of years to come. These atoms I have borrowed have existed from the beginning and will exist in some form being used throughout the future. We are from the stars!

APPENDIX

CONVERSION FACTORS

AMERICAN STANDARD

LENGTH

    1 IN = 2.54 CM

    1 FT = 12 IN = 30.48 CM = 0.3048 M

    1 YD = 36 IN = 91.44 CM = 0.9144 M

    1 MI = 5280 FT = 1.609344 KM

    1 NAUTICAL MILE = 1.15 MI = 6072 FT = 1.85 KM = 1 MINUTE LATITUDE

    1 DEGREE LATITUDE = 69 MI

    1 LIGHT YEAR (LY) = 5.88×10^{12} MI = 9.46×10^{12} KM

AREA

    1 IN^2 = 6.45 CM^2

    1 FT^2 = 144 IN^2 = 929 CM^2 = 0.0929 M^2

    1 YD^2 = 0.836 M^2

    1 MI^2 = 2.59 KM^2 = 640 ACRES

    1 ACRE = 43,560 FT^2 = 208 FT 8.5 IN SQUARE = 4840 YD^2

VOLUME

    1 IN^3 = 16.387 CM^3

    1 FT^3 = 0.028 M^3

    1 YD^3 = 0.765 M^3

    1 MI^3 = 4.166 KM^3

LIQUID VOLUME

    1 PT = 0.473 LITERS = 0.5178 LB

    1 QT = 0.946 LITERS = 2.071 LB

    1 GAL = 3.785 LITERS = 8.286 LB

MASS

    1 OZ = 28.35 GM

    1 LB = 16 OZ = 453.6 GM = 0.4536 KG

DENSITY

    1 LB/IN^3 = 27.47 GM/CM^3

    1 LB/FT^3 = 16.2 KG/M^3

    1 LB/GAL = 1.717 KG/LITER

RATE

    1 FT/SEC^2 = 30.38 CM/SEC^2 = 0.3038 M/SEC^2

    1 MI/HR = 1.61 KM/HR = 4.47×10^{-1} KM/SEC = 0.000447 KM/SEC

    1 KT (NAUTICAL MILES PER HOUR) = 1.15 MI/HR = 1.85 KM/HR

TEMPERATURE

    $F = 1.8 C + 32 = (9/5) C + 32$

PRESSURE

    1 ATMOSPHERE = 14.7 LB/IN^2

CONVERSION FACTORS

METRIC

LENGTH

       1 CM = 0.3937 IN = 0.0328 FT = 0.0109 YD

       1 M = 100 CM = 39.37 IN = 3.28 FT = 1.09 YD

       1 KM = 1000 M = 39,370 IN = 3280 FT = 1090 YD = 0.6213712 M

       NAUTICAL MILE = 1.85 KM = 6072 FT = 1.15 MI = 1 MINUTE LATITUDE

       1 DEGREE LATITUDE = 111 KM

       1 LIGHT YEAR (LY) = 9.46×10^{12} KM = 5.88×10^{12} MI

AREA

       1 CM^2 = 0.155 IN^2

       1 M^2 = 1550 IN^2 = 10.76 FT^2 = 1.2 YD^2

       1 KM^2 = 0.386 MI^2

VOLUME

       1 CM^3 = 0.061 IN^3

       1 M^3 = 35.3 FT^3 = 1.3 YD^3

       1 KM^3 = 0.24 MI^3

LIQUID VOLUME

       1 LITER = 1000 CC = 1 KG = 2.113 PT = 1.057 QT = 0.264 GAL

MASS

       1 GM = 0.035 OZ = 1 CC = 2.2×10^{-3} LB = 0.0022 LB = 1.1×10^{-6} TON

       1 KG = 1000 GM = 1000 CC = 350 GM = 2.1875 LB = 1.1×10^{-3} TON

DENSITY

       1 GM/CM^3 = 0.0364 LB/IN^3

       1 KG/M^3 = 0.062 LB/FT^3

       1 KG/LITER = 0.58 LB/GAL

RATE

       1 CM/SEC^2 = 328×10^{-4} FT/SEC^2 = 0.0328 FT/SEC^2

       1 M/SEC^2 = 3.28 FT/SEC^2

       1 KM/SEC = 0.6214 MI/SEC = 2237 MI/HR

       1 KM/HR = 0.6214 MI/HR

       1 KT = 1.852 KM/HR = 1.15 MI/HR

TEMPERATURE

       C = (F - 32)/1.8 = 5/9(F - 32)

       K = C + 273.15

       C = K - 273.15

PRESSURE

       1 ATMOSPHERE = 760 mm Hg

CONSTANTS

c = speed of light = 299,792,458 m/sec = 186,282.397 mi/sec

g = acceleration = 9.80665 m/sec^2 = 980.665 cm/sec^2 = 32.174 ft/sec^2

G = universal gravitation = 6.67259 x 10^{-11} newton m^2/kg^2

h = Planck's constant = 6.6260755 x 10^{-34} joule sec = 6.626 x 10^{-27} erg sec

m_e = 9.1093897 x 10^{-31} kg

e = electron charge = 1.60217733 x 10^{-19} coulomb

N_A = Avogardro's number = 6.0221367 molecules/mole

ħ = h/2π = 1.054559 x 10^{-27} erg sec

Earth (g) at pole   = 983.217 cm/sec^2

Earth (g) at equator = 978.039 cm/sec^2

$$\Delta g = 5.178 \text{ cm/sec}^2$$

1 Å = 10^{-8} cm = 10^{-10} m

1 nanometer = 10^{-7} cm = 10^{-9} m

10 Å = 1 nanometer

Probability -   5 coins @ 2 sides = 2^5 = 32 combinations

3 dice @ 6 sides = 6^3 = 216 combinations

FRACTIONS - DECIMALS

1/64	0.015625		33/64	0.515625
1/32	0.03125		17/32	0.53125
3/64	0.046875		35/64	0.546875
1/16	0.0625		9/16	0.5625
5/64	0.078125		37/64	0.578125
3/32	0.09375		19/32	0.59375
7/64	0.109375		39/64	0.609375
1/8	0.125		5/8	0.625
9/64	0.140625		41/64	0.640625
5/32	0.15625		21/32	0.65625
11/64	0.171875		43/64	0.671875
3/16	0.1875		11/16	0.6875
13/64	0.203125		45/64	0.703125
7/32	0.21875		23/32	0.71875
15/64	0.234375		47/64	0.734375
1/4	0.25		3/4	0.75
17/64	0.265625		49/64	0.765625
9/32	0.28125		25/32	0.78125
19/64	0.296875		51/64	0.796875
5/16	0.3125		13/16	0.8125
21/64	0.328125		53/64	0.828125
11/32	0.34375		27/32	0.84375
23/64	0.359375		55/64	0.859375
3/8	0.375		7/8	0.875
23/64	0.390625		57/64	0.890625
13/32	0.40625		29/32	0.90625
27/64	0.421875		59/64	0.921875
7/16	0.4375		15/16	0.9375
29/64	0.453125		61/64	0.953125
15/32	0.46875		31/32	0.96875
31/64	0.484375		63/64	0.984375
1/2	0.5		1	1.0

METRIC - INCH

mm	inch
1	0.03937
2	0.07874
3	0.11811
4	0.15748
5	0.19685
6	0.23622
7	0.27559
8	0.31496
9	0.35433
10 = 1 cm	0.39370
11	0.43307
12	0.47244
13	0.51174
14	0.55118
15	0.59055
16	0.62992
17	0.66929
18	0.70886
19	0.78803
20 = 2 cm	0.79740
21	0.82677
22	0.86614
23	0.90551
24	0.94488
25	0.98415
25.4 = 2.54 cm	1.0
26	1.02342
27	1.06297
28	1.10216
29	1.14153
30 = 3 cm	1.18090

VELOCITY

km/sec	mi/sec	mi/hr	km/hr	mi/hr
1	0.6214	2,237	1	0.6214
2	1.2428	4,474	2	1.2428
3	1.8642	6,711	3	1.8642
4	2.4856	8,948	4	2.4856
5	3.1060	11,185	5	3.1060
6	3.7284	13,422	6	3.7284
7	4.3498	15,659	7	4.3498
8	4.9712	17,896	8	4.9712
9	5.5916	20,133	9	5.5916
10	6.214	22,370	10	6.214
20	12.428	44,740	20	12.428
30	18.642	67,110	30	18.642
40	24.856	89,480	40	24.856
50	31.060	111,850	50	31.060
60	37.284	134,220	60	37.284
70	43.498	156,590	70	43.498
80	49.712	178,960	80	49.712
90	55.916	201,330	90	55.916
100	62.14	223,700	100	62.14
200	124.28	447,400	200	124.28
300	186.42	671,100	300	186.42
400	248.56	894,800	400	248.56
500	310.60	1,118,500	500	310.60
600	372.84	1,342,200	600	372.84
700	434.98	1,565,900	700	434.98
800	497.12	1,789,600	800	497.12
900	559.16	2,013,300	900	559.16
1,000	621.4	2,237,000	1,000	621.4
10,000	6,214.0	22,370,000	10,000	6,214.0
100,000	62,140.0	223,700,000	100,000	62,140.0

DISTANCE - VELOCITY - ACCELERATION

DISTANCE	$d = 16t^2$ ft		t = time
VELOCITY	$v = 32\ t$ ft/sec		
ACCELERATION	$a = 32$ ft/sec^2		

TIME (seconds)	DISTANCE (feet)	VELOCITY (ft/sec)	VELOCITY (mi/hr)
1	16	32	21.82
2	64	64	43.64
3	144	96	65.46
4	256	128	87.28
5	400	160	109.10
6	576	192	130.92
7	784	224	152.74
8	1,024	256	174.56
9	1,296	288	196.38
10	1,600	320	218.20
11	1,936	352	240.02
12	2.304	384	261.84
12.44	2,640 (½ mi)	398.08	271.44
13	2,704	416	283.66
14	3,076	448	305.48
15	3,600	480	327.30
16	4,096	512	349.12
17	4,624	544	370.94
18	5,184	576	392.76
18.17	5,280 (1 mi)	581.44	396.47
19	5,776	608	414.58
20	6,400	640	436.40
21	7,056	672	458.22
22	7,744	704	480.04
22.32	7,920 (1 ½ mi)	714.24	487.02
23	8,464	736	501.86
24	9,216	768	523.68
25	10,000	800	545.50
25.69	10,560 (2 mi)	821.08	560.56
26	10,816	832	567.32
27	11,644	864	589.14
28	12,544	896	610.96
28.73	13,200 (2 ½ mi)	919.36	626.89
29	13,456	928	632.78
30	14,400	960	654.60
31	15,376	992	676.42
31.57	15,840 (3 mi)	1,010.24	688.86
32	16,384	1,024	698.24
33	17,424	1,056	720.06
34	18,496 (3 ½ mi)	1,088	741.88
35	19,600	1,120	763.70

NUMBER	EXPONENTIAL	PREFIX	SYMBOL
Septillion	10^{24}	Yotta	Y
Sextillion	10^{21}	Zetta	Z
Quintillion	10^{18}	Eta	E
Quadrillion	10^{15}	Peta	P
Trillion	10^{12}	Tera	T
Billion	10^{9}	Giga	G
Million	10^{6}	Mega	M
Thousand	10^{3}	Kilo	k
Hundred	10^{2}	Hecto	h
Ten	10^{1}	Deka	da
Tenth	10^{-1}	Deci	d
Hundredth	10^{-2}	Centi	c
Thousandth	10^{-3}	Milli	m
Millionth	10^{-6}	Micro	u
Billionth	10^{-9}	Nano	n
Trillionth	10^{-12}	Pico	p
Quadrillionth	10^{-15}	Femto	f
Quintillionth	10^{-18}	Atto	a
Sextillionth	10^{-21}	Zepto	z
Septillionth	10^{-24}	Yocto	y

A femtosecond (10^{-15} of a second) is to one second as one second is to thirty-two million years.

TEMPERATURE

Fahrenheit = 180/100 Celsius + 32 = 9/5 C + 32 = 1.8 C + 32

C = 5/9 (F - 32) = (F - 32)/1.8

Kelvin = C + 273.15

Kelvin (K)	Celsius (C)	Fahrenheit (F)	
373	100	212	Boiling
368	95	203	
363	90	194	
358	85	185	
353	80	176	
348	75	167	
343	70	158	
338	65	149	
333	60	140	
328	55	131	
323	50	122	
318	45	113	
313	40	104	
308	35	95	
303	30	86	
298	25	77	
293	20	68	
288	15	59	
283	10	50	
278	5	41	
273	0	32	Freezing
268	-5	23	
263	-10	14	
258	-15	5	
253	-20	-4	
0	-273.15	-459.67	Absolute Zero

TIME

PERIOD	SECONDS	MINUTES	HOURS
Minute	60	---	---
Hour	3,600	60	---
Day	86,400	1,440	24
Week	604,800	10,080	168
Year	31,557,600	525,960	8,766

SPEED - DISTANCE

Speed of light in a vacuum C	186,282.39937994096 miles/second
	299,792.458 kilometers/second
Speed of light in water @ 77% C	143,437.4475221 miles/second
	230,840.19266 kilometers/second
Speed of electricity - one-third C	62,094.13312647 miles/second
	99,930.81933333 kilometers/second

Distance light travels:

Year	5,878,625,446,655 miles
	9,460,730,472,581 kilometers
Day	16,094,799,306.38 miles
	25,902,068,371.20 kilometers
Hour	670,616,637.7658 miles
	1,079,252,848.8 kilometers
Minute	11,176,943.96276 miles
	17,987,547.48 kilometers
Second	983,571,068.7232 feet
	299,792,458 meters
Micro-second (10^{-6}) (Millionth)	983.57 feet
	299.79 meters
Nano-second (10^{-9}) (Billionth)	11.80284 inches
	0.29979 meters
	29.979 centimeters

GREEK ALPHABET

A	α	Alpha
B	β	Beta
Γ	γ	Gamma
Δ	δ	Delta
E	ε	Epsilon
Z	ζ	Zeta
H	η	Eta
Θ	θ	Theta
I	ι	Iota
K	κ	Kappa
Λ	λ	Lambda
M	μ	Mu
N	ν	Nu
Ξ	ξ	Xi
O	o	Omnicron
Π	π	Pi
P	ρ	Rho
Σ	σ	Sigma
T	τ	Tau
Υ	υ	Upsilon
Φ	φ	Phi
X	χ	Chi
Ψ	ψ	Psi
Ω	ω	Omega

PARSEC

Parsec - measured parallex is one arcsecond.

1 degree = 60 minutes

1 minute = 60 seconds

1 arcsecond = 1/60 x 1/60 = 1/3600 degree

Θ = 1 arcsecond

One Astronomical Unit (AU) equals the mean distance from the Sun to Earth.

AU = 92,955,826.64048 miles
   = 149,597,900 kilometers

Tan Θ = AU/Parsec

Tan 1/3600 degree = Tan 0.000277778 degree

Tan 1 arcsecond = 0.0000048481368

Parsec = AU/Tan Θ = 92,955,826.64048/0.0000048481368
                  = 92.956 x 10^6 / 4.848 x 10^{-6}
                  = 19,174,056,650,260 miles = 19 x 10^{12} miles

Light Year (LY) = 5,878,625,434,682 miles = 5.9 x 10^{12} miles

LY = Parsec/LY = 3.26265 LY/Parsec

AU = Parsec/AU = 206,270 AU/Parsec

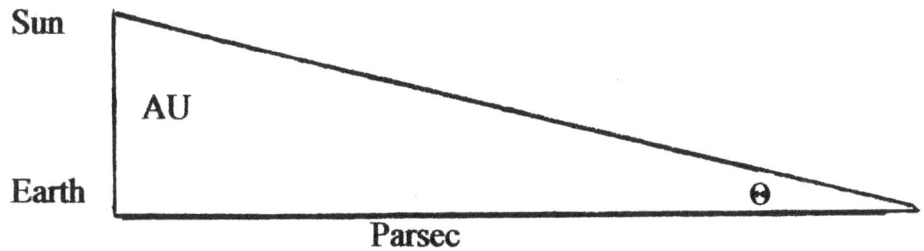

349

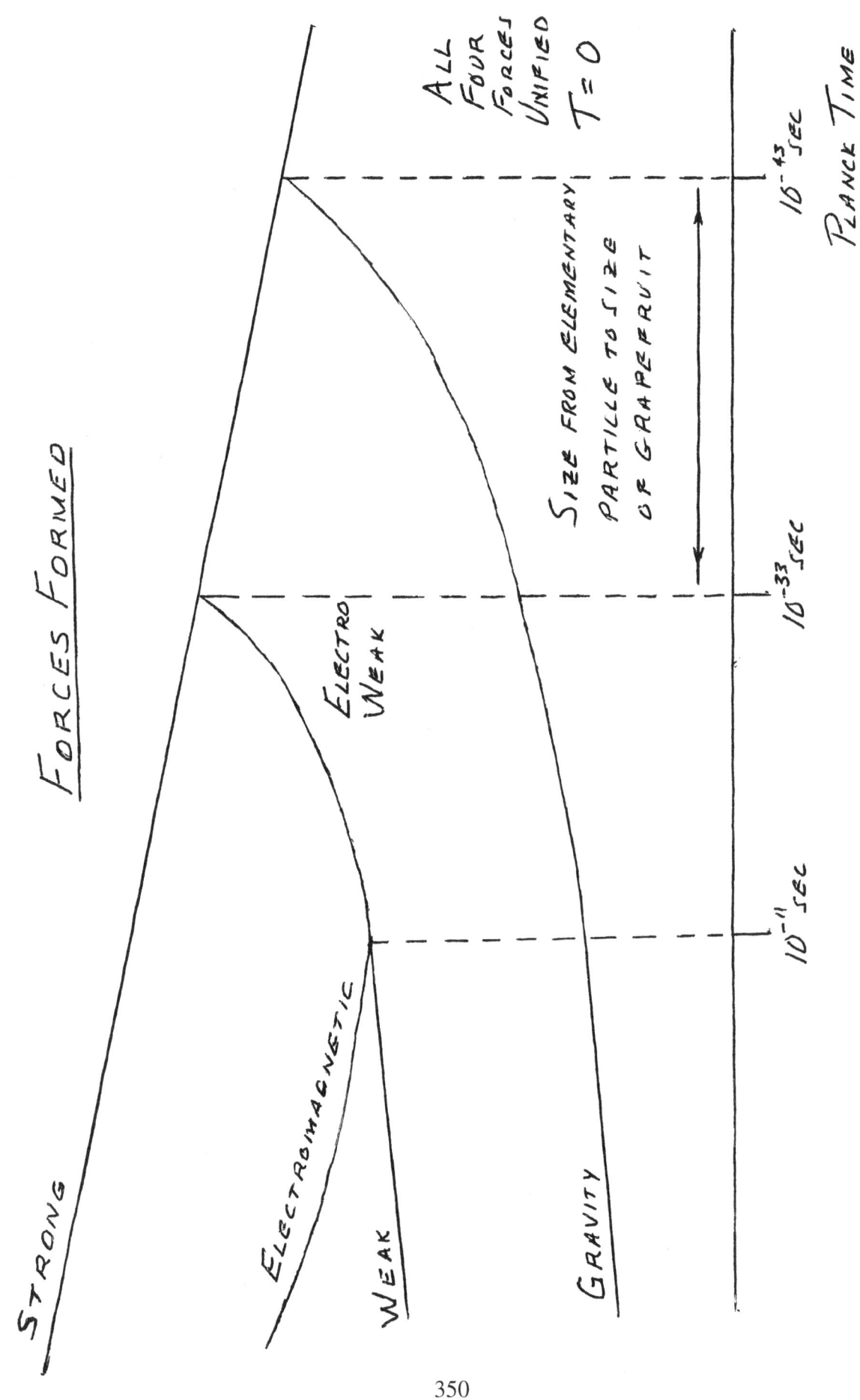

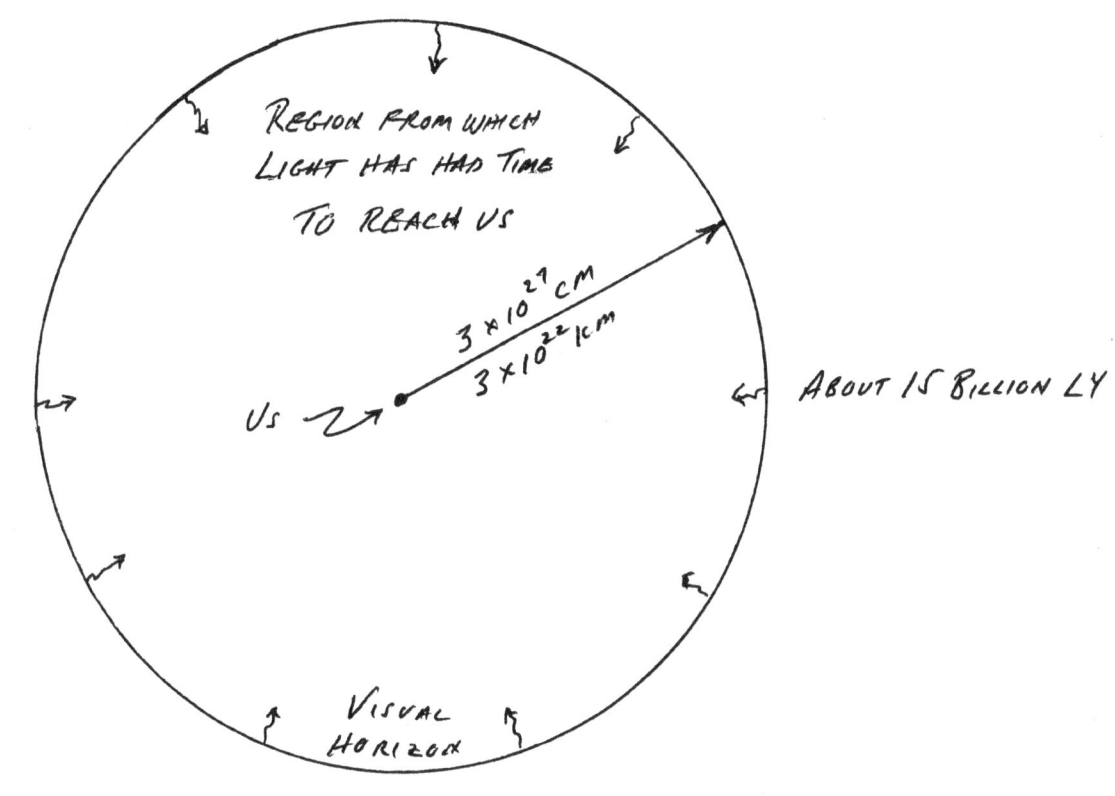

REGION FROM WHICH LIGHT HAS HAD TIME TO REACH US

3×10^{27} CM

3×10^{22} KM

US

ABOUT 15 BILLION LY

VISUAL HORIZON

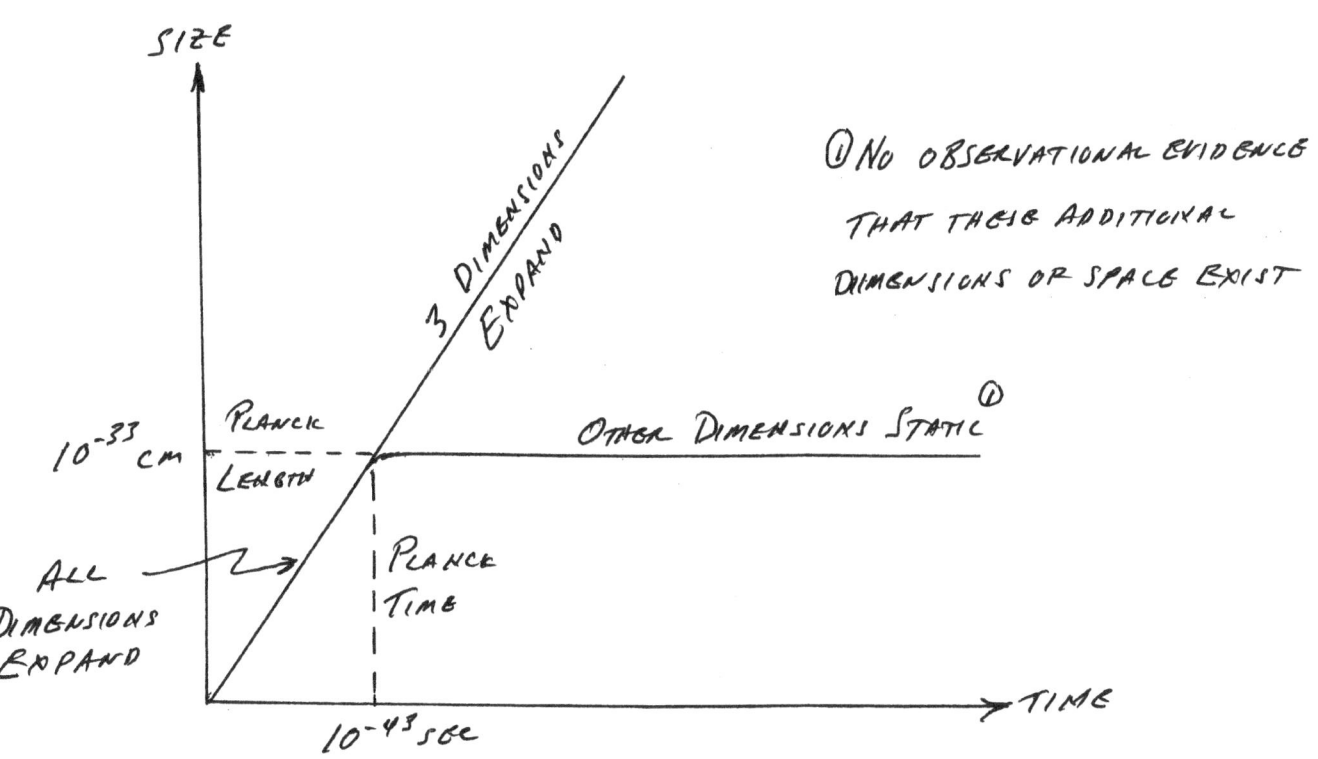

SIZE

3 DIMENSIONS EXPAND

① NO OBSERVATIONAL EVIDENCE THAT THESE ADDITIONAL DIMENSIONS OF SPACE EXIST

10^{-33} CM

PLANCK LENGTH

OTHER DIMENSIONS STATIC ①

ALL DIMENSIONS EXPAND

PLANCK TIME

10^{-43} SEC

TIME

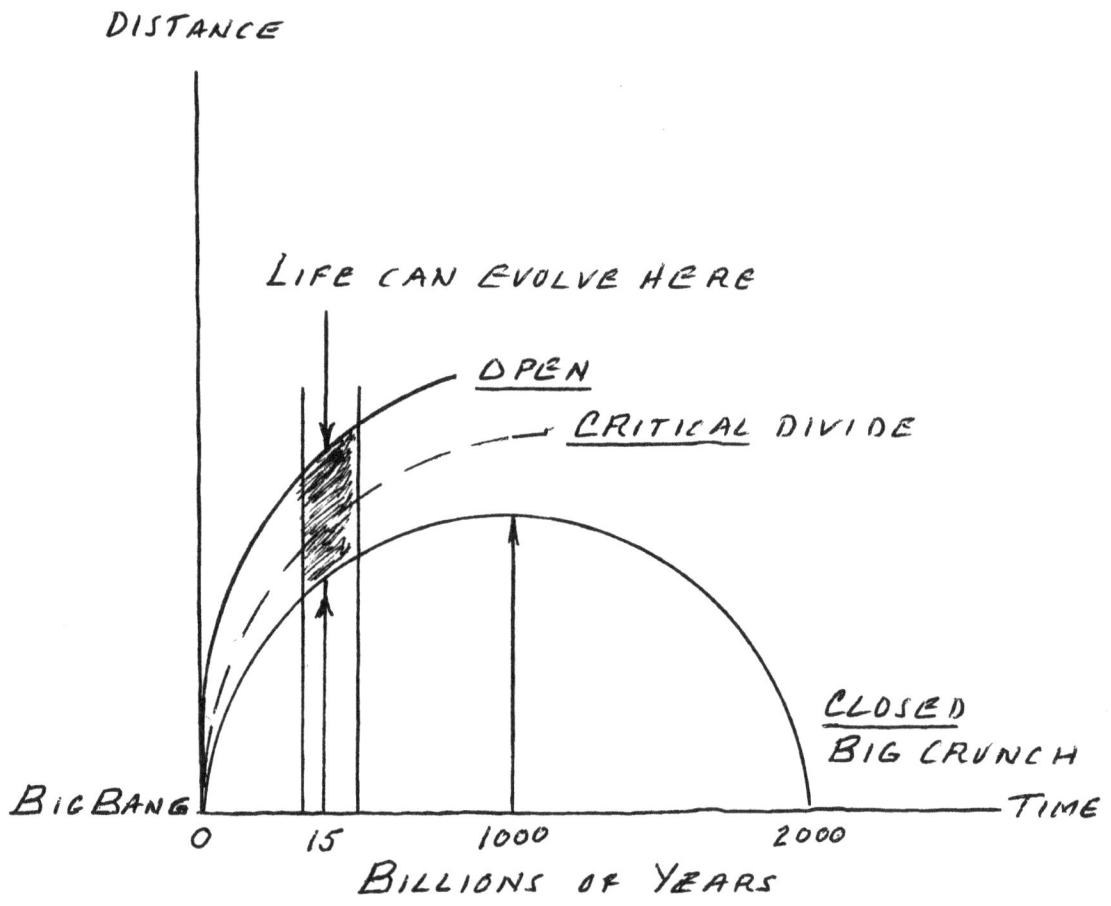

DISTANCE

LIFE CAN EVOLVE HERE

OPEN

CRITICAL DIVIDE

CLOSED
BIG CRUNCH

BIG BANG — Time

0   15   1000   2000

BILLIONS OF YEARS

OPEN – UNIVERSE EXPANSION FASTER THAN
CRITICAL SPEED. GRAVITY CANNOT
GATHER MATERIAL TO FORM GALAXIES
AND STARS. WILL REMAIN DEVOID OF LIFE.

CLOSED – UNIVERSE EXPANSION LESS THAN
CRITICAL SPEED. EXPANSION WILL BE
REVERSED INTO CONTRACTION BEFORE
STARS FORM. NO CHANCE FOR LIFE.

CRITICAL – UNIVERSE APPEARS TO BE NEAR
CRITICAL DIVIDE. GALAXIES AND
STARS FORM. LIFE BEGINS

352

REDSHIFT AND RECESSIONAL VELOCITY

| REDSHIFT | V/C | PRESENT DISTANCE | | LOOK BACK |
		(MPC)	(10^6 ly)	TIME (MY)
0.000	0.000	0	0	0
0.010	0.010	40	129	129
0.025	0.025	98	320	316
0.05	0.049	193	628	613
0.10	0.095	372	1214	1158
0.20	0.180	697	2272	2080
0.25	0.220	844	2753	2473
0.50	0.385	1468	4785	3961
0.75	0.508	1952	6365	4937
1.00	0.600	2343	7638	5619
1.50	0.724	2940	9584	6493
2.00	0.800	3381	11021	7019
3.00	0.882	4000	13038	7606
4.00	0.923	4422	14415	7915
5.00	0.946	4733	15431	8101
10.00	0.984	5587	18214	8454
50.00	0.999	6879	22425	8668
100.00	1.000	7203	23482	8683
Infinity	1.000	7746	25253	8692

$$\text{Redshift} = \frac{\text{Observed wavelengths - true wavelength}}{\text{True wavelength}} = \frac{\text{Recession velocity}}{\text{Speed of light}} = \frac{V}{C}$$

Galaxy location in the Expanding Universe.

MPC = MegaParsec

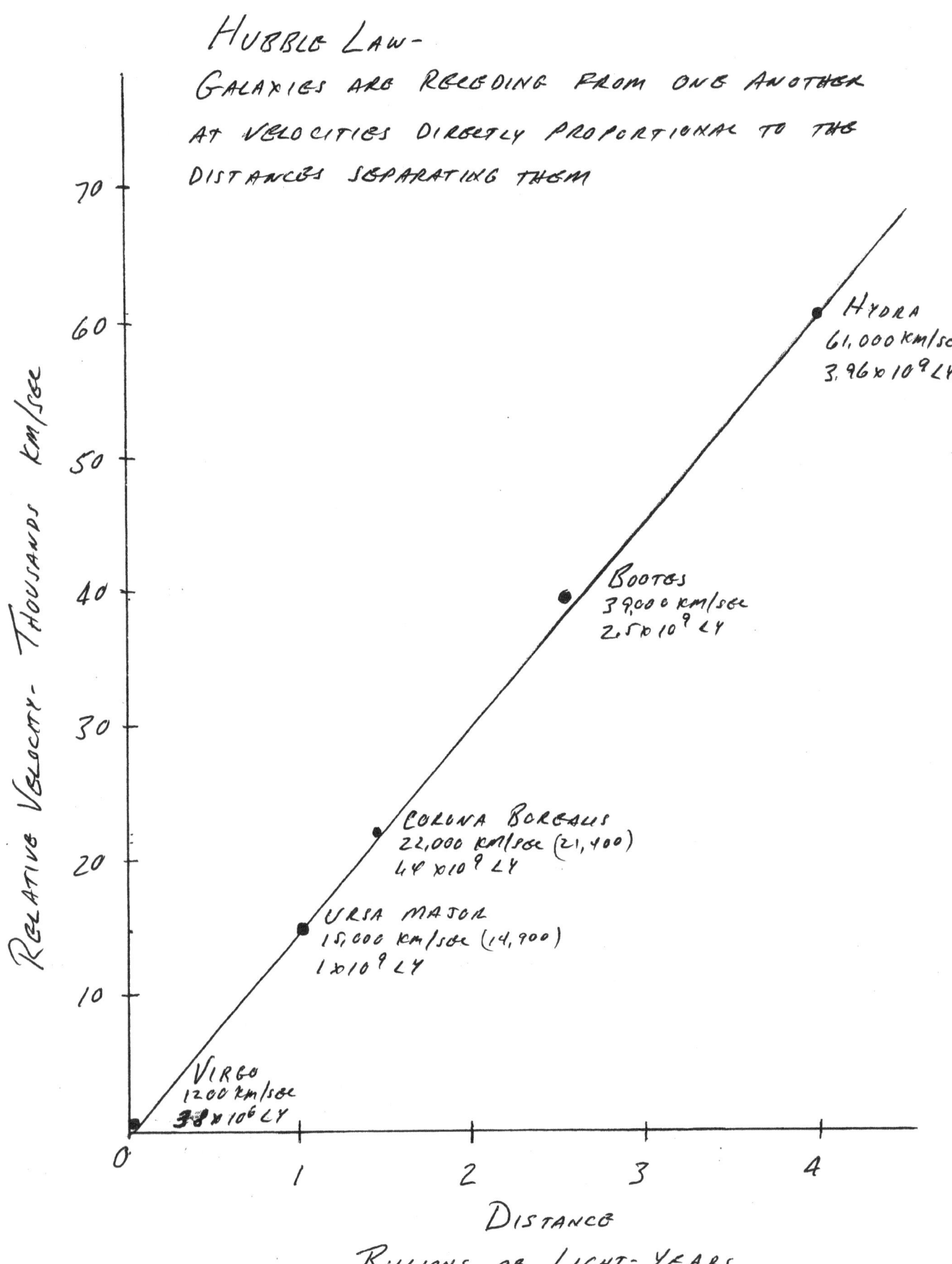

HUBBLE LAW-
GALAXIES ARE RECEDING FROM ONE ANOTHER
AT VELOCITIES DIRECTLY PROPORTIONAL TO THE
DISTANCES SEPARATING THEM

HYDRA
61,000 KM/SEC
3.96×10^9 LY

BOOTES
39,000 KM/SEC
2.5×10^9 LY

CORONA BOREALIS
22,000 KM/SEC (21,400)
1.4×10^9 LY

URSA MAJOR
15,000 KM/SEC (14,900)
1×10^9 LY

VIRGO
1200 KM/SEC
78×10^6 LY

RELATIVE VELOCITY- THOUSANDS KM/SEC

DISTANCE
BILLIONS OF LIGHT-YEARS

354

NEUTRON STAR DIAMETER AND WEIGHT

Neutron Star - 1.4 to 3 Solar Masses (SM)
One Solar Mass = 1.99×10^{30} kg = 2.189×10^{27} tons

Smallest = 1.4 SM = 2.786×10^{30} kg = 3.0646×10^{27} tons

Volume = Mass/Density = V = $\dfrac{M}{D}$ = $\dfrac{2.786 \times 10^{33} \text{ gm}}{2.3 \times 10^{14} \text{ gm/cm}^3}$ = 1.2×10^{19} cm^3

V = 4/3 π r^3 r^3 = $\dfrac{V}{4/3\ \pi}$ = $\dfrac{1.2 \times 10^{19} \text{cm}^3}{4/3\ \pi}$ = 2.86×10^{18} cm^3

r = 1.42×10^6 cm = 14.2 km

Diameter = 28.4 km = 17.6 mi

Weight = (Volume)(Density of nuclear matter)
Grams = $(1.2 \times 10^{19}$ cm$^3)(2.3 \times 10^{14}$ gm/cm$^3) = 2.76 \times 10^{33}$ gm
Tons = $(1.2 \times 10^{19}$ cm$^3)(0.061$in^3/cm$^3)(4.15 \times 10^9$ tons/in$^3) = 3.0378 \times 10^{27}$ tons

Largest = 3 SM = 5.97×10^{30} kg = 6.567×10^{27} tons

Volume = Mass/Density = V = $\dfrac{M}{D}$ = $\dfrac{5.970 \times 10^{33} \text{ gm}}{2.3 \times 10^{14} \text{ gm/cm}^3}$ = 2.6×10^{19} cm^3

V = 4/3 π r^3 r^3 = $\dfrac{V}{4/3\ \pi}$ = $\dfrac{2.6 \times 10^{19} \text{cm}^3}{4/3\ \pi}$ = 6.2×10^{18} cm^3

r = 1.84×10^6 cm = 18.4 km

Diameter = 36.8 km = 22.8 mi

Weight = (Volume)(Density of nuclear matter)
Grams = $(2.6 \times 10^{19}$ cm$^3)(2.3 \times 10^{14}$ gm/cm$^3) = 5.98 \times 10^{33}$ gm
Tons = $(2.6 \times 10^{19}$ cm$^3)(0.061$in^3/cm$^3)(4.15 \times 10^9$ tons/in$^3) = 6.5819 \times 10^{27}$ tons

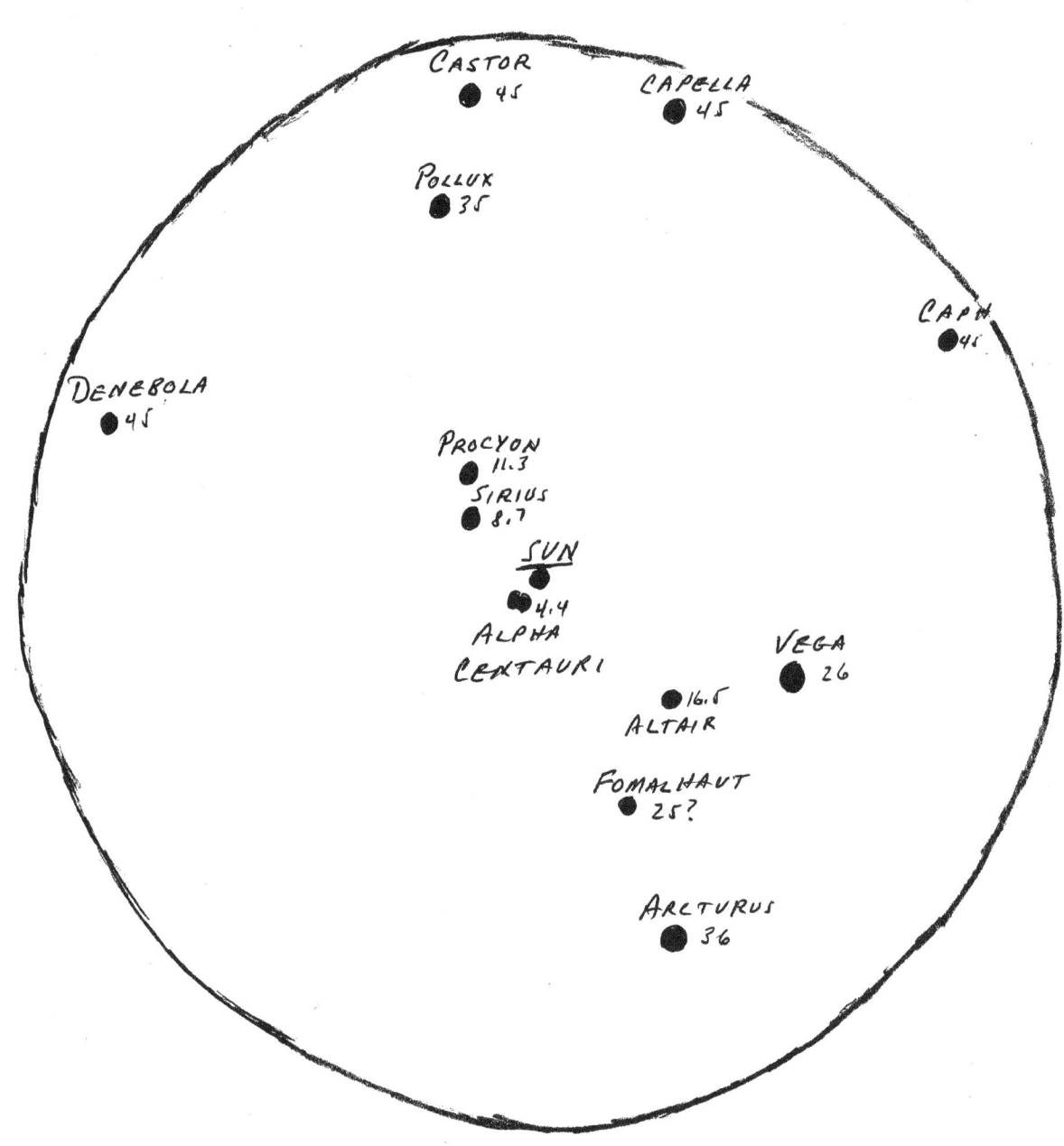

STARS BRIGHTER THAN THE SUN WITHIN
50 LIGHT YEARS OF THE SUN — SPHERICAL

SUN

REGION	INNER RADIUS KM	K TEMPERATURE	DENSITY KG/M^3	DEFINING PROPERTIES
CORE	0	15,000,000	150,000	Energy generated by nuclear fusion.
RADIATOR ZONE	200,000	7,000,000	15,000	Energy transported by electromagnetic radiation.
CONVECTION ZONE	500,000	2,000,000	150	Energy carried by convection.
PHOTO-SPHERE	696,000	5,800	2×10^{-4}	Escaping electromagnetic radiation - the part of the Sun we see.
CHROMO-SPHERE	696,500	4,500	5×10^{-6}	Cool lower atmosphere.
TRANSITION ZONE	698,000	8,000	2×10^{-10}	Rapid temperature increase.
CORONA	706,000	1,000,000	10^{-12}	Hot, low density upper atmosphere.
SOLAR WIND	10,000,000	2,000,000	10^{-23}	Solar material excapes into space and flows outward through the solar system.

SOLAR SYSTEM

OBJECT	ORBIT VELOCITY	DIST TO SUN	YEAR	ROTATION VELOCITY	AXIS ROTATION	ATMOSPHERE	DIAMETER MILES	MASS TONS	ESCAPE VELOCITY	SURFACE TEMPERATURE	MOONS
SUN	480,000 528,000	-	-	29 DAYS EQ 31 DAYS POLE	W-E	H	861,800	2.18×10^{27}	1,274,912	10^6 K	-
MERCURY	107,037	36×10^6	87.969 DAYS	6	W-E	-	3,030	363×10^{18}	9,503	-279.4 F 800.6 F	0
VENUS	78,282	67.19×10^6	224.7	4	E-W	CO_2	7,516	$5,386 \times 10^{18}$	23,254	890.6 F	0
EARTH	66,588	93.3×10^6	365.24	1,037	W-E	N, O	7,922	$6,587 \times 10^{18}$	25,043	-40 F 125 F	1
MOON	1,740	(238,712)	(27.322)	10	W-E	-	2,159	80.85×10^{18}	5,322	-276 F 231.8 F	-
MARS	53,979	142×10^6	687	550	W-E	CO_2	4,219	706×10^{18}	11,234	-225.4 F 80.6 F	2
ASTEROIDS	40,000	250×10^6	-	-	-	-	-	-	-	-	-
JUPITER	29,202	483×10^6	11.862 YR	28,400	W-E	H, He	88,676	2.1×10^{24}	136,396	-153.8 F -99.4 F	28
SATURN	21,689	886×10^6	29.46	22,800	W-E	H, He	74,930	625.24×10^{21}	88,098	-388.88 F	30
URANUS	15,205	1.79×10^9	84.01	5,800	E-W	H, He	32,472	957×10^{21}	46,603	-743.88 F	21
NEPTUNE	12,075	2.8×10^9	164.8	5,750	W-E	H, He	30,678	113.3×10^{21}	41,300	-179.2 F	8
PLUTO	10,599	3.69×10^9 2.75×10^9 minimum 4.58×10^9 maximum	247.69	13	W-E	-	1,984	170×10^{18}	8,165	-369.4 F	1

KEPLER'S THIRD LAW

The square of the planet's obital period is proportional to the cube of its semi-major axis.

$$T^2 = a^3 \qquad T^2/a^3 = 1$$

PLANET	DIST TO SUN (AU)	PERIOD (T)	AU	T	AU³	T²	T²/AU³
Mercury	36 x 10⁶	87.969 days	0.39	0.24	0.059	0.0576	0.98
Venus	67.58 x 10⁶	224.701	0.73	0.62	0.389	0.3844	0.99
Earth	92.9 x 10⁶	365.256	1.0	1.0	1.0	1.0	1.0
Mars	141 x 10⁶	1.88 years	1.52	1.88	3.512	3.534	1.006
Jupiter	485 x 10⁶	11.862	5.22	11.86	142.237	140.66	0.99
Saturn	868 x 10⁶	29.46	9.34	29.46	814.781	867.892	1.065
Uranus	1.8 x 10⁹	84.01	19.38	84.01	7278.826	7057.68	0.97
Neptune	2.8 x 10⁹	164.8	30.14	164.8	27379.767	27159.04	0.99
Pluto	3.65 x 10⁹	247.69	39.29	247.7	60652.134	61355.29	1.01

EARTH ESCAPE VELOCITY – 25,043 MI/HR = 11 KM/SEC

$G = 6.67 \times 10^{-11}$ NEWTON M^2/KG^2

EARTH MASS – $m = 5.98 \times 10^{24} KG$

EARTH RADIUS – $d = 6,350 KM = 6.35 \times 10^6 M$

$$V = \sqrt{\frac{2Gm}{d}} = M/SEC = 10^{-3}\sqrt{\frac{2Gm}{d}} \ KM/SEC$$

$$V = \sqrt{\frac{2(6.67 \times 10^{-11})(5.98 \times 10^{24})}{6.35 \times 10^6}}$$

$$= \sqrt{\frac{80.4 \times 10^{13}}{6.35 \times 10^6}}$$

$$= \sqrt{12.6 \times 10^7}$$

$$= 10^3\sqrt{126}$$

$$= 11,000 \ M/SEC = 11 \ KM/SEC$$

$$NEWTON = \frac{M \ KG}{SEC^2} \qquad\qquad G = \frac{M^3}{KG \ SEC^2}$$

$$G = \left(\frac{M \ KG}{SEC^2}\right)\frac{M^2}{KG^2} = \frac{M^3}{KG \ SEC^2}$$

$$V = \sqrt{\frac{M^3}{KG \ SEC^2} \cdot \frac{KG}{m}} = \sqrt{\frac{M^2}{SEC^2}} = M/SEC$$

Escape Velocity

$$V = 10^{-3}\sqrt{\frac{2Gm}{d}} \quad km/sec$$

$$= 10^{-3}\sqrt{2G\frac{m}{d}}$$

$$= 10^{-3}\sqrt{2(6.67\times10^{-11})\frac{m}{d}}$$

$$= 10^{-3}\sqrt{(13.34\times10^{-11})\frac{m}{d}}$$

$$= 10^{-3}(10^{-5})\sqrt{1.334\frac{m}{d}}$$

$$= 1.155\times10^{-8}\sqrt{\frac{m}{d}} \quad km/sec$$

Orbit Velocity

$$V = 10^{-3}\sqrt{\frac{Gm}{d}}$$

$$= 10^{-3}\sqrt{(6.67\times10^{-11})\frac{m}{d}}$$

$$= 10^{-8}\sqrt{0.667\frac{m}{d}}$$

$$= 0.817\times10^{-8}\sqrt{\frac{m}{d}} \quad km/sec$$

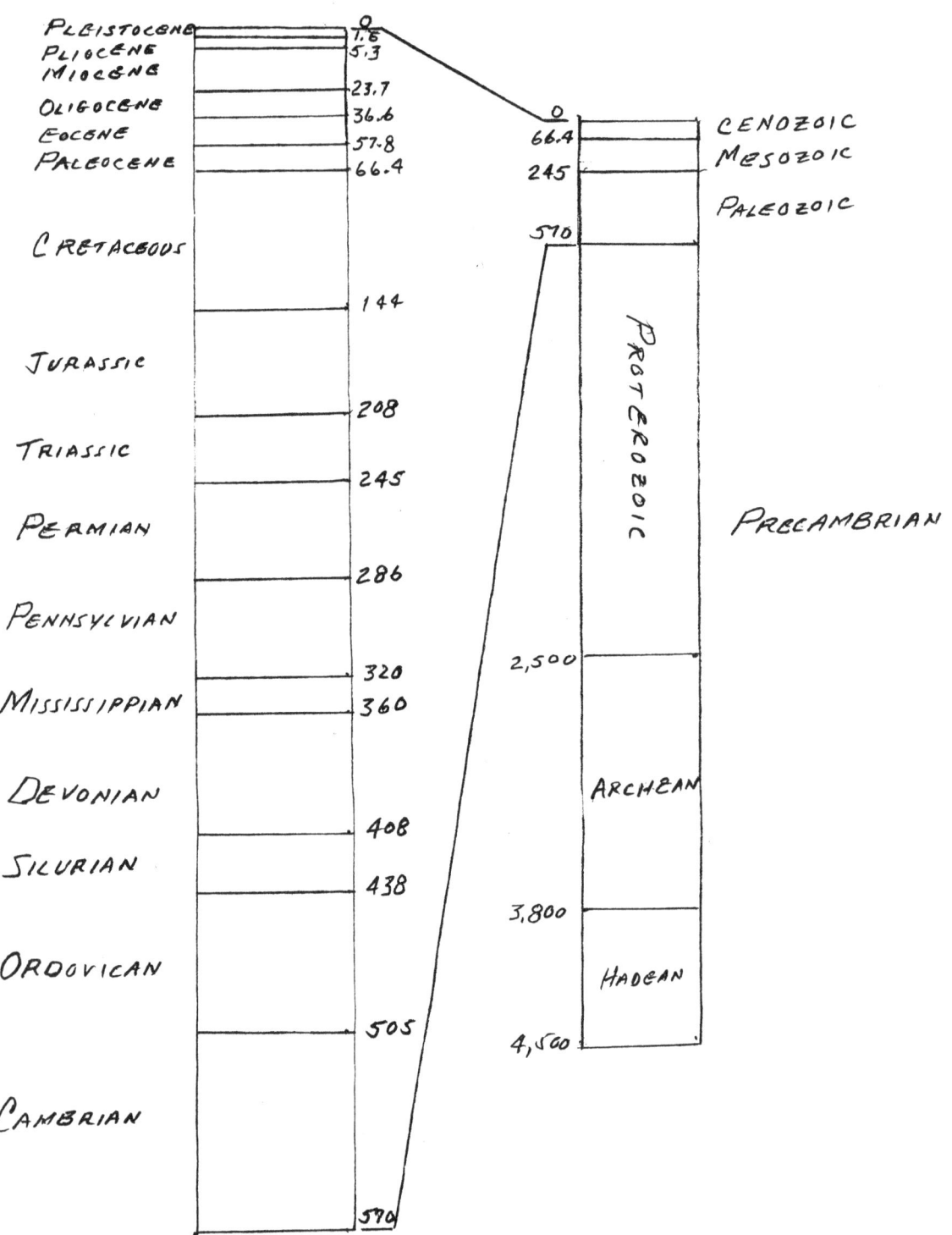

PLEISTOCENE
PLIOCENE
MIOCENE
OLIGOCENE
EOCENE
PALEOCENE

CRETACEOUS

JURASSIC

TRIASSIC

PERMIAN

PENNSYLVIAN

MISSISSIPPIAN

DEVONIAN

SILURIAN

ORDOVICAN

CAMBRIAN

0
1.6
5.3
23.7
36.6
57.8
66.4

144

208

245

286

320
360

408

438

505

570

0
66.4
245

570

2,500

3,800

4,500

CENOZOIC
MESOZOIC
PALEOZOIC

PROTEROZOIC

PRECAMBRIAN

ARCHEAN

HADEAN

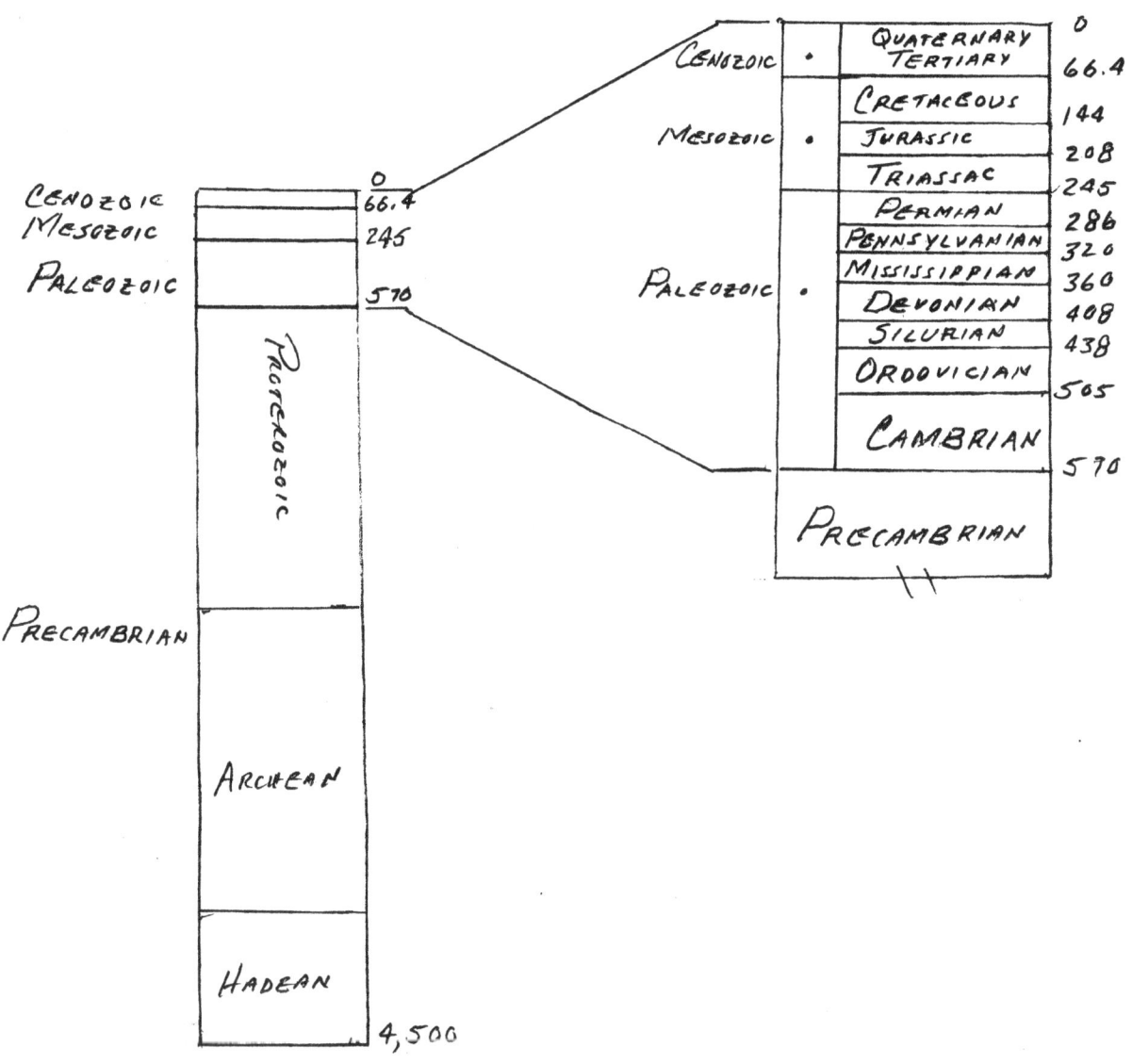

NUCLEI VOLUME WEIGHT

Radius of nucleus $\quad R = (1.2 \times 10^{-13}\text{ cm})A^{1/3} \quad A = \text{mass number}$

Mass of nucleus $\quad M = (1.67 \times 10^{-24}\text{ gm})A$

Volume of a sphere $\quad V = (4/3)\pi r^3$

Density of nuclear matter $= \dfrac{\text{Mass}}{\text{Volume}}$

$$= \frac{(1.67 \times 10^{-24}\text{ gm})A}{(4/3)\pi(1.2 \times 10^{-13}\text{ cm})^3 A}$$

$= 2.3 \times 10^{14}\text{ gm/cm}^3$
$= 8.3 \times 10^{12}\text{ lb/in}^3$
$= 4.15 \times 10^{9}\text{ tons/in}^3$
$= 4.15$ BILLION TONS PER CUBIC INCH

Sphere 1 cm in diameter.

$V = (4/3)\pi r^3 = (4/3)\pi(0.5)^3 = (4/3)\pi(0.125) = 0.52\text{ cm}^3$

$M = DV = (2.3 \times 10^{14}\text{ gm/cm}^3)(0.52\text{ cm}^3) = 1.2 \times 10^{14}\text{ gm}$

$(1.2 \times 10^{14}\text{ gm})(1.1 \times 10^{-6}\text{ tons/gm}) = 1.32 \times 10^{8}\text{ tons} = 132 \times 10^{6}\text{ tons}$

$= 132$ MILLION TONS

Sphere 1 inch in diameter.

$V = (4/3)\pi r^3 = (4/3)\pi(0.5)^3 = (4/3)\pi(0.125) = 0.52\text{ in}^3(16.387\text{ cm}^3/\text{in}^3\,) = 8.52\text{ cm}^3$

$M = DV = (2.3 \times 10^{14}\text{ gm/cm}^3)(8.52\text{ cm}^3) = 19.6 \times 10^{14}\text{ gm}$

$(19.6 \times 10^{14}\text{ gm})(1.1 \times 10^{-6}\text{ tons/gm}) = 21.56 \times 10^{8}\text{ tons} = 2.156 \times 10^{9}\text{ tons}$

$= 2.156$ BILLION TONS

NUCLEI VOLUME QUANTITY

<u>Hydrogen</u>   A = 1.00797

$R = (1.2 \times 10^{-13} \text{ cm})A^{1/3}$ A = mass number
$R = (1.2 \times 10^{-13} \text{ cm})(1.00797)^{1/3} = (1.2 \times 10^{-13} \text{ cm})(1.0026) = 1.203 \times 10^{-13} \text{ cm}$
Diameter = 2.406×10^{-13} cm   Electron Field = 2.406×10^{-9} cm

$V = (4/3)\pi r^3 = V = (4/3)\pi(1.203 \times 10^{-13} \text{ cm})^3 = (4/3)\pi(1.74 \times 10^{-39} \text{ cm}^3)$
   $= 7.29 \times 10^{-39} \text{ cm}^3$

Sphere 1 cm in diameter - number of nuclei = $\dfrac{0.52 \text{ cm}^3}{7.29 \times 10^{-39} \text{ cm}^3} = 71.3 \times 10^{39}$

Sphere 1 inch in diameter - number of nuclei = $\dfrac{8.52 \text{ cm}^3}{7.29 \times 10^{-39} \text{ cm}^3} = 1.17 \times 10^{42}$

<u>Uranium</u>   A = 238.03

$R = (1.2 \times 10^{-13} \text{ cm})A^{1/3}$ A = mass number
$R = (1.2 \times 10^{-13} \text{ cm})(238.03)^{1/3} = (1.2 \times 10^{-13} \text{ cm})(6.2) = 7.44 \times 10^{-13} \text{ cm}$
Diameter = 14.88×10^{-13} cm   Electron Field = 14.88×10^{-9} cm

$V = (4/3)\pi r^3 = V = (4/3)\pi(7.44 \times 10^{-13} \text{ cm})^3 = (4/3)\pi(4.118 \times 10^{-37} \text{ cm}^3)$
   $= 1.725 \times 10^{-36} \text{ cm}^3$

Sphere 1 cm in diameter - number of nuclei = $\dfrac{0.52 \text{ cm}^3}{1.725 \times 10^{-36} \text{ cm}^3} = 301 \times 10^{36}$

Sphere 1 inch in diameter - number of nuclei = $\dfrac{8.52 \text{ cm}^3}{1.725 \times 10^{-36} \text{ cm}^3} = 1.17 \times 10^{39}$

NUCLEI MASS NUMBER AND MASS WEIGHT

	MASS NO.	MASS WT.	CHARGE
Proton	1836.10	1.670×10^{-24} gm	+1
Neutron	1838.63	1.672×10^{-24} gm	0
Up Quark	611.19	0.556×10^{-24} gm	+2/3
Down Quark	613.72	0.558×10^{-24} gm	-1/3

PROTON

Up Quark	611.19	0.556×10^{-24} gm	+2/3
Up Quark	611.19	0.556×10^{-24} gm	+2/3
Down Quark	613.72	0.558×10^{-24} gm	-1/3
Total	1836.10	1.670×10^{-24} gm	+1

NEUTRON

Up Quark	611.19	0.556×10^{-24} gm	+2/3
Down Quark	613.72	0.558×10^{-24} gm	-1/3
Down Quark	613.72	0.558×10^{-24} gm	-1/3
Total	1838.63	1.672×10^{-24} gm	0

NUCLEI MASS NUMBER

A = Up Quark

B = Down Quark

PROTON

$2A + B = 1836.10$
$B = 1836.10 - 2A$

NEUTRON

$A + 2B = 1838.63$

$A + 2(1836.10 - 2A) = 1838.63$
$A + 3672.20 - 4A = 1838.63$
$3A = 1833.57$
$A = 611.19 = $ Up Quark Mass Number

$B = 1836.10 - 2(611.19)$
$\quad = 1836.10 - 1222.38$
$\quad = 613.72 = $ Down Quark Mass Number

NUCLEI MASS WEIGHT

A = Up Quark

B = Down Quark

PROTON

$2A + B = 1.670 \times 10^{-24}$ gm
$B = 1.670 \times 10^{-24}$ gm - 2A

NEUTRON

$A + 2B = 1.672 \times 10^{-24}$ gm

$A + 2(1.670 \times 10^{-24}$ gm - 2A$) = 1.672 \times 10^{-24}$ gm
$A + 3.340 \times 10^{-24}$ gm - 4A $= 1.672 \times 10^{-24}$ gm
$3A = 1.668 \times 10^{-24}$ gm
$A = 0.556 \times 10^{-24}$ gm = Up Quark Mass Weight

$B = 1.670 \times 10^{-24}$ gm - 2(0.556 $\times 10^{-24}$ gm)
  $= 1.670 \times 10^{-24}$ gm - 1.112 $\times 10^{-24}$ gm
  $= 0.558 \times 10^{-24}$ gm = Down Quark Mass Weight

FREQUENCY EMITTED BY AN ELECTRON CIRCLING A NUCLEUS AT A DISTANCE OF $1 A°$

$$f_{osc} = \frac{1}{2\pi} \sqrt{\frac{e^2}{m_e r^3}}$$

$$= \frac{1}{2\pi} \sqrt{\frac{(4.8 \times 10^{-10} esu)^2}{(9.11 \times 10^{-28} gm)(10^{-8} cm)^3}} = 2.5 \times 10^{15} sec^{-1}$$

$e = 4.8 \times 10^{-10} esu$

$r = 1 A° = 10^{-8} cm$

$m_e = 9.11 \times 10^{-28} gm$ — MASS OF ELECTRON

$m_p = 1.67 \times 10^{-24} gm$ — MASS OF PROTON

AN ELECTROMAGNETIC WAVE OF THIS FREQUENCY LIES IN THE ULTRA VIOLET. MOST ATOMS EMIT READILY IN THIS FREQUENCY

$$\lambda = \frac{c}{f} = \frac{3 \times 10^{10} cm/sec}{2.5 \times 10^{15} sec^{-1}} = 1.2 \times 10^{-5} cm$$

$$E = h f = 6.63 \times 10^{-27} ERG\text{-}sec \left(2.5 \times 10^{15} sec^{-1}\right)$$

$$= 16.6 \times 10^{-12} ERG$$

$$\frac{16.6 \times 10^{-12} ERG}{1.6 \times 10^{-12} ERG/ev} = 10 eV$$

$$E = f\left(\frac{4}{10^{15}}\right) eV = 2.5 \times 10^{15} \left(\frac{4}{10^{15}}\right) eV = 10 eV$$

$$= (2.5 \times 10^{15})(4 \times 10^{-15}) eV$$

F = frequency emitted by an electron circling a nucleus at a distance of 10^{-10} centimeters. This orbit would be halfway between the nucleus and the valence electron. The frequency of this wave identifies it as being in the x-ray range.

λ = wavelength
E = energy

$F = 2.5 \times 10^{18} \ sec^{-1}$
$\lambda = 1.2 \times 10^{-8} \ cm$
E = 10,000 eV (ten thousand electron-volts)

F = frequency emitted by an electron circling a nucleus at a distance of 10^{-12} centimeters. The orbit would be just outside the nucleus. The frequency of this wave identifies it as being in the gamma ray range.

$F = 2.5 \times 10^{21} \ sec^{-1}$
$\lambda = 1.2 \times 10^{-11} \ cm$
E = 10,000,000 eV = 10 MeV (ten Million electron-Volts)

Energy - Visible light   1.714 - 3.428 eV
         Ultraviolet   3.428 - 120 eV
         X-Ray   120 - 12,000 eV
         Gamma Ray   12,000 eV - 12,000 MeV

Ultraviolet energy may range to 120 times more powerful than visible light; X-Ray energy is up to 12,000 times more energetic and Gamma Ray energy can be 12 billion times more than visible light. Life cannot exist in frequencies above visible light.

LASER

LASER - Light Amplification by Stimulated Emission of Radiation.

Stimulated Emission - A photon may be of the right frequency to stimulate emission from a particular state without running any danger of being absorbed by the system in that state.

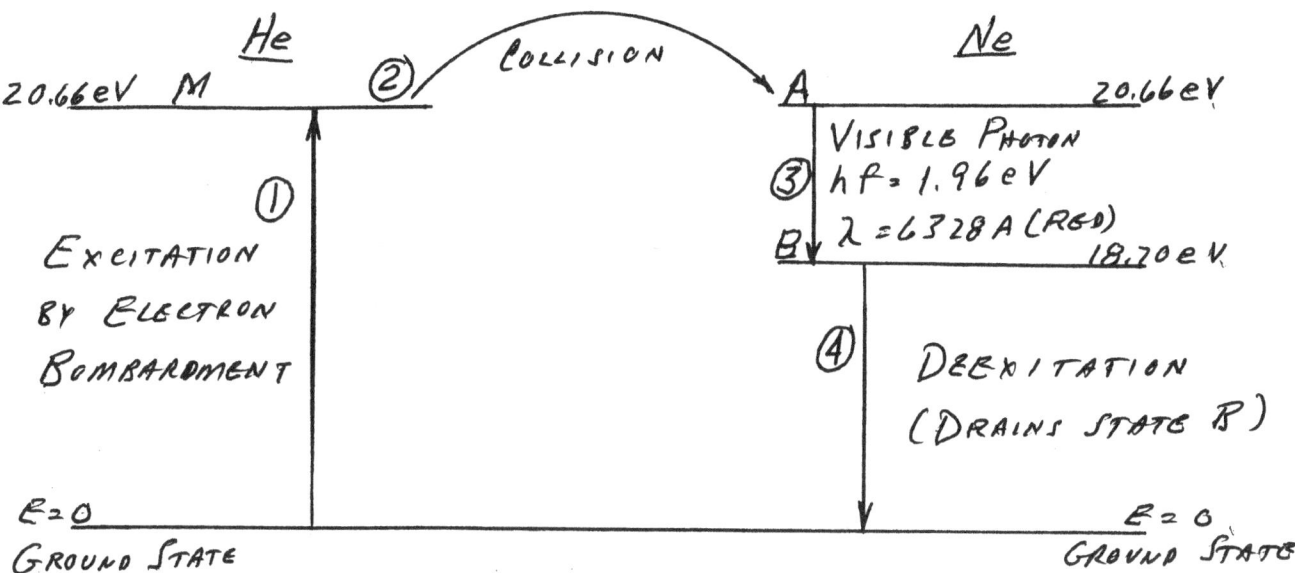

1. Electrons excite helium atoms to metastable state (M).

2. Helium atom transfers its energy to neon, exciting it to state (A).

3. Stimulated emission causes neon atom to jump from state (A) to state (B), emitting photon of red light.

4. In collision with tube walls, neon dissipates the remainder of its excitation energy.

   By this indirect means, more neon atoms find themselves in state (A) than in state (B) at any one time, so that a photon of 1.96 eV is more likely to stimulate emission than to be absorbed. It is due to mirrors that the stimulated emission builds up to a high intensity in one particular direction.

SCHRODINGER'S WAVE EQUATION

PROBABILITY OF THE LOCATION OF AN ELECTRON IN AN ATOM.

$$\left(-\frac{\hbar^2}{2m}\nabla^2 + V\right)\psi = i\,h\,\frac{\partial \psi}{\partial t}$$

PLANCK'S CONSTANT - h 6.6×10^{-27} gm cm²/sec (ERG SEC)

6.6×10^{-34} JOULE-SEC

WAVE FUNCTION - ψ

HEISENBERG'S UNCERTAINTY PRINCIPLE

$$\Delta x \,\Delta p = \hbar$$

Δ UNCERTAINTY OR $h = 6.6255 \times 10^{-27}$ ERG SEC

x POSITION $\hbar = 1.05459 \times 10^{-27}$ ERG SEC

p MOMENTUM

$\hbar$ $h/2\pi$

$$\Delta x\,\Delta V(m) \geq h$$

    V VELOCITY

    m MASS OF PARTICLE

$$(mv = p)$$

PERIODIC TABLE OF THE ELEMENTS

	1A	2A	3B	4B	5B	6B	7B	8	8	8	1B	2B	3A	4A	5A	6A	7A	0
1	1 H 1.008																	2 He 4.003
2	3 Li 6.939	4 Be 9.012											5 B 10.81	6 C 12.01	7 N 14.01	8 O 15.99	9 F 18.99	10 Ne 20.18
3	11 Na 22.99	12 Mg 24.31											13 Al 26.98	14 Si 28.09	15 P 30.97	16 S 32.06	17 Cl 35.45	18 Ar 39.94
4	19 K 39.10	20 Ca 40.08	21 Sc 44.96	22 Ti 47.90	23 V 50.94	24 Cr 51.99	25 Mn 54.94	26 Fe 55.85	27 Co 58.93	28 Ni 58.71	29 Cu 63.54	30 Zn 65.37	31 Ga 69.72	32 Ge 72.59	33 As 74.92	34 Se 78.96	35 Br 79.91	36 Kr 83.80
5	37 Rb 85.47	38 Sr 87.62	39 Y 88.91	40 Zr 91.22	41 Nb 92.91	42 Mo 95.94	43 Tc 99	44 Ru 101.07	45 Rh 102.91	46 Pd 210	47 Ag 107.87	48 Cd 112.40	49 In 114.82	50 Sn 118.69	51 Sb 121.75	52 Te 127.60	53 I 126.9	54 Xe 131.30
6	55 Cs 132.91	56 Ba 137.34	57* La 138.9	72 Hf 178.49	73 Ta 180.95	74 W 183.85	75 Re 186.2	76 Os 190.2	77 Ir 192.2	78 Pt 195.09	79 Au 196.97	80 Hg 200.59	81 Tl 204.37	82 Pb 207.19	83 Bi 208.98	84 Po 210	85 At 210	86 Rn 222
7	87 Fr 223	88 Ra 226	89⊙ Ac 227	104 Rf 253	105 Db 258	106 Sg 259	107 Bh 261	108 Hs 264	109 Mt 266	110 — 272	111 — 272	112 — 277						

* LANTHANIDES

57 La 138.9	58 Ce 140.12	59 Pr 140.91	60 Nd 144.24	61 Pm 145	62 Sm 150.35	63 Eu 151.96	64 Gd 157.25	65 Tb 158.93	66 Dy 162.50	67 Ho 164.93	68 Er 167.26	69 Tm 168.94	70 Yb 173.04	71 Lu 174.97

⊙ ACTINIDES

89 Ac 227	90 Th 232.04	91 Pa 231	92 U 238.03	93 Np 237	94 Pu 244	95 Am 241	96 Cm 247	97 Bk 247	98 Cf 250	99 Es 254	100 Fm 257	101 Md 258	102 No 254	103 Lr 256

373

Actinium	Ac	89	Mendelevium	Md	101
Aluminum	Al	13	Mercury	Hg	80
Americium	Am	95	Molybdenum	Mo	42
Antimony	Sb	51	Neodymium	Nd	60
Argon	Ar	18	Neon	Ne	10
Arsenic	As	33	Neptunium	Np	93
Astatine	At	85	Nickel	Ni	28
Barium	Ba	56	Niobium	Nb	41
Berkelium	Bk	97	Nitrogen	N	7
Beryllium	Be	4	Nobelium	No	102
Bismuth	Bi	83	Osmium	Os	76
Bohrium	Bh	107	Oxygen	O	8
Boron	B	5	Palladium	Pd	46
Bromine	Br	35	Phosphorus	P	15
Cadmium	Cd	48	Platinum	Pt	78
Calcium	Ca	20	Plutonium	Pu	94
Californium	Cf	98	Polonium	Po	84
Carbon	C	6	Potassium	K	19
Cerium	Ce	58	Praseodymium	Pr	59
Cesium	Cs	55	Promethium	Pm	61
Chlorine	Cl	17	Protactinium	Pa	91
Chromium	Cr	24	Radium	Ra	88
Cobalt	Co	27	Radon	Rn	86
Copper	Cu	29	Rhenium	Re	75
Curium	Cm	96	Rhodium	Rh	45
Dubnium	Db	105	Rubidium	Rb	37
Dysprosium	Dy	66	Ruthenium	Ru	44
Einsteinium	Es	99	Rutherfordium	Rf	104
Erbium	Er	68	Samarium	Sm	62
Europium	Eu	63	Scandium	Sc	21
Fermium	Fm	100	Seaborgium	Sg	106
Fluorine	F	9	Selenium	Se	34
Francium	Fr	87	Silicon	Si	14
Gadolinium	Gd	64	Silver	Ag	47
Gallium	Ga	31	Sodium	Na	11
Germanium	Ge	32	Strontium	Sr	38
Gold	Au	79	Sulfur	S	16
Hafnium	Hf	72	Tantalum	Ta	73
Hassium	Hs	108	Technetium	Tc	43
Helium	He	2	Tellurium	Te	52
Holmium	Ho	67	Terbium	Tb	65
Hydrogen	H	1	Thallium	Tl	81
Indium	In	49	Thorium	Th	90
Iodine	I	53	Thulium	Tm	69
Iridium	Ir	77	Tin	Sn	50
Iron	Fe	26	Titanium	Ti	22
Krypton	Kr	36	Tungsten	W	74
Lanthanum	La	57	Uranium	U	92
Lawrencium	Lr	103	Vanadium	V	23
Lead	Pb	82	Xenon	Xe	54
Lithium	Li	3	Ytterbium	Yb	70
Lutetium	Lu	71	Yttrium	Y	39
Magnesium	Mg	12	Zinc	Zn	30
Manganese	Mn	25	Zirconium	Zr	40
Meitnerium	Mt	109			

METALLIC		METALLIC RADIOACTIVE	RADIOACTIVE	NONMETALLIC	GASEOUS
Al 13	Mo 42	Ac 89	At 85	B 5	Ar 18
Sb 51	Nd 60	Am 95	Bk 97	Br 35	Cl 17
As 33	Ni 28	Fm 100	Bh 107	C 6	F 9
Ba 56	Nb 41	Fr 87	Cf 98	Eu 63	He 2
Be 4	Os 76	Np 93	Cm 96	Ge 32	H 1
Bi 83	Pd 46	Pu 94	Db 105	I 53	Kr 36
Cd 48	Pt 78	Po 84	Es 99	Lu 71	Ne 10
Ca 20	K 19	Ra 88	Hs 108	P 15	N 7
Ce 58	Re 75	Tc 43	Lr 103	Pr 59	O 8
Cs 55	Rh 45	U 92	Mt 109	Rb 37	*Rn 86
Cr 24	Ru 44		Md 101	Se 34	Xe 54
Co 27	Sm 62		No 102	Si 14	
Cu 29	Sc 21		Pm 61	S 16	
Dy 66	Ag 47		Pa 91	Tm 69	
Er 68	Na 11		*Rn 86	Yb 70	
Gd 64	Sr 38		Rf 104		
Ga 31	Ta 73		Sg 106		
Au 79	Te 52				
Hf 72	Tb 65				
Ho 67	Tl 81				
In 49	Th 90				
Ir 77	Sn 50				
Fe 26	Ti 22				
La 57	W 74				
Pb 82	V 23				
Li 3	Y 39				
Mg 12	Zn 30				
Mn 25	Zr 40				
Hg 80					

*Rn - Radon - classified both as radioactive and gaseous.

ELEMENT DESCRIPTIONS

ACTINIUM - Ac - 89 - A radioactive metallic element found in uranium ores and used as a source of alpha rays.

ALUMINUM - Al - 13 - A silvery-white, ductile metallic element used to form many hard, light corrosion-resistant alloys.

AMERICIUM - Am - 95 - A white metallic radioactive element used as a radiation source in research.

ANTIMONY - Sb - 51 - A metallic element used in a wide variety of alloys; with lead in battery plates, and in paints, semiconductors and ceramic products.

ARGON - Ar - 18 - A colorless, odorless, inert gaseous element consisting approximately one percent of the Earth's atmosphere. Used in electric lamps.

ARSENIC - As - 33 - A highly poisonous metallic element used in insecticides, weed killers, solid-state devices and various alloys.

ASTATINE - At - 85 - A highly unstable radioactive element used in medicine as a radioactive tracer.

BARIUM - Ba - 56 - A soft, silvery-white metal, used to deoxidize copper, in various alloys and in rat poison.

BERKILIUM - Bk - 97 - A synthetic radioactive element.

BERYLLIUM - Be - 4 - A high-melting, lightweight, corrosion-resistant, rigid, steel-gray metallic element used as an aerospace structural material, as a moderator and reflector in nuclear reactors, and in a copper alloy used for springs, electrical contacts and nonsparking tools.

BISMUTH - Bi - 83 - A white, crystalline, brittle metallic element used in alloys to form sharp castings for objects subject to high temperatures and various low-melting alloys for fire-safety devices.

BOHRIUM - Bh - 107 - An artificially produced radioactive element.

BORON - B - 5 - A soft, brown, amorphous or crystalline nonmetallic element used in flares, nuclear reactor control elements, abrasives and hard metallic alloys.

BROMINE - Br - 35 - A heavy, corrosive, reddish-brown, nonmetallic liquid element used in gasoline antiknock mixtures, fumigants and photographic chemicals.

CADMIUM - Cd -48 - A soft, bluish-white metallic element used in low-friction alloys, solders, dental amalgams and nickel-cadmium storage batteries.

CALCIUM - Ca - 20 - A silvery metallic element that occurs in bone, shells, limestone and gypsum and forms compounds used to make plaster, quicklime, cement and metallurgic and electronic materials.

CALIFORNIUM - Cf - 98 - A synthetic radioactive element produced in trace quantities by helium isotope bombardment of curium.

CARBON - C - 6 - A naturally abundant nonmetallic element that occurs in many inorganic and all organic compounds, exists in amorphous, graphitic, and diamond allotropes, and is capable of chemical self-bonding to form a large number of chemically, biologically and commercially important long-chain molecules.

CERIUM - Ce - 58 - A lustrous, iron-gray, malleable metallic element used in various metallurgical and nuclear applications.

CESIUM - Cs - 55 - A soft, silvery-white ductile metal, liquid at room temperature, the most electropositive and alkaline of the elements, used in photoelectric cells.

CHLORINE - Cl - 17 - A highly irritating, greenish-yellow gaseous element, used in water purification, as a disinfectant, a bleaching agent, and in making chloroform and carbon tetrachloride.

CHROMIUM - Cr - 24 - A lustrous, hard, steel-gray metallic element for hardening steel alloys, for producing stainless steels and for use in corrosion-resistant decorative platings.

COBALT - Co - 27 - A hard, brittle metallic element, used for magnetic alloys, high-temperature alloys, and glass and ceramic pigments.

COPPER - Cu - 29 - A ductile, malleable, reddish-brown metallic element that is an excellent conductor of heat and electricity and is used for electrical wiring, water piping, and corrosion-resistant parts.

CURIUM - Cm - 96 - A silvery, metallic synthetic radioactive element.

DUBNIUM - Db - 105 - An artificially produced radioactive element.

DYSPROSIUM - Dy - 66 - A soft, silvery metal used in nuclear research.

EINSTEINIUM - Es - 99 - A synthetic element first produced by neutron irradiation of uranium in a thermonuclear explosion.

ERBIUM - Er - 68 - A soft, malleable, silvery element, used in metallurgy and nuclear research and to color glass and porcelain.

EUROPIUM - Eu - 63 - A silvery-white, soft element used to absorb neutrons in research.

FERMIUM - Fm - 100 - A synthetic transuranic metallic element.

FLUORINE - F - 9 - A pale-yellow, highly corrosive, poisonous gaseous element, the most electronegative and reactive of all the elements.

FRANCIUM - Fr - 87 - A highly unstable radioactive metallic element.

GADOLINIUM - Gd - 64 - A silvery-white, malleable, ductile metallic element, used to improve the high-temperature characteristics of iron, chromium and related metallic alloys.

GALLIUM - Ga - 31 - A rare metallic element used in semiconductor technology and as a component of various low-melting alloys.

GERMANIUM - Ge - 32 - A brittle, crystalline, gray-white semiconducting element widely used as a semiconductor and as an alloying agent and catalyst.

GOLD - Au - 79 - A soft, yellow, corrosion-resistant, highly malleable and ductile metallic element that has many uses.

HAFNIUM - Hf - 72 - A brilliant, silvery, metallic element used in nuclear reactor control rods and in the manufacture of tungsten filaments.

HASSIUM - Hs - 108 - An artificially produced radioactive element.

HELIUM - He - 2 - A colorless, odorless, tasteless inert gaseous element used to provide lift for balloons and as an inert component of various artificial atmospheres.

HOLMIUM - Ho - 67 - A relatively soft, malleable, stable metallic element.

HYDROGEN - H - 1 - A colorless, highly flammable, gaseous element used in producing synthetic ammonia and methanol, in petroleum refining, as a reducing atmosphere and in oxyhydrogen torches and rocket fuel.

INDIUM - In - 49 - A soft, malleable, silvery-white metallic element used as a silver-plating for mirrors and in making transistors.

IODINE - I - 53 - A lustrous, grayish-black, corrosive, poisonous element having radioactive isotopes. Used as tracers in thyroid disease and therapy, and in compounds as germicides, antiseptics and dyes.

IRIDIUM - Ir - 77 - A very hard and brittle, exceptionally corrosion-resistant, whitish-yellow metallic element used to harden platinum and in high-temperature materials, electrical controls and wear-resistant bearings.

IRON - Fe - 26 - A silvery-white, lustrous, malleable, ductile, magnetic or magnetizable metallic element used alloyed in many important structural materials.

KRYPTON - Kr - 36 - A whitish, inert gaseous element used chiefly in gas-discharge lamps and fluorescent lamps.

LANTHANUM - La - 57 - A soft, silvery-white metallic element used in glass manufacture and in lighting.

LAWRENCIUM - Lr - 103 - A synthetic radioactive element having isotopes with mass numbers 255 through 260.

LEAD - Pb - 82 - A soft, bluish-white, dense metallic element used in solder and type metal, bullets, radiation shielding and paints.

LITHIUM - Li - 3 - A soft, silvery, highly reactive metallic element. Used as a heat transfer medium, in thermonuclear weapons and in alloys.

LUTETIUM - Lu - 71 - A silvery-white rare-earth element used in nuclear technology.

MAGNESIUM - Mg - 12 - A light, silvery, moderately hard metallic element used in structural alloys, pyrotechnics, flash photography and incendiary bombs.

MANGANESE - Mn - 25 - A gray-white, brittle metallic element, alloyed with steel to increase such properties as strength, hardness and wear resistance.

MEITNERIUM - Mt - 109 - An artificially produced radioactive element.

MENDELELIUM - Md - 101 - A radioactive transuranium element of the actnide series.

MERCURY - Hg - 80 - A silvery-white poisonous metallic element used in thermometers, barometers, vapor lamps and batteries and in the preparation of chemical pesticides.

MOLYBDENUM - Mo - 42 - A hard, gray, metallic element used to toughen alloy steels.

NEODYMIUM - Nd - 60 - A bright silvery rare-earth metallic element, used for coloring some glass in in some lasers.

NEON - Ne - 10 - An inert gaseous element occurring in the atmosphere to the extent of eighteen parts per million, used in display and television tubes.

NEPTUNIUM - Np - 93 - A silvery, metallic, naturally radioactive element.

NICKEL - Ni - 28 - A silvery, hard, ductile, metallic element used in alloys, in corrosion-resistant surfaces and batteries, and for electroplating.

NIOBIUM - Nb - 41 - A silvery, soft, ductile metallic element used in steel alloys, arc welding and superconductivity research.

NITROGEN - N - 7 - A nometallic element consisting nearly four-fifths of the air by volume, occurring as a colorless, ordorless, almost inert gas, in various minerals and in all proteins.

NOBELIUM - No - 102 - A synthetic radioactive element produced in trace amounts.

OSMIUM - Os - 76 - A bluish-gray, hard, metallic element, used in platinum alloys, as a catalyst and in making pen points and instrument pivots.

OXYGEN - O - 8 - A colorless, tasteless, odorless gaseous element constituting 21 percent of the atmosphere by volume and essential to most combustion and combustive processes.

PALLADIUM - Pd - 46 - A soft, ductile, steel-white, tarnish-resistant, metallic element alloyed for use in electric contacts, jewelry, nonmagnetic watch parts and surgical instruments.

PHOSPHORUS - P - 15 - A highly reactive, poisonous nonmetallic element used in fireworks, safety matches, incendiary shells, fertilizers, steel and glass.

PLATINUM - Pt - 78 - A silver-white, corrosive-resistant metallic element used in electrical components, electroplating, jewelry, dentistry and as a catalyst.

PLUTONIUM - Pu - 94 - A naturally, radioactive, silvery metallic element used as a reactor fuel and in nuclear weapons.

POLONIUM - Po - 84 - A naturally radioactive metallic element, occurring in minute quantities as a product of radium disintegration and produced by bombarding bismuth or lead with neutrons.

POTASSIUM - K - 19 - A soft, silver-white, light, highly reactive metallic element found in or converted to a wide variety of salts and in fertilizers and soaps.

PRASEODYMIUM - Pr - 59 - A soft, silvery, malleable, ductile rare-earth element, used to color glass yellow and in metallic alloys.

PROMETHIUM - Pm - 61 - A radioactive rare-earth element.

PROTACTINIUM - Pa - 91 - A rare radioactive element chemically similar to uranium.

RADIUM - Ra - 88 - A rare brilliant-white, luminescent, highly radioactive metallic element used in radiotherapy, as a neutron source and as a constituent of luminescent paints.

RADON - Rn - 86 - A colorless, radioactive, inert gaseous element formed by disintegration of radium and used in radiotherapy.

RHENIUM - Re - 75 - A rare, dense, silvery-white metallic element with a high melting point used for electrical contacts and with tungsten for high-temperature thermocouples.

RHODIUM - Rh - 45 - A hard, durable, silvery-white metallic element that is used to form high-temperature alloys with platinum and is plated on other metals to produce a corrosion-resistant coating.

RUBIDIUM - Rb - 37 - A soft, silvery-white, acid-resistant alkali element used in photocells and in the manufacture of vacuum tubes.

RUTHENIUM - Ru - 44 - A hard, white, acid-resistant metallic element used to harden platinum and palladium and in nonmagnetic wear-resistant alloys.

RUTHERFORDIUM - Rf - 104 - An artificially produced radioactive element.

SAMARIUM - Sm - 62 - A silvery or pale-gray metallic rare-earth element used in laser materials, in infrared absorbing glass and as a neutron absorber.

SCANDIUM - Sc - 21 - A silvery-white, lightweight metallic element found in various rare minerals.

SEABORGIUM - Sg - 106 - An artificially produced radioactive element.

SELENIUM - Se - 34 - A nonmetallic element resembling sulfur, used as a semiconductor and in xerography.

SILICON - Si - 14 - A nonmetallic element occurring extensively in the earth's crust in silica and silicates, used in glass, semiconducting devices, concrete, brick, refractories, pottery and silicones.

SILVER - Ag - 47 - A lustrous white, ductile, malleable metallic element, highly valued for jewelry and tableware and widely used in coinage, photography, dental and soldering alloys, electrical contacts and printed circuits.

SODIUM - Na - 11 - A soft, light, extremely malleable silver-white metallic element used in the production of a wide variety of industrially important compounds.

STRONTIUM - Sr - 38 - A soft, silvery, easily oxidized metallic element used in pyrotechnic compounds and various alloys.

SULFUR - S - 16 - A pale-yellow nonmetallic element used in gunpowder, rubber vulcanization, the making of insecticides and pharmaceuticals and in preparing industrial chemicals.

TANTALUM - Ta - 73 - A very hard, heavy gray metallic element used to make electric light-bulb filaments, lightning arresters, nuclear reactor parts and some surgical instruments.

TECHNETIUM - Tc - 43 - A silvery-gray, radioactive metallic element, used as a tracer and to eliminate corrosion in steel.

TELLURIUM - Te - 52 - A brittle silvery-white metallic element, used to alloy stainless steel and lead, in ceramics and, in the form of bismuth telluride, in thermonuclear devices.

TERBIUM - Tb - 65 - A soft silvery-gray metallic rare-earth element, used in electronics and as a laser material.

THALLIUM - Tl - 81 - A soft, malleable, highly toxic metallic element, used in rodent and ant poisons and low-melting glass.

THORIUM - Th - 90 - A silvery-white metallic element used in magnesium alloys.

THULIUM - Tm - 69 - A bright silvery rare-earth element, one isotope of which is used in small portable medical x-ray units.

TIN - Sn - 50 - A malleable, silvery metallic element used to coat other metals to prevent corrosion and in numerous alloys, as soft solder, pewter-type metal and bronze.

TITANIUM - Ti - 22 - A strong, low-density, highly corrosion-resistant, lustrous white metallic element used in alloys requiring low weight, strength and high-temperature stability.

TUNGSTEN - W - 74 - A hard, brittle, corrosion-resistant gray to white metallic element used in high-temperature structural materials and electrical elements, notably lamp filaments requiring thermally compatible glass-to-metal seals.

URANIUM - U - 92 - A heavy silvery-white radioactive metallic element used in research, nuclear fuels and nuclear weapons.

VANADIUM - V - 23 - A bright white soft ductile metallic element, used in rust-resistant high-speed tools, as a carbon stabilizer in some steels and as a catalyst.

XENON - Xe - 54 - A colorless, odorless, highly unreactive gaseous element found in minute quantities in the atmosphere.

YTTERBIUM - Yb - 70 - A soft bright silvery rare-earth element used as an x-ray source for portable irradiation devices, in some laser materials and in some special alloys.

YTTRIUM - Y - 39 - A silvery metallic element used to increase the strength of magnesium and aluminum alloys.

ZINC - Zn - 30 - A bluish-white, lustrous metallic element used to form a wide variety of alloys including brass, bronze, various solders and in galvanizing iron and other metals.

ZIRCONIUM - Zr - 40 - A lustrous grayish-white, strong, ductile metallic element used chiefly in ceramic and refractory compounds, as an alloying agent and in nuclear reactors.

GLOSSARY

ABSOLUTE ZERO - 0 Kelvin, -273.15^0 Celsius, -459.7^0 Fahrenheit - the lowest temperature theoretically possible, achieved when all heat has been removed from a body and all electron activity has ceased.

ACCELERATION - the rate of change of velocity of a moving object.

ACCRETION - gradual growth of bodies, such as planets, by the accumulation of other, smaller bodies.

ACCRETION DISK - flat disk of matter spiraling down onto the surface of a star or black hole. Often, the matter originated on the surface of a companion star in a binary system.

ALPHA PARTICLE - helium nucleus - 2 protons, 2 neutrons

AMINO ACIDS - organic molecules which form the basis for building the proteins that direct metabolism in living creatures.

AMPLITUDE - the maximum deviation of a wave above or below the zero point.

ANALOG - representing a value by an infinitely varying system.

ANGULAR MOMENTUM - measures a rotational property of motion.

ANODE - positive electrode that attracts negative particles.

ANTHROPOLOGY - The study of humankind, extant and extinct, from an all-encompassing holistic approach.

ANTIMATTER - matter made of particles with identical mass and spin as those of ordinary matter, but with opposite charge and quantum properties.

ANTIPARTICLE - see ANTIMATTER.

ANTI-PROTON - antimatter with a negative charge. A man-made particle.

APHELION - the point on the elliptical path of an object in orbit about the Sun that is most distant from the Sun.

APOGEE - furtherest distance to the object being orbited.

ARCHAEOLOGY - the study of the human past.

ARCSECOND - 1/3600 of a degree - 60 seconds times 60 minutes of a degree.

ARTIFACT - any object used or manufactured by humans.

ASTHENOSPHERE - layer of the Earths's interior, just below the lithosphere, over which the surface plates slide.

ASTRONOMICAL UNIT (AU) - the average distance of the Earth from the Sun. 149,603,500 kilometers or 92,963,614.9 miles.

ASTRONOMY - the science that studies the natural world beyond the Earth.

ASTROPHYSICS - the science that studies the physics and chemistry of extra-terrestrial objects. The alliance of physics and astronomy, which began with the advent of spectroscopy, made it possible to investigate what celestial objects are and not just where they are.

ATMOSPHERIC PRESSURE - 14.7 pounds per square inch at sea level.

ATOM - the fundamental unit of a chemical element. An atom consists of a nucleus, containing protons and neutrons; and electrons, which occupy shells that surround the nucleus and are centered on it.

AU - see ASTRONOMICAL UNIT

AZTEC - the last major pre-Columbian civilization in Mesoamerica; its capital, Tenochtitlan, lies under the modern Mexico City.

BARYON - massive elementary particle with half-integral (½) spin that experiences the strong nuclear force and consist of three quarks. Nucleons - protons and neutrons, lambda, sigma and omega particles are baryons.

BASALT - a volcanic rock that is dark in color and usually quite fluid in the molten state.

BETA DECAY - a neutron, when isolated, becomes unstable and decays in twelve to fifteen minutes into a proton, an electron and a neutrino.

BETA PARTICLE - high energy photons beyond the x-ray region. From the nucleus, it is called a gamma ray.

BETELGEUSE - a red giant star with a diameter of 1.1 billion kilometers, or 682 million miles.

BIG BANG - the event that cosmologists consider the beginning of the Universe, in which all matter, space and radiation in the Universe came into being.

BINARY STAR SYSTEMS - a system which consists of two stars in orbit about their common center of mass, held together by their mutual gravitational attraction. Most stars are found in binary star systems.

BIOSPHERE - the living portion of the Earth that interacts with all other geologic and biologic processes.

BLUE SHIFT - a shift in the frequency of a photon toward the higher energy level. Appears as approaching.

BOSON - elementary particle with integer spin (0, 1, 2) that do not obey the Pauli Exclusion Principle. They include the photons and the W and Z particles, carriers of the electromagnetic and the electroweak (gravitational) forces respectively.

BROWN DWARF - a substellar object that is below the minimum mass required for nuclear fusion reactions to occur in its core.

BYA - billion years ago - 1,000,000,000.

CAMBRIAN PERIOD - 570 MYA - 504 MYA

CARBONIFEROUS PERIOD - 365 MYA - 290 MYA

CATHODE - negative electrode that attracts positive particles.

CATHODE RAY - beam of high energy, high speed electrons.

CELESTIAL MECHANICS - the study of motions of gravitationally interacting objects, such as planets and stars.

CELESTIAL SPHERE - imaginary sphere surrounding Earth, to which all objects in the sky were considered to be attached.

CENOZOIC ERA - 65 MYA to present.

CENTER OF GRAVITY - the point in which two or more objects in which balance of the system occurs. In the Earth-Moon system the center of gravity is a point 2,900 miles from the center of the Earth in the direction of the Moon. Centers of gravity may be in the space between the objects.

CENTER OF MASS - the average position in space of a collection of massive bodies, weighed by their masses. In an isolated system this point moves with constant velocity, according to Newtonian mechanics.

CENTRIPETAL FORCE - force needed to keep an object in a circular path.

CERN - European Center for Nuclear Research, Geneva, Switzerland.

CHANDRASEKHAR LIMIT - the maximum mass, approximately 1.4 times the mass of our Sun (1.4 SM) above which an object cannot support itself by electron degeneracy pressure. This is the maximum mass of a white dwarf.

CHIP - popular name for an integrated circuit, a complete electronic circuit etched onto a piece of semiconductor material, commonly silicon.

CHLOROPHYL - converts electromagnetic solar energy into chemical energy.

CLASSICAL ARCHAEOLOGY - the study of Old World Greek and Roman civilizations.

CLIMATE SCIENCE - includes the fields of astronomy, physics, oceanography, geology and meterology.

COHERENT LIGHT - two or more waves of light with the same frequency and a constant phase difference.

COULOMB BARRIER - electromagnetic zone of resistance surrounding protons (or other electrically charged particles) that tend to repel other protons (or other particles of like charge).

COMA - the diffuse envelope that covers the nucleus of a comet - together called the head - formed by the gases and dust driven out of the nucleus by solar radiation.

COMPOUND - two or more elements in combination, such as salt - sodimum chloride (NaCl).

CONSTELLATION - a human grouping of stars in the night sky into a recognizable pattern.

CORE - the central region of Earth, surrounded by the mantle. The central region of any planet or star. The Earth's core has a diameter of 2,600 kilometers (1,612 miles) and an outer core diameter of 7,000 kilometers (4,340 miles).

CORONA - the tenuous outer atmosphere of the Sun, which lies just above the chromosphere, and, at great distances, turns into the solar wind.

COSMIC BACKGROUND RADIATION - the blackbody radiation, now mostly in the microwave band, which consists of relic photons left over from the very hot, early phase of the big bang.

COSMIC RAYS - very energetic charged particles. Ninety percent are protons (hydrogen nucleus), nine percent helium nuclei and one percent are electrons. Slow moving ones originate in the Sun. Fast moving ones originate in distant galaxies. Supernovas also originate cosmic rays. These rays are caught in Earth's magnetosphere and Van Ellen belts on their way to Earth. These rays travel at the speed of light and collide with atoms in our bodies, plants, animals, etc., can cause genetic mutations and are a source of biological evolution.

COSMOLOGICAL PRINCIPLE - the principle that there is no center to the Universe. The Universe is everywhere isotropic on the largest scales, from which it follows that it is homogeneous.

COSMOLOGY - the science concerned with discerning the structure and composition of the Universe as a whole. Combines astronomy, astrophysics, particle physics and a variety of mathematical approaches including geometry and topology.

COVALENCE - two elements combined into a compound by the sharing of electrons.

CRATER - bowl-shaped depression on the surface of a planet or moon, resulting from a collision with interplanetary debris.

CRETACEOUS PERIOD - 140 MYA - 65 MYA.

CRUST - layer of Earth which contains the solid continents and the seafloor.

DARK GLOBULE - interstellar dust and reflection nebula. Dark particles in the sky that block stellar radiation (photons) from distant sousrces.

DE BROGLIE, LOUIS - proposed in 1924 that a particle could be described by a wave whose wavelenght was determined by its momentum.

DE BROGLIE WAVELENGTH - Plank's Constant divided by momentum. All bodies - electrons, protons, planets, etc. - have a wavelength that is related to their momentum. Particles have wave properties. Under proper conditions, every particle will produce an interference or diffraction pattern.

DECCAN VOLCANOES - obliterated the surface of India 65 million years ago, eliminating all life on the island. Depths of 8,000 feet (2,000 meters) were added covering 200 thousand square miles (518 thousand kilometers). Also known as the Deccan Traps.

DECLINATION - the difference between true and magnetic north.

DENSITY - a measure of the compactness of the matter within an object. It is computed by dividing the mass by the volume of the object. Units are kilograms per cubic meter or grams per cubic centimeter.

DEVONIAN PERIOD - 413 MYA - 365 MYA.

DIFFRACTION - any deviation of wave motion away from straight-line propagation.

DIGITAL - representing a value as a limited set of discrete signals, usually on/off pulses of electricity.

DIODE - device that allows an electric current to flow in one direction only.

DISTANCE TO LIGHTNING STRIKE - the number of seconds between when the lightning strikes and the time you hear the thunder, divided by five, equals the number of miles to the point of the lightning strike.

DNA - molecular structure and chemical process of life.

DYNAMICS - the branch of mechanics that deals with the motion and equilibrium of systems under the action of forces, usually from outside the system.

EAST AFRICAN RIFT - the split will create a new sea in the continent of Africa.

ECLIPTIC - the apparent path of the Sun, relative to the stars on the celestial sphere, over the course of a year. The great circle of the celestial sphere which is cut by the plane containing the orbit of the Earth.

ELECTROMAGNETIC FORCE (EMF) - fundamental force of nature that acts on all electrically charged particles. Classical electromagnetics is based on Maxwell's and Faraday's equations and on the theory of quantum electrodynamics (QED).

ELECTROMAGNETIC RADIATION - radiation consisting of electromagnetic waves.

ELECTROMAGNETIC WAVE - a wave produced by the acceleration of an electric charge and propagated by the periodic variation of intensities of perpendicular electrical and magnetic fields.

ELECTROSTATIC UNIT - esu.

ELECTRON - light elementary particle with a negative electrical charge. Electrons are found in shells surrounding the nuclei of atoms. Their interactions with the electrons of neighboring atoms create the chemical bonds that link atoms together as molecules and compounds.
Mass - 1, spin - ½.
Charge - $e = 4.8033 \times 10^{-10}$ esu $= 1.6022 \times 10^{-19}$ coulomb.
Mass - $m = 9.1096 \times 10^{-28}$ gm.
Energy - $e = mc^2 = 0.5110$ MeV.

ELECTRON'S NEUTRINO - V_e - zero mass, spin ½, neutral charge.

ELECTRON ORBITAL - two electrons, one has spin up and the other has spin down, are allowed in each orbital according to the Pauli Exclusion Principle.

ELECTRON VOLT (eV) - 1 eV = 1.6×10^{-12} erg; 1.6×10^{-19} joules.

ELECTROVALENCE - two elements combined into a compound by the electrostatic attraction of the atom - sodium chloride (NaCl) - positive ion attracted to a negative ion to produce salt.

ELECTROWEAK THEORY - theory demonstrating links between the electromagnetic and the weak nuclear forces. Indicates that in the high energies that characterized the early Universe, electromagnetism and the weak force functioned as a single electroweak force. This is also known as the Weinberg-Salam theory.

ELEMENT - matter made up of one particular atom. The number of protons in the nucleus of the atom determines which element it represents.

ELLIPSE - an oval having both ends alike.

ENTROPY - a measure of the amount of disorder. Natural systems tend to proceed toward a state of greater disorder.

EOCENE EPOCH - 54 MYA - 38 MYA.

EON - on the geologic time scale, the longest unit of time, comprised of several eras.

EPICENTER - the surface area over the hypocenter of an earthquake.

EPOCH - an interval of geologic time longer than an age and shorter than a period.

ERA - on the geologic time scale, the unit of time below eon; composed of several periods.

ESCAPE VELOCITY - the minimum velocity that an object must have in order to escape from the surface of a planet or moon's gravitational attraction.

ETA - a particle with mass of 1074, spin 0, charge neutral: a meson.

EVENT - a point in four-dimensional space-time; a location in both space and time.

EVENT HORIZON - the surface, or effective boundary, of a black hole. No material or light can escape from within the event horizon. The radius is approximately three kilometers times the mass of the black hole, given in solar units.

EXCITED STATE - the state of an atom when one of its electrons is in a higher energy orbital than the ground state. Atoms can become excited by absorbing a photon of a specific energy, or by colliding with a nearby atom.

EXCLUSION PRINCIPLE - the rule that no two fermions can occupy the same quantum state. Also known as Pauli's Exclusion Principle.

EXTRAPOLATE - to infer an unknown from something that is known.

EXTRATERRESTRIAL - any object not on or part of Earth.

FAULT - a braking of crustal rocks caused by Earth's movement.

FERMILAB - a proton accelerator named after Enrico Fermi. Located in Chicago.

FERMION - particles with half-integral spin which obey the exclusion principle.

Leptons	Hadrons	
electrons	Baryons	Mesons
muon	proton	pion
tauon	neutron	kaon
neutrinos (3)		

FIBONACCI SERIES - an arithmetic operation in which each succeeding unit is equal to the total of the preceding two: 1, 1, 2, 3, 5, 8, 13, 21, 34, 55, ...

FIELD - a region of space marked by a physical property, as gravitational or electromagnetic force or fluid pressure, having a determinable value at every point in the region.

FISSION - a nucleus of U-235, after absorbing a neutron to become an excited nucleus of U-236, might break into fragments and emit several neutrons which enter other U-235 nuclei to continue a chain reaction.

FLINTS - skeletons of protozoan radiolarians are made of silica and sink to fossilize into flints.

FLUID DYNAMICS - an interaction between two fluids. In climate science these would be the ocean and the atmosphere.

FOCUS - see HYPOCENTER.

FOSSIL - the remains or traces of an organism preserved from the geologic past. Buried rapidly in the absence of oxygen or bacteria to prevent decay. Turns to stone.

FREQUENCY - the number of wave crests passing any given point in a given period of time.

FUSION - mechanism of energy generation in the core of the Sun, in which light nuclei are fused into heavier ones - hydrogen into helium - releasing energy in the process.

GALACTIC CENTER - the center of the Milky Way or any other galaxy. The point about which the disk of a spiral galaxy rotates.

GALAXY - gravitationally bound collection of a large number of stars. The Sun is a star in the Milky Way galaxy.

GALAXY CLUSTER - a collection of galaxies held together by their mutual gravitational attraction. Can be 100 million light years in diameter.

GAMMA DECAY - emission of high energy photons from the nucleus whose frequency is between the x-ray and cosmic ray frequencies.

GAMMA PARTICLE - high energy photons beyond the x-ray region.

GEOTHERMAL - the generation of hot water or steam by hot rocks in the Earth's surface.

GLACIAL TILL - glacial drift composed of an unconsolidated, heterogeneous mixture of clay, sand, gravel and boulders.

GLUON - quanta that carry the strong nuclear force. Like photons, vector bosons and gravitons - the carriers respectively of electromagnetism, the weak force and gravitation - gluons and massless bosons. Consequently, for simplicity's sake, some physicists lump together all the force-carrying quanta under the name "gluon."

GONDWANA (LAND) - 570 MYA - 240 MYA. Contained present-day Antarctica, South America, Africa, Arabia, Madagascar, India, Australia and New Zealand.

GRADIENT - rate at which temperature or pressure changes.

GRANITE - composed of quartz, feldspar and mica.

GRAVITATIONAL FORCE - fundamental force of nature, generated by all particles that possess mass. Interpreted by means of Newton's mechanics or by the general theory of relativity.

GRAVITATIONAL MASS - manifests itself only in the presence of gravitational force.

GRAVITON - zero mass, spin 2 with a neutral charge.

GRAVITY - the attractive effect that any massive object has on all other massive objects. The greater the mass of the object, the stronger is its gravitational pull.

GREENHOUSE EFFECT - the partial trapping of solar radiation by a planetary atmosphere, similar to the trapping of heat in a greenhouse. Carbon dioxide transmits visible light coming in from the Sun, but absorbs infrared radiation that rises from the surface of the Earth and holds this heat in the atmosphere instead of reflecting it back into space.

GROUND STATE - an atom with the nuclei and electrons located in their normal shells. The lowest energy state that an electron can have with an atom.

HADRONS - elementary particles that are influenced by the strong nuclear force. There are two types of hadrons: mesons, have zero spin; and baryons, which have spin ½ or 3/2. Composed of quarks.

HALF-LIFE - the time it takes for half of a given quantity of radioactive material to decay.

HAMILTONIAN - named after the Irish physicist William Rowan Hamilton. Begins quantization of a mechanical system.

HERTZ - HZ - see FREQUENCY.

HOLOCENE EPOCH - 10,000 years ago to present.

HOMO ERECTUS - the ancestor of modern humans who evolved from Australopithecus, contained a brain about two-thirds the size of contemporary humans and established a number of flake tool traditions.

HOMOGENEITY - the property of a geometry in which all points are equivalent. The surface of an unmarked sphere is homogeneous; the surface of a circle is not, the edge points are different from the points on the cube face. Uniform throughout.

HOT SPOTS - tectonic plates pass over holes in the mantle of the Earth and magma is pushed through, causing volcanoes to form. Hot spots formed the Hawaiian Island chain, the Emperor Island chain and Yellowstone Park.

HUBBLE'S CONSTNANT - 75 kilometers per second per megaparsec. Each megaparsec is equal to 3.26 million light years. Expansion is the same in the Universe to any observer, there is no common center, all objects are moving away from each other equally proportionally.

HUBBLE'S LAW - velocity of recession of a galaxy is proportional to the distance of the galaxy from Earth. $V = HD$. A galaxy twice as far away recedes at twice the speed.

HYDROCARBON - compounds that contain carbon and hydrogen only.

HYDROSPHERE - layer of the Earth which contains the liquid oceans and accounts for roughly seventy percent of Earth's total surface area.

HYPERBOLIC PARABALOID - saddle shaped object.

HYPERSONIC - velocity in excess of five times the speed of sound.

HYPOCENTER - also called FOCUS - the poiint of origin of earthquakes.

IGNEOUS - rock that comprises the earth's crust formed of hardened magma.

IGNEOUS ROCK - produced by volcanic action.

INDETERMINANCY PRINCIPLE - quantum precept indicating that the position and trajectory of a particle cannot both be known with perfect exactitude.

INERTIA - the tendancy of an object to continue in motion at the same speed and in the same direction, unless acted upon by a force.

INERTIAL MASS - a permanent property of matter. Equivalent to gravitational mass. The force required to accelerate mass sixteen feet per second would equal the weight of the mass on Earth.

INFLATION - short period of unchecked cosmic expansion early in the history of the Universe. During inflation, the Universe expanded by a factor of 10^{50}.

INFRARED - the region of the electromagnetic spectrum just outside the visible range, corresponding to light of slightly longer wavelength than that of red light.

INTEGRATED CIRCUIT - electronic circuit that can contain millions of components etched on a small wafer of semiconductor material.

INTERFERENCE - the ability of two or more waves to interact in such a way that they either reinforce or cancel each other. Waves in phase interfere constructively - giving a point of illumination. Waves out of phase interfere destructively - giving a point of darkness. Used to measure wavelength.

INTERPLANETARY SPACE - the space between the objects in the solar system.

INTERPOLATE - information follows a predictable pattern - 10, ?, 20, 25, 30, ... (15 is between 10 and 20).

ION - an atom or molecule that has lost one or more electrons and has a positive electrical charge.

IONIZED - state of an atom that has at least one electron removed.

ISOTOPE - nuclei containing the same number of protons but different number of neutrons. Most elements can exist in several isotopic forms. A common example of an isotope is deuterium, which differs from normal hydrogen by the presence of a neutron in the nucleus.

ISOTROPY - the property of uniformity in all directions. The property of a geometry of being the same in all directions. The surface of an unmarked sphere is isotropic; the surface of a cylinder is not isotropic.

JOVIAN - relating to the planet Jupiter. Commonly used in referring to all large-sized objects.

JOVIAN PLANET - one of the four giant outer planets of the solar system which resemble Jupiter in physical and chemical composition. Saturn, Uranus and Neptune are Jovian planets.

JURASSIC PERIOD - 210 MYA - 140 MYA.

KAON - mass 966.4, spin zero, K⁻ positive; mass 974.3, spin zero, K^0 neutral - a meson.

KINEMATICS - the branch of mechanics that deals with pure motion without reference to the masses or forces involved in it.

KINETIC ENERGY - the energy of a body or a system with respect to the motions of the body or of the particles in the system.

KINETICS - the branch of mechanics that deals with the action of forces in producing or changing the motion of masses.

LAGRANGIAN - a mathematical expression named after the French physicist Joseph Lagrange.

LAMBDA - particle, baryon, mass 2183.1, spin ½, neutral charge.

LASER BEAM SPREAD - use a big telescope backward. If wavelength is a micron (one millionth of a meter) and the aperture is a meter - the spread would be one part in a million. You could go a million miles and the beam would only be a mile across.

LAURASIA - the northern half of Pangaea consisted of what is now North America, Greenland, Europe and most of Asia. 570 MYA - 240 MYA.

LENGTH CONTRACTION - an apparent contraction of the length of an object in motion relative to a given observer, caused by the Lorentz transformation from one frame to another.

LEPTON - dimensionless and do not participate in the strong force: electron, muon and neutrino. Not composed of quarks.

LIGHT CONE - the surface representing all possible paths of light which arrive at or depart from a particular event.

LIMESTONE - skeletons of the planktonic protozoa known as foraminifera are made of carbonates ($CaCO_3$ - calcium carbonate), and sink to the seabeds to form chalk that transforms to limestone.

LINE SPECTRA - light emitted by matter. Consists of a definite set of discrete wavelengths.

LITHOSPHERE - Earth's crust and a small portion of the upper mantle that makes up the Earth's plates. This layer of the Earth undergoes tectonic activity.

LORENTZ CONTRACTION - see LENGTH CONTRACTION.

LORENTZ TRANSFORMATION - the transformation, valid for all relative velocities, which describes how to relate coordinates and observations in one inertial frame to those in another such frame.

LY - light year. Distance light travels in one year - 5,878,625,446.655 miles.

MAGIC WRIST - STM and robotic manipulator which can feel an atom.

MAGMA - a molten rock material generated within the Earth. It is the constituent of igneous rocks, including volcanic eruptions.

MANTLE - the part of the Earth below the crust and above the core, composed of dense iron-magnesium rich rocks.

MARBLE - a hard limestone transformed by heat and pressure.

MASS - the measure of the amount of matter in an object. Inertial mass indicates the object's resistance to changes in its state of motion. Gravitational mass indicates its response to the gravitational force. In the general theory of relativity, gravitational and intertial mass are revealed to be aspects of the same quantity.

MAYA, CLASSIC - major pre-Columbian civilization in Middle America; the Classic Maya period, characterized by the use of the long-count calendar, extends from 300 - 900.

MECHANICS - the branch of physics that deals with the action of forces on bodies and with motion; comprised of kinetics, statics and kinematics.

MEGALITHIC - of or pertaining to the Middle Neolithic period; characterized by the presence of large stone monuments.

MERIDIAN - a line that extends from pole to pole. On Earth, these are the lines that run form the North Pole to the South Pole.

MESOAMERICA - the area in Central America in which various Classic and Post-classic civilizations developed; including the Olmec, Teothuacan, Aztec and Maya.

MESOLITHIC - the Middle Stone Age; a period of transition from hunting and gathering to agriculture, featuring settlements based on broad-spectrum wild resource exploration.

MESONS - see HADRONS . Consist of two quarks with zero spin. Pions, kaons, and eta.

MESOZOIC ERA - 245 MYA - 65 MYA.

METALS - elements with a valence shell of no more than three electrons.

METEORITE - any part of a meteoroid that survives passage through the atmosphere and lands on the surface of the Earth.

MIDDLE PALEOLITHIC - generic grouping for those cultures that flourished between about 35,000 and 100,000 years ago; the best known such tradition is termed Mousterian, the culture carried by Neanderthals.

MILLION ELECTRON VOLTS - MeV

MIOCENE EPOCH - 26 MYA - 5 MYA.

MOLECULE - a tightly bound collection of atoms held together by the electromagnetic fields of the atoms. The smallest unit of a chemical compound. Molecules, like atoms, emit and absorb photons at specific wavelengths.

MUON - a negatively charged particle with a positively charged antiparticle with a spin of 1/2 and a mass of 206.77. Same as an electron, but more massive. A short-lived particle.

MUON'S NEUTRINO - zero mass, 1/2 spin with a neutral charge.

MYA - million years ago

NEANDERTHAL - Homo Sapiens neanderthalensis. An extinct species existing about 100,000 to 50,000 years ago and predominately associated with Mousterian culture in the Stone Age.

NEBULA - a general term used for any "fuzzy" patch on the sky, either light or dark. May be either a mass of luminous gas or a cluster of stars very far away from the solar system. A cloud of dust in a galaxy, such as the Orion Nebula.

NEOLITHIC - the New Stone Age, during which self-sufficient agriculture developed; characterized by the use of polishing and grinding stones and the origin of ceramics.

NEURON - a cell with specialized processes that is the fundamental functional unit of the nervous tissue. The complexity of the human brain depends on the number of interconnections (neurons). The brain consists of at least ten billion neurons, each a little computer in itself.

NEUTRINO - electrically neutral, massless particles that respond to the weak nuclear force but not the strong nuclear or electromagnetic forces. Neutrinos have energy, momentum, angular momentum, and either electron-family number or muon-family number. Free neutron (beta decay) releases a proton, electron and a neutrino. Over 8×10^{28} pass through Earth each second and 8×10^{12} pass through each person each second.

NEUTRON - electrically neutral, massive particle formed in the nucleus of atoms. Each neutron is composed of one *up* quark and two *down* quarks. Its mass is 939.6 MeV, slightly more than that of the proton. Stable within the nucleus, the neutron, if isolated, decays with a half-life of fifteen minutes (beta decay) into a proton, electron and neutrino.

NEUTRON STAR - a very dense star composed of neutrons having a diameter of about thirty kilometers (eighteen miles).

NEW WORLD - the Western Hemisphere - North and South America and the neighboring islands.

NON-METALS - elements with a valence shell of five or more electrons.

NOVA - a star that suddenly increases enormously in brightness, then slowly fades back to its original luminosity. These are the result of explosions on the surface of faint white-dwarf stars, caused by matter falling onto their surfaces from debris of a larger binary companion.

NUCLEI - (1) the central part of atoms, composed of protons and neutrons (each consisting of quarks) and contain nearly all of each atom's mass. (2) the central region of a galaxy.

NUCLEONS - protons and neutrons, the constituents of atomic nuclei.

NUCLEUS - a central part around which other parts are grouped: core. Examples: atoms, comets.

OBLATE SPHEROID - a sphere which is flattened at the poles - Earth.

OLIGOCENE EPOCH - 38 MYA - 26 MYA.

OMEGA - a baryon with a mass of 3276, spin ½ with a negative charge.

ORDOVICAN PERIOD - 504 MYA - 441 MYA.

ORGANIC CHEMISTRY - the chemistry of carbon compounds.

OZONE LAYER - layer of Earth's atmosphere at an altitude of twenty to fifty kilometers (twelve to thirty miles) where incoming ultraviolet solar radiation is absorbed by oxygen, ozone and nitrogen in the atmosphere.

PALEOCENE EPOCH - 65 MYA - 54 MYA.

PALEOMAGNETISM - the study of ancient, or fossilized, magnetism.

PALEONTOLOGY - branch of geology devoted to the study of ancient life based on fossils.

PALEOZOIC ERA - 570 MYA - 245 MYA.

PANGAEA (ALL EARTH) - focused mainly around the South Pole but stretched practically to the North Pole joining all the land into one huge continent. It began to break up in the Jurassic Period some 200 MYA.

PARSEC - the distance which a star must lie in order that its measured parallax is exactly one arcsecond, equal to 206,270 AU (astronomical units).

PARTICLES - fundamental units of matter and energy. All may be classed as fermions, which have a half-integral spin and obey the exclusion principle; and bosons, which have an integral spin and do not obey the exclusion principle. The term particle is metaphoric, in that all subatomic particles all show aspects of wave-like behavior.

PARTICLE PHYSICS - the branch of science that deals with the smallest known structures of matter and energy. As their experimental investigation usually involves the application of considerable energy, particle physics overlaps with high-energy physics.

PAULI EXCLUSION PRINCIPLE - see EXCLUSION PRINCIPLE.

PERIGEE - closest distance to object being orbited.

PERIHELION - point closest to object being orbited.

PERIOD - a division of geologic time longer than an epoch and included in an era.

PERMIAN PERIOD - 290 MYA - 245 MYA.

PHANEROZOIC ERA - 570 MYA to present.

PHOTOELECTRIC EFFECT - a light beam shining upon a metal plate causes an electrical current to flow. When a photon of sufficient energy strikes a metal surface, it ejects an electron; these liberated electrons constitute the observed electric current.

PHOTON - the quanta of electromagnetic force. The name comes from the fact that light is a form of electromagnetism. Photons have zero mass and can travel infinite distances. Spin 1, neutral charge.

PHOTON ENERGY - one electron volt - 1 eV.

PHYSICS - the science that deals with matter, energy, motion and force.

PION - a meson, mass 273.1, spin zero; pi⁻ - positive charge; pi⁻ - negative charge.

PLANCK CONSTANT - (h) fundamental constant of quantum mechanics. $h = 6.6255 \times 10^{-27}$ gm cm²/sec (erg sec).

PLANCK LENGTH - 10^{-33} centimeters. The distance light has traveled in Planck's time.

PLANCK TIME - 10^{-43} seconds after the big bang.

PLASMA - a gas consisting of free protons and electrons; a gas in which the electrons have been stripped from their nuclei.

PLATE TECTONICS - the motions of regions of the Earth's lithosphere which drift with respect to one another. It is also known as continental drift. The surface of the Earth is divided into many plates that are in continual motion which results in earthquakes, volcanic action, continental drift and other long-reaching effects.

PLEISTOCENE EPOCH - 5 MYA - 1.6 MYA.

POLARIZATION - wave can be caused to vibrate in a particular direction perpendicular to its direction of polarization.

POLARIZE - transverse waves can be caused to vibrate in a particular direction perpendicular to its direction of propagation.

POSITRON - an elementary particle having the same mass and spin as an electron (e⁻) but having a positive (e⁻) charge equal in magnitude to that of the electron; the antiparticle of the electron.

POTENTIAL ENERGY - kinetic energy at rest. The energy of a body or a system with respect to the position of a body or the arrangement of the particles of the system.

PRECAMBRIAN - 4.5 BYA - 570 MYA.

PRECESSION - the slow change in the direction of the axis of a spinning object, caused by some external influence, usually gravity.

PRIMORDIAL - belonging to or typical of the earliest stage of development.

PROLATE SPHEROID - a sphere elongated at the poles; examples are a nucleus and a football.

PROTON - a massive particle with positive electric charge found in the nuclei of atoms. Composed of two *up* quarks and one *down* quark. The mass is 938.3 MeV, slightly less than that of the neutron.

PULSARS - rapidly rotating neutron stars. Matter is so strongly condensed that the entire star acts like a giant atomic nucleus constructed wholly of neutrons. The size would be so small that it would not shade New York City. A thimbleful would weigh a billion tons. A white dwarf weighs about one ton per thimbleful. The magnetic field is a trillion times stronger than Earth.

PERTURBATION - a disturbing condition.

PYROCLASTIC - the fragmented ejecta released explosively from a volcanic vent.

QCD - quantum chromodynamics - quantum number, "color" for quarks.

QED - quantum electrodynamics - electrical charge in the affairs of electrons.

QUANTA - fundamental *units* of energy.

QUANTUM - fundamental *unit* of energy.

QUANTUM CHROMODYNAMICS - see QCD - the quantum theory of the strong nuclear force, which it envisions as being conveyed by quanta called gluons. The name derives from the assignment of a quantum number called color (red, green, blue) to distinguish how quarks function in response to the strong force.

QUANTUM ELECTRODYNAMICS - see QED - the quantum theory of electromagnetic force, which it envisions as being carried by quanta called photons.

QUANTUM FLUCTUATIONS - the small variations that must be p resent in a quantum field due to the uncertainty principle.

QUANTUM LEAP - the disappearance of a subatomic particle, an electron, at one location and its simultaneous reappearance at another.

QUANTUM MECHANICS - motion of atomic structure. Energy transfers from matter to radiation is quanitized. Planck's constant determines the entire scale of the submicroscopic world - the energy of photons, the spin of the particles and the size of atoms. Albert Einstein related Plank's constant (h) to the energy carried by electromagnetic radiation.

QUANTUM PHYSICS - physics based upon the quantum principle, that energy is emitted not as a continium but in discrete units. It is a statistical theory.

QUANTUM SPACE - vacuum with the potential to produce virtual particles.

QUANTUM TUNNELING - a quantum leap through a barrier.

QUARKS - fundamental particles from which all hadrons are made. According to the theory of quantum chromodynamics, protons, neutrons and their higher-energy cousins are composed of a trio of quarks, while the mesons are each made of one quark and one antiquark. Quarks are held together by the strong nuclear force. Quarks are not found in isolation in nature today. Six flavors: up, down, strange, charmed, bottom and top. Up, charmed and top quarks have a +2/3 charge; down, strange and bottom quarks have a - 1/3 charge. Dimension of each is 10^{-20} cm.

QUATERNARY PERIOD - 1.6 MYA to present.

RADIATION - a way in which energy is transferred from place to place in the form of a wave. Light is a form of electromagnetic radiation.

RADIOACTIVITY - the spontaneous decay or disintegration of an unstable atomic nucleus. This phenomenon is usually accompanied by the emission of ionizing radiation - alpha particles, beta particles, gamma rays.

RED DWARF - a dying star that has a light reddish color. It is of low mass and has a surface temperature of 2000 degrees Celsius.

RED GIANT - a star that is in the early stages of dying. It begins an expansion process as the hydrogen fuel is depleted during its later stage of evolution. Its diameter increases many times as it expands searching for hydrogen atoms in space. Luminosity increases due to the increased surface area. When our Sun reaches this stage of life, its edge will have increased to Earth's orbit, engulfing Mercury, Venus and Earth as it reaches its maximum diameter.

REDSHIFT - a shift in the frequency of a photon toward a lower energy. Appears to be receding. As you lower the frequency of visible light, it moves toward the red end of the spectrum.

REFLECTION - the angle of incidence equals the angle of reflection.

REFRACTION - the deflection of a wave passing from one medium to another. This is caused by the difference in the speed of the wave in the two media.

RELATIVITY, GENERAL THEORY - Einstein's theory of electrodynamics of moving systems gravitationally.

RELATIVITY, SPECIAL THEORY - Einstein's theory of gravitational force associated with light and its universal characteristics.

RERADIATE - photons emitted from an atom as its electron returns to its ground state from its excited state. An example is a photon of light hitting a mirror entering the electron field of the outer atom causing an electron to move to a higher energy level, its *excited* state, and emitting a photon as it returns to its *ground* state.

RIFT - the center of an extensional spreading center, where continental or oceanic plate separation occurs.

SCALE OF THE UNIVERSE - the observable radius: 10^{26} meters = 10^{23} kilometers = 6×10^{22} miles.

SCANNING TUNNEL MICROSCOPE - see STM

SCIENTIFIC METHOD - the set of rules used to guide science based on an idea that scientific "laws" be continually tested, and replaced if found inadequate.

SEDIMENTARY ROCK - deposited by debris over time.

SEISMIC WAVE - a wave that travels outward from the site of an earthquake through the Earth.

SEMICONDUCTOR - a material with electrical conductivity somewhere between that of a metal and that of an insulator.

SETI - search for extraterrestrial intelligence.

SIDEREAL DAY - the time needed for a star on the celestial sphere to make one complete rotation in the sky.

SIDEREAL MONTH - the time required for the Moon to complete one trip around the celestial sphere.

SIDEREAL TIME - the measurement of time based on the rotation of the Earth with respect to the stars.

SIDEREAL YEAR - the time required for the constellations to complete one cycle around the sky and return to their starting points, as seen from a given point on Earth. (SIDEREAL TIME)

SIGMA - a baryon with three possible charges. Positive - mass 2327.7, spin ½; negative - mass 2343.3, spin ½; neutral - mass 2333.7, spin ½.

SILURIAN PERIOD - 441 MYA - 413 MYA.

SOLAR DAY - the interval between two successive meridian passages of the Sun.

SOLAR MASS (SM) - 1.99×10^{30} kilograms (2.19×10^{27} tons).

SOLAR UNIT - see SOLAR MASS.

SOLAR WIND - an outward flow of fast moving charged particles from the Sun. Consist of plasma - free protons and free electrons which have been stripped from the nucleus of hydrogen atoms. Speed is 900 thousand to 1.62 million miles per hour.

SPECTROGRAPH - a device, usually based on a finely etched grate that performs the function of a prism, for breaking up light into its constituent parts and making a photographic or electronic record of the resulting spectrum.

SPECTROSCOPY - scientific investigation of an object by studying its spectrum.

SPECTRUM - a record of the distribution of matter or energy (light) by wavelength. Spectrum can be studied to learn the chemical composition and motion of stars and planets.

SPHERE - a three dimensional round object - a ball.

SPIN - particles of spin ½ make up the matter of the Universe. Spin 0, 1 and 2 are forces between matter particles. May exist in some circumstances as real particles. These appear to us as waves of light or gravitational waves. Spin 0 is like a dot - looks the same from every direction. Spin 1 is like an arrow - looks different form different directions. Spin 2 looks like a double-headed arrow - looks the same as if turned 180 degrees. Spin ½ turns through two revolutions. $\hbar = h/2\pi = 1.055 \times 10^{-27}$ erg sec.

STATICS - the branch of mechanics that deals with bodies at rest or forces in equilibrium.

STM - scanning tunneling microscope. Uses the barrier tunneling phenomenon of quantum mechanics to produce photographs in which individual atoms can be seen.

STRATA - layered rock formations; also called beds.

STRATIFICATION - a pattern of layering in sedimentary rock, lava flows or water or materials of different composition or density/

STRATOSPHERE - the portion of Earth's atmosphere lying above the troposphere, extending up to an altitude of forty to fifty kilometers (almost thirty miles).

STRONG NUCLEAR FORCE - fundamental force of nature that binds quarks together, and holds nucleons (composed of quarks) together as the nuclei of atoms. Portrayed in quantum chromodynamics as conveyed by quanta called gluons.

SUBDUCT, SUBDUCTED, SUBDUCTION - plate being drawn under or overridden by another in plate tectonic action.

SUBDUCTION ZONE - an area where the oceanic plate dives below a continental plate into the asthenosphere. Ocean trenches are the surface expression of the subduction zone.

SUPERCLUSTER - grouping of several clusters of galaxies into a larger, but not necessarily gravitationally bound, unit.

SUPERCONDUCTIVITY - the disappearance at a very low temperature of all electrical resistance - 23 Kelvin. Without resistance, electrical energy cannot be converted into heat, and therefore, no power is wasted. Once the voltage is applied, the current will continue to flow indefinitely.

SUPERNOVA - an immense stellar explosion which can increase a star's intrinsic brightness by as much as a billion times.

SUPERPOSITION - waves, fields, coexist in space without influencing each other.

SUPERSONIC - velocity in the excess of the speed of sound.

SWAG - scientific wild-assed guess.

TAIL OF A COMET - formed of molecules and very finely divided dust particles driven out from the coma of the comet by the force of the solar wind.

TECTONICS - in geology, the history of the larger features of the Earth (rock formations and plates) and the forces and movements that produce them.

TEMPERATURE IN SPACE - the area around Earth is 250 degrees Fahrenheit in the sunlight to minus 250 degrees Fahrenheit in the unlit area.

TEOTIHUACAN - a major urban center that flourished during the classic period of Mexican pre-history, from roughly 300 BC to 900. Located about twenty-five miles north of modern Mexico City.

TERRESTRIAL - anything that is of Earth.

TERRESTRIAL PLANET - the rocky planets of Mercury, Venus, Earth and Mars.

TERTIARY PERIOD - 65 MYA - 1.6 MYA.

TETHYS SEA - the sea between Laurasia and Gondwana which became the Mediterranean Sea after Laurasia and Gondwana squeezed together forming Pangaea.

THERMODYNAMICS, FIRST LAW - the heat added to a system equals an increase in internal energy plus external work done by the system.

THERMODYNAMICS, SECOND LAW - heat will never flow from a cold body to a hot body.

THUNDER - a violent explosion of air heated by lightning.

TRANSISTOR - a semiconductor device that regulates a current passing through it.

TRANSIT OF VENUS - Venus passes directly in front of the Sun - June 7, 2004 and June 5, 2012.

TRENCH - a depression on the ocean floor caused by subduction.

TRIASSIC PERIOD - 245 MYA - 210 MYA.

TROPOSPHERE - the portion of Earth's atmosphere from the surface to about fifteen kilometers (nine miles).

TROPICAL YEAR - the time interval between two successive passages of the Sun through the vernal equinox: the calendar year, or 365.2422 mean solar days.

ULTRAVIOLET - region of electromagnetic spectrum just outside the visible range, corresponding to the wavelengths slightly shorter than that of blue light.

UPWELLING - the process by which warm, less dense surface water is drawn away from along the shore by offshore currents and replaced by cold, denser water brought up from the subsurface (coming from below).

VERNAL EQUINOX - the point at which the ecliptic intersects the celestial equator, the Sun having a northerly motion (March 21).

VIRTUAL PARTICLES - short-lived particles that arise from a vacuum. Their existence is permitted by the indeterminacy principle.

W PARTICLE - massive bosons thought to have been abundant in the early Universe when the unified electroweak force was manifest. Have integer spin.

WATCH - a portable timepiece. The term comes from the shipboard practice of dividing up the day into six watches of four hours each.

WAVE - a pattern that repeats itself cyclically in both time and space. Waves are characterized by the velocity with which they move, their frequency and their wavelength.

WAVELENGTH - the length from one point on a wave to a point where it is repeated exactly in space, at a given time.

WEAK NUCLEAR FORCE - fundamental force of nature that governs the process of radioactivity. It is currently accounted for by the electroweak theory.

WEIGHT - a gravitational force experienced by a body.

WHITE DWARF - an approximately Earth-sized star that does not have a source of nuclear energy in its interior.

X-RAYS - high frequency electromagnetic waves emitted by the excitation of the innermost (high-energy) orbital electrons of atoms.

XI - baryons with negative and neutral charges. Negative - mass 2585.5, spin ½; neutral - mass 2573, spin ½.

Z PARTICLE - bosons which are carriers of the electroweak force and have integer spin.

ZEEMAN EFFECT - atoms in a magnetic field emit a spectral pattern slightly different from the pattern emitted without a field due to intrinsic magnetism of atoms.

BIBLIOGRAPHY

The Almagest - Ptolemy
America in 1492 - Alvin M. Josephy, Jr.
Archaeology - David Hurst Thomas
Astronomy - Michael Hoskin
Astronomy From A to Z - Charles Schweighauser
Astronomy Today - Eric Chaisson and Steve McMillan
Basic Physics - Kenneth W. Ford
Billions and Billions - Carl Sagan
Biography of a Planet - Chet Raymo
The Breakup of Pangaea - Robert S. Dietz and John C. Holden
A Brief History of Time - Stephen W. Hawking
The Changing Atmosphere - John Firor
Civilization Past and Present - Wallbank, Taylor, Bailkey, Jewsbury, Lewis and Hackett
Comet - Carl Sagan and Ann Druyan
Coming of Age in the Milky Way - Timothy Ferris
Conceptual Physics - Paul G. Hewitt
Concerning the Two New Sciences - Galileo Galilei
The Cosmic Blueprint - Paul Davies
Cosmic Discoveries - Martin Hewitt
Discovering Relativity for Yourself - Sam Lilley
Egypt in the Age of the Pyramids - Guillemette Andrew
The Electromagnetic Spectrum - Franklyn M. Branley
Encyclopedia Britannica
Epitome of Copernican Astronomy - Johannes Kepler
Essentials of Geology - Fredrick K. Lutgens and Edward J. Tarbuck
Evolution of the Earth - R. H. Dott, Jr. and R. L. Battan
Foundations of Modern Cosmology - John F. Hawley and Katherine A Holcomb
The Founders of the Western World - Michael Grant
The Genius of China - Robert Temple
Genius Talk - Denis Brian
Geological Society of America
God and the New Physics - Paul Davies
Great Books of the Western World - Encyclopedia Britannica, Inc.
The Great Dying - Kenneth J. Hsu
The Harmonics of the World - Johannes Kepler
A History of Civilization - Fernand Brandel
Journal of Geophysical Union
Journal of Geology
Journey to the Stars - Robert Jastrow
The Large Scale Structure of Space Time - Stephen W. Hawking
The Last Three Minutes - Paul Davies
Let the Sea Make a Noise - Walter A. McDongall
Mars - Our Future on the Red Planet - Robert M. Powers

The Matter Myth - Paul Davies and John Gribbin
The Moment of Creation - Big Bang Physics - James S. Trefil
Monsters in the Sky - Paolo Maffei
NASA
The Nature of Space Time - Stephen W. Hawking and Roger Penrose
NOAA
On the Loadstone and Magnetic Bodies - William Gilbert
On the Revolution of the Heavenly Spheres - Nicholas Copernicus
The Origin of the Universe - John W. Barrow
Other Worlds - Paul Davies
The Paranormal - Anthony North
Physical Geology - Robert J. Foster
Physics for Scientists and Engineers - Fishbane, Gasiorowicz and Thornton
Pioneering Space - James E. Oberg and Alcestis R. Oberg
Plate Tectonics - John Ericson
Quantum Chromodynamics - W. Greiner and A Schafer
Quantum Electrodynamics - W. Greiner and J. Reinhardt
Quantum Leap - Costanza
Quantum Mechanics in Curved Space - J. Audretsch and Venzo DeSabbata
Quantum Mechanics, Symmetries - Greiner Muller
Reading the Mind of God - James Trefil
The Ring of Truth - Philip and Phyllis Morrison
Carl Sagan's Universe
The Search for Alexander - Robin Lane Fox
The Search for Extraterrestrial Intelligence - Thomas R. McDonnough
Six Easy Pieces - Richard P. Feynman
Space Warps - John Gribbin
The Story of Astronomy - Lloyd Motz and Jefferson Hane Weaver
The Sumerians - C. Leonard Woolley
Taming the Atom - Hans Christian Von Baeyer
The Three Big Bangs - Phillip M. Dauber and Richard A. Muller
Three Hundred Years of Gravitation - Stephen W. Hawking and Werner Israel
The Time Before History - Colin Tudge
Timescale: An Atlas of the Fourth Dimension - Nigel Calder
U. S. Coast and Geodetic Society
U. S. Geological Society
The Universe Story - Brian Swimme and Thomas Gerry
The Way Science Works
Woods Hole Oceanographic Institution
The World Book Encyclopedia

www.ingramcontent.com/pod-product-compliance
Lightning Source LLC
Chambersburg PA
CBHW081103170526
45165CB00008B/2311